CHEMICAL ENGINEERING METHODS AND TECHNOLOGY

HELIUM

OCCURRENCE, APPLICATIONS AND BIOLOGICAL EFFECTS

CHEMICAL ENGINEERING METHODS AND TECHNOLOGY

Additional books in this series can be found on Nova's website under the Series tab.

Additional e-books in this series can be found on Nova's website under the e-book tab.

CHEMICAL ENGINEERING METHODS AND TECHNOLOGY

HELIUM

OCCURRENCE, APPLICATIONS AND BIOLOGICAL EFFECTS

SIMON B. DAVIS
EDITOR

nova publishers
New York

For permission to use material from this book please contact us:
Telephone 631-231-7269; Fax 631-231-8175
Web Site: http://www.novapublishers.com

Library of Congress Cataloging-in-Publication Data

Helium : occurrence, applications and biological effects / editor, Simon B. Davis.
pages cm
Includes bibliographical references and index.
ISBN: 978-1-62808-612-6 (hardcover)
1. Helium. 2. Helium--Industrial applications. 3. Helium--Physiological effect. I. Davis, Simon B.
TP245.H4H45 2013
665.8'22--dc23

2013028317

Published by Nova Science Publishers, Inc. † New York

CONTENTS

PREFACE

Helium is the second most abundant element in the solar system after the hydrogen atom and also the simplest of all atoms. In this book the authors present topical research in the study of the occurrence, applications and biological effects of helium. Topics discussed include the discontinuous yielding, stress drops and strength of metals in liquid helium; the use of helium with oxygen at pressure as a breathing gas mixture; the physics and chemistry of helium clusters and droplets; the interaction of helium atoms and ions with matter; postseismic leakage of mantle and crustal helium from seismically active regions; weak superfluidity of the nonstationary Josephson effect in He-4 and possible applications; and clean energy from Helium-3 fueled muon-catalyzed aneutronic fusion.

Chapter 1 - The mechanical behavior of metals under load in liquid helium is studied. The mechanisms and manifestations of low-temperature discontinuous yielding (jerky flow) are considered. The accompanying thermal effects are evaluated by analyzing energy conversion and deformation behavior. Criteria for new limiting states that occur during discontinuous yielding are established.

A two-parameter (temperature and stress) criterion of instability of plastic deformation of metals is formulated. It can be used to calculate the maximum critical temperature and the critical stress at given temperature. The influence of the loading conditions and design/technology factors on the discontinuous yielding characteristics and mechanical properties of some steels and alloys at a temperature of 4 K was studied experimentally. Prestrain, as well as stress concentrators in combination with low rigidity of the structure or supercritical strain rates may cause fracture of a structural member. A method for the experimental simulation of a strain jump at room temperature is developed and shown to be capable of reproducing and studying effects similar to those observed in liquid helium. A nonlinear multiparameter mathematical model of a strain jump is set up.

The dependence of strain jumps on the mechanical characteristics of the material, the dynamic characteristics of the testing machine, rated strain rate, and scale effect is established numerically for the widest possible ranges of these factors. A method for determining ultimate stresses with allowance for the low-temperature hardening of metals and the effects of discontinuous yielding is proposed. Recommendations on how to improve the tensile test method for metals at temperatures below 20 K are formulated.

Chapter 2 - The use of compressed air or nitrogen-oxygen mixture in diving induce from 30 metres of sea water several neurological and psychomotor disturbances due to the increase of partial pressure of nitrogen and to its narcotic potency called nitrogen narcosis. To avoid these disturbances, nitrogen has been replaced by helium, an inert gas with a low narcotic potency in relationship with its very low lipid solubility. With helium-oxygen synthetic gas mixture, depths higher than 50 metres or pressure higher of 0.5 MPa have been reached up to 6.1MPa. However, from 1 MPa physiological and neurophysiological changes different from the symptoms of nitrogen narcosis have been reported. In man, these changes consist of electroencephalographic modifications from 1MPa, occurrence of tremor and sleep disturbances from around 2MPa, myoclonia and dysmetria from 4MPa, and in experimental animals, epileptic seizures for pressure higher than 8MPa. Theses disturbances have been regrouped in a high pressure nervous syndrome (HPNS), as helium did not seem implicated in these disruptions. Indeed, if helium has narcotic properties, the signs and symptoms will be similar to or worse than 0.9MPa compressed air at about 4.1MPa. However owing to antagonism of the weak narcotic effect of helium by the increased pressure (called the pressure reversal effect) there is no helium narcosis. However, in the light of results obtained with narcotic gases added to helium oxygen mixture, such as nitrogen or hydrogen, mood changes and some somesthetic hallucinations reported at pressure higher than 4-5 MPa could be in relationship with a narcotic effect of helium rather than a pressure effect. Moreover hallucinatory behaviours have been reported in monkeys in helium oxygen mixture for pressure of 8MPa and above.

Neuropharmacological and neurochemical studies (microdialysis and voltammetry) performed with helium-oxygen mixture at pressure have demonstrated changes in amino-acid (gama-amino-butyric acid and glutamate) and monoamine (dopamine and serotonin) neurotransmission regulations at the level of basal ganglia, brain structures implicated in the regulation of the motor, loco-motor and cognitive processes.

Chapter 3 - Many microscopic properties of the only natural superfluid element helium have only become available within the last 18 years through a number of novel experiments. The diffraction of matter-waves of small helium clusters have established the unusually large bond distance (5.2 nm) and extremely weak binding energy (10^{-7} eV) of the dimer and other exotic properties of small helium clusters. The high resolution spectroscopy of closed shell chromophore molecules attached to small clusters or embedded inside large superfluid droplets have revealed that as a result of superfluidity molecules rotate virtually without friction and that only a hand-full of He atoms are required. These and a number of other experiments demonstrate that helium droplets are the most gentle and coldest of all cryomatrices with enormous potential for a wide range of future applications.

Chapter 4 - The electron loss and electron capture cross sections, equilibrium target thickness, energy loss and ranges are studied experimentally for helium atoms and helium ions with velocities $(2.6 - 20)\times10^8$ cm/s in various targets. The energy, angular and charge distributions of helium ions and atoms reflected from a metal surface at grazing incidence are measured at the energy 250 and 300 keV. On the bases of experimental data some effects of ion interaction with matter are discussed theoretically (nonmonotonic dependence of range, energy loss, and charge-changing cross sections on nuclear charges of the colliding particles, density effects in gaseous and solid targets etc.). The experimental values are given by figures and by tables in appendix

Chapter 5 - It is well known that mantle degassing does not occur homogeneously over the Earth's surface. The elevated $^3He/^4He$ ratios found in volcanic regions and tectonically active areas are higher than the atmospheric values. The anomalous helium isotopic ratios have been interpreted to transfer mantle volatiles into the crust by processes or mechanisms such as magmatic intrusion, continental underplating and lithospheric rifting. This study was undertaken to elucidate the geographic distribution of $^3He/^4He$ ratios around seismically active regions in Japan, using helium isotope data obtained from gas samples. Several case studies suggest that there seem to be a significant trend of high ^3He emanations around the source regions following inland earthquakes, resulting in leakage of mantle volatiles through crustal pathways (faults) due to more frequent development of higher permeability pathways and/or upwelling of mantle fluids through the ductile lower crust. The other case is significant decrease of $^3He/^4He$ ratios found in groundwaters before and after a large inland earthquake. Episodic faulting could release accumulated crustal (radiogenic) helium from host rocks, or enhance the transfer of mantle volatiles through permeable fault zones, such that subsequent fluid flow proximal to the source region could then explain the spatio-temporal variations in $^3He/^4He$ ratios.

Chapter 6 - Unusual occurrences of nonstationary Josephson effect in superfluid He-4 are considered. Nonstationary Josephson effect is analyzed in the context with parametric amplifying of acoustic oscillations on kinetic nonlinearity of superfluid Josephson element. Mechanism of quantum fluctuations compression, following parametric rating, is taken into account. Appearance of Josephson fluxon in superfluid medium parted by a wall with narrow channels serving as weak links is analyzed by analogy with Josephson vortexes in the case of weak superconductivity. Possible applications of the effects above are discussed. Parametric amplifying of acoustic oscillations on conditions of quantum fluctuations compression is proposed to be used for making a high frequency ($f >> 1000$ Hz) gravitational waves detector. In this case if one have in mind that the efficiency of gravitational waves generation by a quadrupole radiator is proportional to the sixth power of frequency it seems interesting to use such a detector in a gravitational analog of Hertz experiment. Since superfluid Josephson vortexes are of quantum nature, it may be possible to model entangled states, which underline the matter wave q-bit system.

Chapter 7 - Helium-3 is a stable isotope of helium, terrestrially scarce, but with an estimated 6.5x10^8kg contained within the lunar regolith. Despite the difficulties inherent in commercial exploitation of these remote reserves, research progresses into its potential for use in fusion. It has now been posited that by employing existing and developing technologies, an energy efficient, helium-3 fuelled, muon-catalyzed fusion (μCF) reactor is able to be constructed. This reactor, which avoids the problems associated with maintenance of the extremely high temperatures normally required to catalyze fusion, is also aneutronic, offering total freedom from the need to deal with radioactive materials. The key processes of this radiation free, helium-3 powered μCF reactor are discussed, including a review of some of the technologies that are required, the issue of alpha-sticking in the reaction area, and the predicted muon-wave that simulations have suggested would partially negate the alpha-sticking while providing the primary agency for the predicted energy efficiency.

In: Helium: Occurrence, Applications and Biological Effects ISBN: 978-1-62808-612-6
Editor: Simon B. Davis

Chapter 1

METALS IN LIQUID HELIUM: DISCONTINUOUS YIELDING, STRESS DROPS, AND STRENGTH

E. V. Vorob'ev, V. A. Strizhalo and T. V. Anpilogova
G. S. Pisarenko Institute for Problems of Strength, National Academy of Sciences of Ukraine, Ukraine

ABSTRACT

The mechanical behavior of metals under load in liquid helium is studied. The mechanisms and manifestations of low-temperature discontinuous yielding (jerky flow) are considered. The accompanying thermal effects are evaluated by analyzing energy conversion and deformation behavior. Criteria for new limiting states that occur during discontinuous yielding are established.

A two-parameter (temperature and stress) criterion of instability of plastic deformation of metals is formulated. It can be used to calculate the maximum critical temperature and the critical stress at given temperature. The influence of the loading conditions and design/technology factors on the discontinuous yielding characteristics and mechanical properties of some steels and alloys at a temperature of 4 K was studied experimentally. Prestrain, as well as stress concentrators in combination with low rigidity of the structure or supercritical strain rates may cause fracture of a structural member. A method for the experimental simulation of a strain jump at room temperature is developed and shown to be capable of reproducing and studying effects similar to those observed in liquid helium. A nonlinear multiparameter mathematical model of a strain jump is set up.

The dependence of strain jumps on the mechanical characteristics of the material, the dynamic characteristics of the testing machine, rated strain rate, and scale effect is established numerically for the widest possible ranges of these factors. A method for determining ultimate stresses with allowance for the low-temperature hardening of metals and the effects of discontinuous yielding is proposed. Recommendations on how to improve the tensile test method for metals at temperatures below 20 K are formulated.

INTRODUCTION

The substantial progress in power engineering, astronautics, medicine, physical research, and other fields has become possible due to the wide use of the cryogenic technology based on liquefied gases. Among them, liquid helium stands out as a gas with the lowest boiling point (about 4 K); therefore, only this coolant can ensure the stable operation of superconducting magnetic systems. They made it possible to create essentially new medical diagnostic equipment, efficient high-power generators, magnetic energy storage devices, particle accelerators, fusion reactors, etc. Load-carrying members of such systems are distinguished by especially extreme operating conditions, including high mechanical stresses, strong electromagnetic and other physical fields, and ultralow temperatures.

At cryogenic temperatures below 30 K, however, metallic materials exhibit qualitatively different deformation mechanisms: the plastic deformation process loses stability and becomes discrete in both time and space, though the loading process remains continuous. This effect is called discontinuous yielding or jerky flow of metals. It is manifested as serrations on stress–strain curves with stress drops reaching 60% of the initial level for some metals. There is extensive literature on this phenomenon, but its strength aspects have received a little attention [1] (revue).

In designing engineering facilities intended to operate at cryogenic and especially ultralow temperatures, the problem arises of how to make the full utilization of the strength reserves of structural materials because it is necessary to minimize the consumption of expensive coolants and materials and to raise the electromagnetic efficiency of superconducting magnetic systems. Low-temperature hardening is known to be a considerable strength reserve in metals. However, the existing methods for evaluating and using this effect when defining ultimate stresses fail to describe the unstable deformation (discontinuous yielding) of structural alloys.

The abrupt change from elastic deformation to local yielding should be regarded as a sharp drop in resistance to the load or a sudden loss of the load-carrying capacity of a structural member.

Therefore, an essential aspect in the low-temperature strength design of cryogenic engineering facilities is the description of the discontinuous yielding of metals. This problem can be resolved by conducting an integrated study of this effect and developing criteria and methods for the evaluation of limiting states of materials during unstable deformation.

Such research has been carried out at the G. S. Pisarenko Institute for Problems of Strength, National Academy of Sciences of Ukraine.

Among other things, the following important questions have been answered: will the effect occur in given conditions? how will certain factors affect the magnitude of strain jumps and the mechanical characteristics of the material? is a failure possible during or after a strain jump? The answers provided at the stage of strength design of structures will ensure their operational reliability and allow formulating recommendations on improvement of the standard tensile test method for metals. In what follows, we will discuss some results of such a study.

Mechanisms and Manifestations of the Discontinuous Yielding in Metals

The weak serrations observed on the stress–strain curves of some alloys at room and high temperatures are known to be caused by specific deformation mechanisms. These mechanisms include mechanical twinning, deformation polymorphism, deformation recrystallization, and dynamic strain aging and underlie the Portevin–Le Chatelier effect, which accompanies, for example, the deformation of aluminum alloys.

Load drops are not great in these cases. It should be noted that microheterogeneity or microinstability is typical of the plastic deformation of metals and alloys, up to individual dislocations that occur and move irregularly.

Therefore, the stress–strain curves of either polycrystalline or monocrystalline samples recorded at high resolution are serrated rather than smooth curves.

Unlike these effects, low-temperature discontinuous yielding is universal. As temperature is decreased, the plastic deformation of most metals, metal composites, some metallic glasses, and nonmetallic materials (such as ionic crystals NaCl and LiF) become macroscopically unstable.

Serrations are observed in metals with any lattice and any grain size, in tension and compression, above and below the λ-point at which helium becomes superfluid, with and without contact with liquid helium, in normal and superconducting states. In different materials, discontinuous yielding sets in at different temperatures in a range from 30 K to absolute zero. Discontinuous yielding occurs or intensifies when:

- the temperature is decreased;
- the strain rate is decreased;
- the purity of metal is reduced, or the content of alloying elements is increased;
- the prestrain is increased;
- superconducting state changes into normal state.

As the temperature of almost all steels and alloys is decreased, the shape of their stress–strain curves changes from smooth to wavy, and then to serrated (Figure 1).

The magnitude of strain and stress discontinuities monotonically increases, and each discontinuity is accompanied by necking of the specimen. The number N of discontinuities ranges from 1000 in aluminum monocrystals to 100 and more in aluminum alloys, to several tens in austenitic steels, and to one or two in some titanium alloys (Figure 1).

Moreover, in pure FCC metals, jerky flow usually occurs at later stages of deformation and may alternate with smooth flow, their amplitude of discontinuities ($\Delta\sigma_d / \sigma_0$) being 1–10%. In BCC metals, serrations immediately follow the yield point and reach high amplitude. It increases not only with the strain hardening, but also with the strength of metal or alloy.

The total number of occurrences of discontinuous yielding, its behavior, amplitude, and number of discontinuities are strongly dependent on strength, which, in turn, depends on "internal" factors such as the type of lattice, initial structural state, alloying elements and impurities, grain size and on external factors such as loading conditions (temperature, stress mode, loading rate, rigidity of testing machine) and design/technology factors (prestrain, surface finish, dimensions and shape of specimens, stress concentrators, etc.).

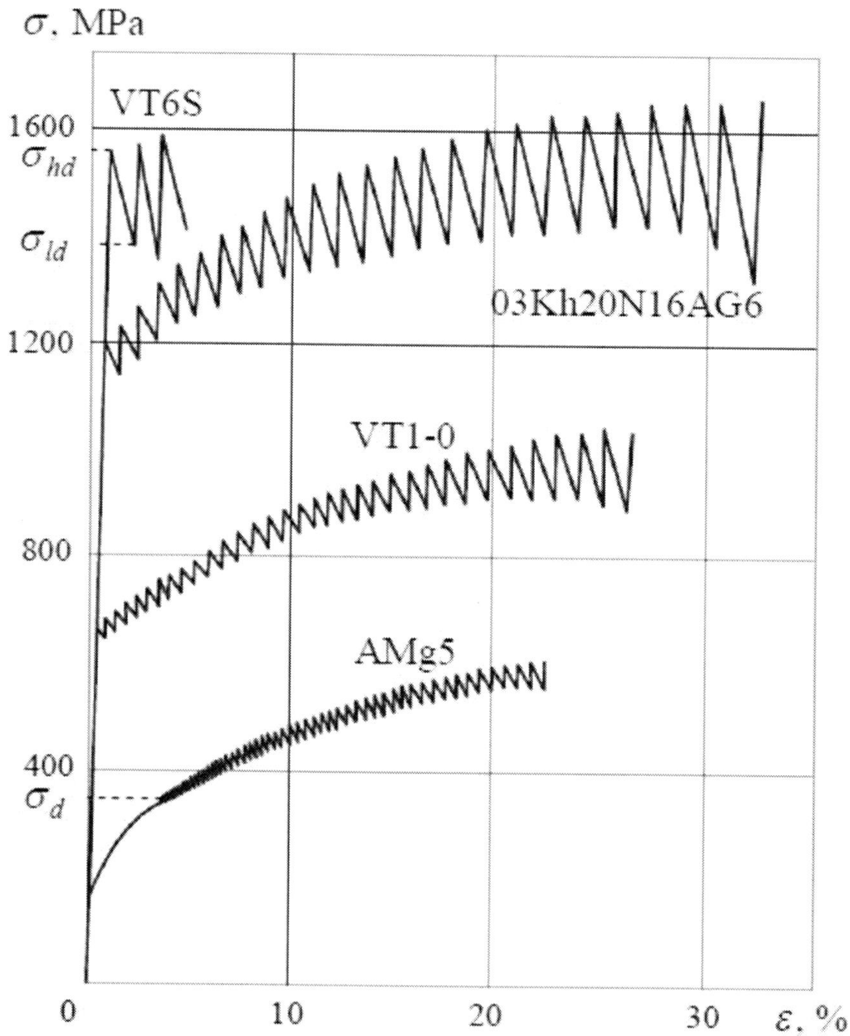

Figure 1. Stress–strain curves for various materials at a temperature of 4 K.

Discontinuous yielding can be either suppressed or intensified by varying these factors; therefore, they should be "controlled" to ensure the required structural strength of cryogenic-engineering products.

Besides pure mechanical effects such as multiple necking, low-temperature discontinuous yielding is accompanied by nonmechanical phenomena. One is the radiation of sound waves, their amplitude and frequency being different in different materials, which means that they can be identified from their acoustic spectrum. For example, the regularly discontinuous deformation of an aluminum monocrystal produces a clock-tick sound. The sound of AMg5 alloy amplifies, while the discontinuous deformation of high-strength steels and alloys sounds as thuds. These sounds are due to the internal processes in the specimen and the high dynamic loads on the specimen–machine system caused by strain jumps. Another, more fundamental effect is the electric pulses generated by strain jumps [2]; i.e., a catastrophic local flow generates an electric current in the specimen.

This phenomenon is explained by the entrainment of conduction electrons by quickly moving dislocations, which may be regarded as inverse electroplasticity. Moreover, even a weak electric (as well as magnetic) pulse is capable of initiating a strain jump [3]. These phenomena indicate the unique nature of the deformation of metals near absolute zero.

The basic discontinuous-yielding models employ the adiabatic heating hypothesis [4] and the dislocation stress (athermal) hypothesis (catastrophic breakdown of barriers by dislocation pileups [5] and initiation of multiple dislocation sources [6]).

The adiabatic hypothesis is based on the concept of the thermomechanical instability of the deformation process that occurs at very low temperatures, when the heat capacity of metals drops almost to zero. In such conditions, even a small amount of heat generated by plastic deformation is sufficient to easily heat up the lattice. The heat is localized in a small volume of the specimen, causing a significant increase in temperature. Because of the sharp decrease in resistance to deformation, a local plastic flow develops at a great rate in adiabatic conditions (there is no sufficient time for heat transfer).

Numerous experiments (see, e.g., [4, 7]) confirm that a strain jump is accompanied by a sharp rise in temperature. The critical temperature T_c above which discontinuities do not occur was discovered in [4]. For example, $T_c = 15$ K for Al–2Mg alloy.

The dislocation hypothesis [5] assumes that the lattice possesses high geometric heterogeneity. At a certain very high stress level, dislocation pileups formed in a crystal no longer bypass barriers, but rather break through them. In local regions, this stress is close to the theoretical shear strength. It is believed that this also leads to the spontaneous generation of dislocations near the barrier. However, this hypothesis fails to explain some experimental facts such as critical temperature, relationship between discontinuity amplitude and local heating, etc. There is also no explanation to the fact that dislocation pileups break through barriers synchronously over the entire specimen.

We may propose the following description for the starting and synchronizing mechanisms of discontinuous deformation. Let us consider a rod of cross-sectional area F stretched by a force directed along the rod axis z. Since the stress state of the rod is linear and homogeneous, the axial macrostrains are almost identical in different cross-sectional zones. Therefore, discontinuous yielding can be initiated by the starting mechanism in any section of the specimen. As indicated above, discontinuous deformation can be started up by a thermal, mechanical, or electromagnetic pulse. The first strain jump may immediately follow the elastic region, as well as the subsequent discontinuities occurring after the respective elastic stages (Figure 1). The occurrence of discontinuous yielding may be schematized as follows. After the stress reaches certain quite high level $\sigma = \sigma_0$ in some rather small cross-sectional zone of the rod, major dislocation pileups break through barriers. Dislocations first move in an over-barrier manner, but later their motion becomes thermally activated. The first discontinuity of deformation occurs within a small expanding area δF where the temperature undergoes a discontinuity δT. Such a flow is possible at the expense of redistribution of elastic, plastic, and thermal strains in the yielding zone, which is accompanied by an increase in the stress by $\delta\sigma_z = f(k_T, \bar{\alpha}, \delta T, E, \theta)$, where k_T is the thermal softening coefficient, $\bar{\alpha}$ is the average thermal expansion coefficient, E is the elastic modulus, and θ is the strain hardening modulus.

Making some simplifications and assuming that $\varepsilon_0 = \varepsilon_{p_z} + \varepsilon_{T_z} + \varepsilon_{e_z}$, where $\varepsilon_0 = \sigma_0 / E$, and ε_{p_z}, ε_{T_z}, and ε_{e_z} are plastic, thermal, and elastic strains, respectively, we obtain $\varepsilon_{p_z} \approx k_T \delta T / (E + \theta)$, $\delta\sigma_z = \dfrac{k_T \delta T \delta F}{(1 + \theta / E) F} + \dfrac{\overline{\alpha} \delta T E \delta F}{F} \approx \dfrac{k_T \delta T \delta F}{F}$. The magnitude of a discontinuity is estimated to be on the order of 1–10 Pa. The parameter F represents the scale effect (no discontinuous yielding in large-scale specimens).

A stress drop $\delta\sigma_z$ generates an elastic wave and acts as an initiator and synchronizer of the plastic deformation of the whole rod (multiple breakdowns of barriers by dislocation pileups over the entire volume). Then comes the thermoactivated stage during which intensive heating in one of the sections coinciding with the initial zone causes the flow to localize. This pattern of low-temperature discontinuous yielding is generally in agreement with the hypothesis on the combination of the dislocation-stress and adiabatic-heating mechanisms [1].

THERMAL EFFECTS AND ENERGY CONVERSION DURING DISCONTINUOUS YIELDING

Dislocation motion is a dissipative process in which the major portion of the plastic work is converted into heat. This portion is characterized by a coefficient β and usually considered constant. However, modern studies show that the proportion of plastic work converted into heat monotonically increases to its maximum during the deformation of metals. For example, the radiometric data in [8] suggest a strong dependence of the coefficient β on strain and challenge the classical assumption that it has a constant value of approximately 0.95. At low temperatures, this dependence remains, and the major portion of mechanical work dissipates as heat [9].

The other, athermal portion $(1-\beta)$ goes into structural transformation, formation of various defects, i.e., increases the lattice energy.

Here we deal with an inverse relationship: as the plastic strain increases, the capacity of metals to accumulate energy on microdefects reaches saturation. As a result, the maximum of $(1-\beta)$ corresponds to small initial plastic strain and monotonically decreases with further deformation. The strain-hardening modulus θ is known to be an integral quantitative indicator of structural transformation in metals during plastic deformation. This modulus also monotonically decreases to its minimum $\theta = \sigma_b$ (the true stress corresponding to the ultimate strength $\sigma_в$) at the true strain ε_b. Therefore, the assumption [10] that the athermal portion of the work is proportional to the area of the so-called hardening triangle, a part of the area under the stress-strain curve, appears to be physically consistent. Using this assumption and some simplifications, we obtain

$$\beta = (1 + k_\beta \theta / \sigma)^{-1} \tag{1}$$

where k_β is the coefficient of proportionality; θ and σ are the current values of the hardening modulus and true stress.

The expression for $\beta = f(\varepsilon)$ can be derived from either the constitutive equation, or the strain-hardening law that fits the stress–strain curve, or the upper envelope of serrations on the curve of true stress versus true strain. If such is the equation

$$\sigma = A\varepsilon^m \tag{2}$$

where A and m are constants, then $\theta = d\sigma / d\varepsilon = \sigma m / \varepsilon$, and formula (1) reduces to $\beta = (1 + k_\beta m / \varepsilon)^{-1}$.

If $m = \varepsilon_b$, then

$$\beta = (1 + k_\beta \varepsilon_b / \varepsilon)^{-1} \tag{3}$$

If $\beta_{max} = 0{,}95$ and $\varepsilon = \varepsilon_b$, then k_β = 0.05. Formula (3) is valid for any strain-hardening law describing the convex curve $\sigma = f(\varepsilon)$ because it follows directly from (1) when $\sigma = \sigma_b$. The graph of (3) shows that β increases monotonically and intensively, reaching the value 0.4 even at small strain.

The temperature amplitude experimentally measured during discontinuous yielding ranges from 7 to 60 K [4, 7, 11]. Sensors are placed in axial holes or on the surface of the specimen. The measured temperature of the sensor itself is always much lower than the true temperature because the sensors are slow-responding and, thus, measure the integral temperature of the whole specimen in the conditions of intensive heat transfer. Nevertheless, such measurements allow us not only to experimentally confirm the thermal nature of discontinuous yielding, but also to study the accompanying thermal effects. To this end, we tested 12Kh18N10T steel and AMg6 aluminum alloy in tension at an initial temperature of 4 K in liquid helium [7]. The specimens had a diameter of 4 mm and a gage length of 30 mm. The temperature was measured with a sensor with a response of 1×10^{-2} s/K. The sensor in a Teflon capsule was placed on the surface of the specimen at its middle. The temperature variation and the stress–strain curve were recorded synchronously. In the test, the specimen was loaded in a stress-controlled manner at a rate of $0.28 \times 10^{-3}\ \mathrm{s}^{-1}$. Plastic strain was accumulated only through discontinuous yielding. There were 60 strain jumps during loading of the steel specimens. Note that the temperature discontinuities lag behind the load drops. This lag is due to the very low heat conductivity of metals (low heat-transfer rate) and the slow response of the temperature sensors. The time it takes the sensor to respond to a detectable temperature jump is 0.2 s, which is much longer than the duration of a strain jump. The temperature varies discontinuously, the last jump reaching 20 K. There were 74 discontinuities during the deformation of the AMg6 alloy, the maximum temperature jump being 7 K. The ratio of these temperatures for the steel and aluminum alloy is approximately the same as the ratio of the stresses acting in these materials. Moreover, a common pattern is an increase of the temperature jump with increase in the starting stress and in the magnitude of stress drops. Thus, the higher the strength of the material, the greater the temperature jumps.

Noteworthy are the small temperature jumps in the absence of discontinuous yielding (the corresponding region of the stress–strain curve is smooth). This means that strain microjumps occur even before the discontinuous yielding becomes to be detected. The boundary of this effect is unclear if there is the stage of homogeneous deformation.

Though it is impossible to measure the temperature directly in the zone of plastic flow, it can be quite accurately calculated from experimental data [10, 12]. As an example, let us calculate this temperature for steels 12Kh18N10T, 03Kh20N16AG6, 03Kh13AG19, titanium alloy 3M, and aluminum alloy AMg5. We tested standard fivefold specimens (Ø 3 mm) in tension in liquid helium. The loading rate was 1.6×10^{-3} s^{-1}; the rigidity of the testing machine was 14.5 MN/m. The 03Kh20N16AG6 steel was examined in most detail.

The discontinuous yielding of structural steels and alloys manifests itself on tensile stress-strain curves as alternating regions of elastic deformation and plastic flow accompanied by abrupt drops in load, $\Delta P_d = P_0 - P_d$, and in stress, $\Delta\sigma_d = \sigma_0 - \sigma_d$, where P_0 and P_d (σ_0 and σ_d) are the loads (stresses) corresponding to the beginning and the end of a strain jump. For the first discontinuity, these are the stresses σ_{hd} and σ_{ld} (Figure 1). The stored elastic energy of the testing machine some of which is converted into work of deformation is expressed as $W = P_0^2/2C$, where C is the rigidity of the specimen–machine system. The temperature jump ΔT_m can be determined from the energy-balance equation describing the conversion of the work of deformation into thermal energy:

$$\beta w = \bar{c}\Delta T_m, \tag{4}$$

where w is the specific work of deformation; $\bar{c}$ is the volumetric heat capacity averaged over the temperature range $T_0...(T_0 + \Delta T_m)$. If the deformation is homogeneous, i.e., $\varepsilon_d = \Delta l_d / l$, where l is the length of the plastic-flow zone and Δl_d is the length jump, then $w \approx \sigma_0\varepsilon_d$ or, more precisely, $w = (P_0^2 - P_d^2)/2CV$, where V is the deformed volume. With $\Delta l_d = \Delta P / C$, the work of deformation can be expressed as $\Delta W = \Delta l_d (P_0 + P_d)/2$. Using the expression for enthalpy, we can recast Eq. (4) into an easy-to-use form:

$$\beta(P_0 + P_d)\Delta l_d / 2V = \int_{T_0}^{T_0+\Delta T_m} c\,(T)dT \tag{5}$$

Equations (4) and (5) have been derived assuming adiabatic conditions. This assumption is valid because these equations describe a so rapidly ($\tau_D = 10^{-2}...10^{-4}$ s) developing process that heat transfer can be neglected.

The heat capacity of a unit volume or mass of metals and other crystalline bodies remains approximately constant over a wide temperature range (from room temperature to melting point). As temperature decreases, the heat capacity abruptly drops beginning from the Debye temperature T_D.

At very low temperatures tending to absolute zero, the heat capacity of metals tends to zero as well. As the temperature is reduced from room temperature to 4 K, the heat capacity decreases by several orders of magnitude.

At $T < T_D$, specific heat capacity can be expressed as

$$c = \rho(k_e T + k_p T^3) \tag{6}$$

where ρ is the density of the material; k_e and k_p are the electron and phonon contributions to heat capacity, respectively. Only the latter component is usually taken into account in calculations, which is unacceptable for some metals, including aluminum and its alloys. The densities ρ of the steels and alloys under consideration are very different (Table 1).

This causes the difference of their volumetric heat capacities, which exhibit similar temperature dependences.

A difference between the heat capacities of titanium and aluminum alloys greater than 5% is observed only at temperatures above 150 K, while the heat capacities of all the austenitic steels examined (03Kh20N16AG6, 12Kh18N10T, 03Kh13AG19) are very similar in the temperature range 0–300 K and the curves $c = f(T)$ are located with a narrow band. At temperatures close to absolute zero, these curves for steels are fit by the following empirical formula derived from (6) [13]:

$$c = \rho(0{,}46T + 3{,}8 \cdot 10^{-4} T^3) \tag{7}$$

Table 1. Material characteristics at 4 K

Characteristic	Material				
	12Kh18N10T	03Kh20N16AG6	03Kh13AG19	3M	AMg5
$E \times 10^{-5}$, MPa	2.25	2.30	2.28	1.22	0.83
ρ, kg/m^3	7800	7800	7800	4500	2660
ΔT_{max}, K	195	185	156	168	61,5
σ_B, MPa	1740	1640	1280	1350	521
N	29	26	16	19	108

To calculate the temperature after adiabatic heating, the function $c = f(T)$ was fitted by cubic polynomials over a wide temperature range (0–200 K), with the polynomial coefficients calculated with a routine program.

The values of P_0, P_d, Δl_d were determined from the stress–strain curve, and V from the formula

$$V = \pi \int_0^{l_k} y^2(x)dx, \tag{8}$$

where l_k is the measured length of the neck; $y(x)$ is the generatrix shape function; x is the current coordinate varying from 0 to l_k. If $y(x)$ is a harmonic function (cosinusoid), expression (8) yields $V = \pi l_k (3r_b^2 + 2r_b r_l + 3r_l^2)/8$, where r_b and r_l are, respectively, the maximum and minimum radii of the neck. It is obvious that the temperature corresponding to the last discontinuity is maximum, all the previous discontinuities being characterized by smaller P_0 and Δl_d and, hence, smaller ΔW and ΔT_m. Table 1 summarizes the values of $\Delta T_m = \Delta T_{\max}$ calculated with formula (5) for $\beta = 0.95$ and for the last (Nth) discontinuity. It was established that, first, β has a considerable effect on the temperature jump and, second, the greater the jump number n, the higher the temperature ($\Delta T_n / \Delta T_{\max} \approx 0{,}8$ even at $n/N = 0.5$).

The rigidity of the specimen–machine system was 12, 10, and 9 MN/m for the steels, titanium and aluminum alloys, respectively. This difference is due to the different elastic modules of the materials (Table 1).

Of interest is the effect of the rigidity and initial temperature on the temperature jump. The 03Kh20N16AG6 specimens were tested at C_1=12 MN/m and C_2=8.5 MN/m at temperatures of 4 and 20 K in, respectively, liquid helium and helium vapor [12]. The rigidity of the system was changed by using a small-diameter tie. The results of the tests and calculations are collected in Table 2.

It can be seen that a change in rigidity by 40% does not practically affect the temperature jump, but with increase in the initial temperature and, hence, heat capacity, $\Delta T_{\max}$ decreases by more than 20 K. The result ($\Delta T_{\max}$ = 60–65 K) obtained by Basinski [4] for Al–Mg alloy is similar to that for the AMg5 alloy in Table 1.

As follows from (4) or from the equation $\beta \sigma_0 \varepsilon_d = \bar{c}\ \Delta T$, the higher the strength of a material, the more it heats up. We will now show that an increase in the load leads to an increase in the local strain. This results in a smaller number and greater magnitude of stress drops because $\Delta \sigma_d = K \varepsilon_d$, where K is the effective modulus of the specimen–machine system. It is now possible to linearly relate the maximum temperature jump at T_0 = 4 K to the strength of the material:

$$\Delta T \approx k_{\Delta T}\sigma_{\text{B}} \tag{9}$$

where $k_{\Delta T}$=0.118 K/MPa. The values produced by this formula differ from more exact calculated values by no greater than 6%. Applying formula (9) to nanocrystalline niobium ($\sigma_{0.2}=1810\,\text{MPa}$, $\sigma_{\text{B}}=2000$ MPa [14]), we obtain $\Delta T_{\max}$ = 236 K, which differs from the exact value by 4.9%.

Thus, during discontinuous yielding, the temperature in the plastic-flow zone reaches rather high levels and manifold exceeds the rated temperature; i.e., the deformation is nonisothermal. The thermal instability of the plastic deformation of metals at very low temperatures accompanied by the thermal effects described above is the main cause of discontinuous yielding.

INSTABILITY CRITERION FOR PLASTIC DEFORMATION. INSTABILITY DOMAIN

The criterion-based evaluation of discontinuous yielding involves deriving a system of inequalities that defines the conditions in which either the plastic deformation loses stability if discontinuous yielding sets in after the stage of homogeneous plastic deformation or limiting states occur if the strain jump immediately follows the stage of elastic deformation. In both cases, the criterion is thermal instability or adiabatic deformation [4]. Let us consider the former case of discontinuous yielding. The first step is to set up a system of relations that defines the conditions in which the effect is observed. The serrations on the stress–strain curve recorded at the coolant temperature T_0 and strain rate $\dot{\varepsilon}_n$ originate from a certain point whose coordinates are the true stress σ_c and the true strain ε_c (plastic component), which may be considered as critical parameters. The latter quantity defines the strain-based criterion $\varepsilon/\varepsilon_c \geq 1$. Moreover, the strain-hardening modulus $\theta = d\sigma/d\varepsilon$ reaches the critical level θ_c at this point. Discontinuous yielding occurs when the following system of inequalities holds [15]: $T/T_c \leq 1$; $c/c_c \leq 1$; $\dot{\varepsilon}/\dot{\varepsilon}_c \leq 1$; $\varepsilon/\varepsilon_c \geq 1$; $\sigma/\sigma_c \geq 1$; $\theta/\theta_c \leq 1$. These are partial criteria of discontinuous yielding: temperature, adiabatic, kinetic, strain-base, stress-based, and strain/stress-based criteria, respectively. Here the subscript "c" stands for "critical."

The critical parameters are limited by some maximum values $c_{c_{\max}}$, $T_{c_{\max}}$, $\varepsilon_{c_{\max}}$, $\sigma_{c_{\max}}$. It is clear that $\varepsilon_{c_{\max}} = \varepsilon_b$ and $\sigma_{c_{\max}} = \sigma_b$, where ε_b and σ_b are the true strain and the true stress corresponding to the ultimate strength of the material. The value θ_c cannot exceed the elastic modulus E, while the minimum value $\theta_{c_{\min}} = \sigma_b$ is reached at the upper point of the stress–strain curve because $d\sigma / d\varepsilon = \sigma$. The value σ_c cannot be less then the yield strength, while $\varepsilon_{c_{\min}} = 0$.

Table 2. Discontinuous-yielding and mechanical characteristics of 03Kh20N16AG6 steel

C,	T_0,	$\sigma_{0.2}$,	σ_B,	δ,	ψ,	N	$\Delta l_{d_{\max}}$,	$\Delta P_{d_{\max}}$,	ΔW,	$\Delta T_{\max}$,
MN/m	K	MPa	MPa	%	%		mm	kN	J	K
12	4	1190	1640	35	43	26	0.42	3.3	4.1	185
	20	1180	1620	39	49	30	0.36	2.2	3.5	163
8.5	4	1190	1610	39	45	18	0.57	3.5	5.5	190
	20	1170	1580	38	51	20	0.53	3.0	5.1	169

Notation: $\sigma_{0.2}$ is the offset yield strength, δ is the elongation at rupture, ψ is the contraction at rupture.

To select an integrated criterion of discontinuous yielding, it is necessary to identify the parameters governing the process by analyzing their interrelations. The relationship among the thermal parameters follows from the temperature dependence of heat capacity (6), while the relationship among the mechanical parameters can be established from the strain-hardening law. With (2), the strain-hardening modulus is expressed as $d\sigma / d\varepsilon = \sigma m / \varepsilon$. These equations yield $\sigma_c = A\varepsilon_c^m$; $\varepsilon_c = (\sigma_c / A)^{1/m}$; $\theta_c = \sigma_c m / \varepsilon_c$.

In our case, however, the following strain-hardening model is the most adequate for a number of materials:

$$\sigma = \sigma_y + A\varepsilon^m \tag{10}$$

where σ_y is the yield strength (the values of A and m are different from those given above). It defines the stress range where discontinuous yielding occurs according to $\sigma_c \geq \sigma_y$. Then we have $\sigma_c = \sigma_y + A\varepsilon_c^m$, $\varepsilon_c = \left[(\sigma_c - \sigma_y)/A\right]^{1/m}$, $\theta_c = (\sigma_c - \sigma_y)m / \varepsilon_c$; i.e., these critical parameters can be determined by experimentally finding one of them at a given temperature, namely the discontinuous yield strength σ_d (Figure 1). Its true value $\bar{\sigma}_d$ corresponds to the critical stress σ_c. Since the technical aspect of the effect is of interest, to determine the parameter σ_d, it is necessary to isolate the region of regular discontinuous yielding by excluding the region of so-called quasihomogeneous plastic deformation (domain of uncertainty). It is within this region where microdiscontinuities occur accompanied by small jumps in temperature [7]. The boundary of the regular region can be found from the criterion $\Delta P_d = \Delta P_{\text{nor}}$ (where ΔP_{nor} is the significance level) for the first detected load drop. This criterion defines the condition in which deformation loses stability. For the practicable scales of tensile stress–strain curves of metallic materials, the level $\Delta P_{nor} = 0.01P_0$, where P_0 is the initial load, appears to be optimal.

As will be shown below, temperature may be strongly dependent on the strain rate and some other factors, such as the specimen radius, that govern the heat-transfer conditions. For the sake of simplicity, we assume that the strain rate is small and $T \approx T_0$ because heat transfer to liquid helium occurs very intensively, in the nucleate boiling regime.

Finally, only two parameters (mechanical and thermal) remain, and we select an integrated criterion based on stress and temperature: $\sigma / \sigma_c(T_c) \geq 1$.

The critical values of thermal and mechanical parameters are also interrelated: the point on the stress–strain curve at which serrations appear shifts to the right with increasing temperature; i.e., the higher the T_c and c_c, the higher the σ_c and ε_c and the lower the θ_c. Since θ abruptly drops at the early stages of deformation, the critical stress in high-strength alloys may be close to the offset yield strength, as is the case in steels and titanium alloys at a temperature of 4 K (Figure 1). If we select a stress-based criterion, then it is necessary to derive an analytic formula that relates σ_c and ε_c and describes a limiting curve on the σ, T -plane. This formula was deduced in [16, 17]:

$$\sigma_c \beta_c(\sigma_c) = \frac{\rho}{c_0}\left(k_e T_c + k_p T_c^3\right)\beta_0 \sigma_{c_0} \tag{11}$$

where $\sigma_{c_0} = \overline{\sigma}_d$, β_0 corresponds to σ_{c_0}, c_0 is the heat capacity of metal at temperature T_0, $\beta_c(\sigma_c)$ is the function showing how β varies with σ. It can be derived from (1) and (10):

$$\beta(\sigma) = \left[1 + k_\beta \sigma_b (\sigma_b - \sigma_y)^{\frac{1-m}{m}} / \sigma(\sigma - \sigma_y)^{\frac{1-m}{m}}\right]^{-1} \tag{12}$$

The function $\beta_c(\sigma_c)$ has a similar expression. The value β_0 can be found from (12) when $\sigma = \sigma_{c_0} = \overline{\sigma}_d$ and $\sigma_y = \sigma_{0.2}$.

Unlike Eq. (11), which can only be solved numerically, the inverse problem (given σ_c, find T_c) can be solved analytically. Therefore, it is convenient to use the integrated criterion $T/T_c(\sigma_c) \leq 1$ and the associated equality:

$$\rho(k_e T_c + k_p T_c^3)/(c_0 \beta_c \sigma_c) = 0 \tag{13}$$

here β_c is a coefficient calculated from formula (12) for given σ_c. Transforming (13) to the form

$$T_c^3 + pT_c + q = 0 \qquad (14)$$

we find its real positive root $T_c = \sqrt[3]{-q/2+\sqrt{Q}} + \sqrt[3]{-q/2-\sqrt{Q}}$, where $Q=(p/3)^3+(q/2)^2$, $p = k_e / k_p$, $q = -c_0\beta_c\sigma_c / \rho\beta_0\sigma_{c_0} k_p$.

Formulas (11) and (13) include only two material constants: heat capacity and discontinuous yield strength. On the σ, T-plane, they describe some curve that is the lower boundary of the instability domain of plastic deformation.

Let us validate formula (13) against test data for AMg5 aluminum alloy and 03Kh20N16AG6 steel. The following parameter values have been obtained in static tension tests on standard fivefold specimens with a gage diameter of 3 mm at T_0= 4 K and $\dot{\varepsilon}_n = 1.6\times10^{-3}$ s^{-1}: $\sigma_{0.2}$= 175 MPa, $\sigma_{\text{в}}$= 521 MPa, δ= 35.5%, ε_b= 0.304, σ_b= 706 MPa, σ_d=286 MPa, $\overline{\sigma}_d$= 422 MPa (AMg5) and $\sigma_{0.2}$= 1190 MPa, $\sigma_{\text{в}}$= 1640 MPa, δ = 35%, ε_b= 0.30, σ_b= 2210 MPa, σ_d=1200 MPa, $\overline{\sigma}_d$= 1206 MPa (03Kh20N16AG6). For the temperature dependence of the heat capacity of the AMg5 alloy in the range 0–15 K, we obtained the coefficients k_e=0.495 J/kg K^2 and $k_p = 1.06\times10^{-4}$ J/kg K^4 from available tabulated data, and ρ=2660 kg/m^3. Using the strain-hardening model (10) and formula (13) with $\sigma_c = \sigma_b$, we get $T_{c_{\max}}$ = 8.6 K. Using formula (7) for 03Kh20N16AG6 steel, we obtain ρ=7800 kg/m^3 and $T_{c_{\max}}$ = 27.2 K.

The tests conducted at T_0= 20 K are in agreement with calculations: no discontinuous yielding is observed in the AMg5, whereas the tensile stress–strain curve of the 03Kh20N16AG6 is almost entirely serrated.

The temperature dependence of the critical stress and the instability domains calculated with (13) are shown on the same scale in Figure 2, illustrating the effect of the strength of the material on discontinuous yielding. The curvature of the curve for AMg5 is small because $k_p T^3 << k_e T$ in the initial temperature range 0–12 K.

Thus, we have derived a discontinuous-yielding criterion that allows quite accurate calculation of the critical parameters, unlike other criteria used to make either qualitative or order-of-magnitude estimation.

Influence of the Loading Conditions and Design/Technology Factors on the Mechanical Properties and Characteristics of Alloys at a Temperature of 4 K

How discontinuous yielding is manifested depends in a complicated fashion on the elastic, inertial, and kinetic characteristics of the specimen–machine (part–structure) system and other factors such as loading conditions and design/technology features. In what follows, we will discuss experimental data on the effect of these factors on the plastic deformation, strength, and fracture of alloy specimens, considering that discontinuous deformation is a stage-wise process. In the general case, the stages are subcritical homogeneous deformation, loss of stability of deformation at the discontinuous yield point, a strain jump, and what occurs immediately after it.

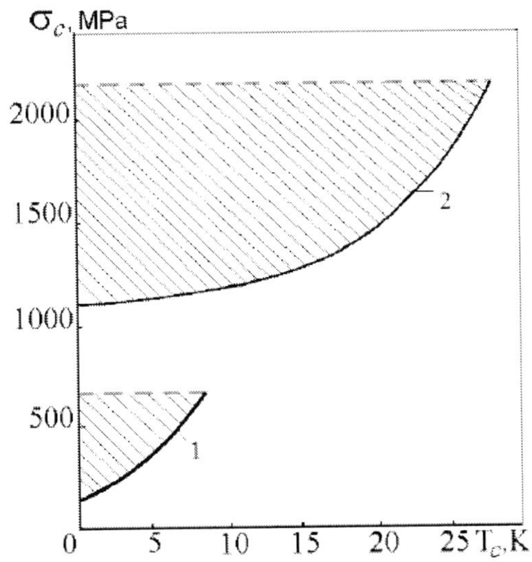

Figure 2. Temperature dependence of critical stress and instability domains for alloy AMg5 (1) and steel 03Kh20N16AG6 (2). The dashed lines represent the levels $\sigma_c = \sigma_b$; the instability domains are hatched.

Effect of the Rigidity of the Testing Machine

At the first stage of plastic deformation, which is subcritical and characterized by intensive strain hardening, the specimen–machine system is in equilibrium and its rigidity does not affect the deformation behavior of the material, as in the elastic case. Note that the stress–strain curves of the same alloy recorded in normal conditions on testing machines that differ in rigidity may not coincide only in the zone of sharp yield point and on the descending section when the hardening condition $dP/dt > 0$ fails.

Low-temperature discontinuous yielding is seen as serrations on the entire (or most of) the stress–strain curve.

Therefore, the effect of rigidity becomes very strong during a strain jump. In the process, elastic energy is converted into the work of plastic deformation and the kinetic energy of moving parts. The amount of elastic energy converted into the work of deformation can be estimated as

$$\Delta W = (P_0^2 - P_c^2)/2C \tag{15}$$

As is seen, this energy is inversely proportional to rigidity or directly proportional to compliance. The less the rigidity of the system, the more elastic energy is expended and the greater the elongation Δl_d and strain ε_d.

It is obvious that the greater the magnitude of each strain jump, the fewer there are such discontinuities because each time deformation localizes in new areas of the specimen. This is experimentally confirmed in [18]. Cylindrical specimens of steels 03Kh20N16AG6, 12Kh18N10T and titanium α-alloy 3M were tested in tension in liquid helium. The rigidity C_m of the testing machine and, hence, the rigidity C of the specimen–machine systems were changed with a special fixture [19]. It is a housed Belleville spring pack linked into the force transmission chain of the testing machine. By changing the number and arrangement of springs in the pack, it is possible to vary the rigidity of the testing machine from 14.5 to 2.0 MN/m.

A decrease in rigidity leads not only to an increase in the magnitude of elongation jumps, but also to a considerable decrease in the elongation at rupture δ and in the relative number of jumps N/δ, and practically does not affect the magnitude of the stress drop $\Delta\sigma_d$.

The parameters δ and N are related as $\delta \geq \sum_{i=1}^{N} \Delta l_{d_i} / L$, where L is the gage length of the specimen. Therefore, the increase in the magnitude of elongation jumps is offset by the decrease in the number of jumps, which is one of the causes of decrease in the elongation δ. As the rigidity is reduced from 12 to 5.7 MN/m, the value of δ for 03Kh20N16AG6 decreases by a factor of 1.4.

The other mechanical characteristics change as well. The contraction at rupture ψ increases from 43 to 48%, while the ultimate strength decreases a little. The mechanical characteristics of the 12Kh18N10T steel and 3M titanium alloy change in a similar manner. Note that a decrease in rigidity does not practically affect the yield strength of all the materials.

In testing machines of high rigidity, the rupture of steel and alloy specimens occurs within a next elastic region, after the reserve of ductility has been completely exhausted. For example, the 03Kh20N16AG6 steel ruptured in this way when the rigidity of the testing machine was 12 or 8.5 MN/m. If a material exhibits linear elastic behavior up to fracture, then it is quasibrittle fracture. After the rigidity is reduced to 5.7 MN/m, this steel fails by adiabatic shear, i.e., during a strain jump. This is another cause of the decrease in the elongation with rigidity. Such failure may occur when the material is weakly strain-hardened and stores a great amount of elastic energy characterized by critical rigidity at given critical stress.

The adiabatic shear failure of the 12Kh18N10T steel occurs at the final stage of deformation even at $C = 12$ MN/m because σ_0 is large and θ is small.

Thus, the rigidity of the specimen–machine (part–structure) system is one of the basic factors related to the static loading conditions. This factor determines the stored elastic energy that is periodically converted into the work of discontinuous deformation, thereby exerting a considerable effect on the discontinuous yielding, mechanical characteristics, and type of failure of materials.

Since, however, the stored elastic energy at the beginning of a discontinuity, $W = \sigma_0^2 F^2 / 2C$, is strongly dependent on the dimensions of the specimen, the analysis of the effect of rigidity will be more complete if combined with the scale effect. The standards on testing materials at cryogenic temperatures should be improved to include the rigidity of testing machines as a controlled parameter.

Effect of Strain Rate

Let us consider the case where discontinuous yielding sets in after the stage of homogeneous plastic deformation [20]. How much the specimen is heated depends on the heating rate, which, in turn, depends on the rated strain rate $\dot{\varepsilon} = \dot{\varepsilon}_n$. The energy balance at this stage is described by the equation

$$\beta V_p \sigma \, d\varepsilon = V_p c(T) dT + qS(T - T_0) dt, \tag{16}$$

where q is the coefficient of heat transfer to the environment; V is the gage volume of the specimen; S is the gage surface area of the specimen. Their ratio $\frac{S}{V} = \frac{2\sqrt{1+\varepsilon}}{R_0}$ changes insignificantly with increasing strain. After thermal equilibrium has been achieved (the heating rate has become equal to the heat-transfer rate), the temperature $T = T_0 + \Delta T_h$ of the specimen no longer rises, i.e., $dT / dt = 0$, where ΔT_h is the temperature jump during homogeneous deformation. Using this condition and Eq. (16), we obtain the rated strain rate:

$$\dot{\varepsilon} = \Delta T_h q S / V \beta \sigma. \tag{17}$$

If $T_0 + \Delta T_h$ is equal to the limit temperature $T_{c_{\max}}$ for discontinuous yielding, then $\dot{\varepsilon} = \dot{\varepsilon}_{ch} = \dot{\varepsilon}_{c_2}$. It is the upper or second critical strain rate. Setting $T_{c_{\max}}$ = 20–30 K, q = 200 W/m^2·K (film boiling), $\sigma = \sigma_{0.2}$, and β =0.9 (which is close to maximum), we obtain the following values of $\dot{\varepsilon}_{ch}$ for Ø3-mm specimens of steels 03Kh20N16AG6, 12Kh18N10T, and titanium alloy 3M, respectively: $(4.0-6.5)\times10^{-3}$; $(4.8-7.7)\times10^{-3}$; $(4.6-7.5)\times10^{-3}$ s^{-1}.

The experimental stress–strain curves of these materials have sections with no serrations when the strain rate is equal to $(5-6)\times10^{-3}$ s^{-1}, which is in agreement with calculation. Similar experimental results were obtained for 12Kh18N10T steel at a temperature of 4 K in [7], where low-strength aluminum alloy AMg6 was also examined, for which $T_{c_{\max}}$ = 10 K. The deformation of this alloy becomes almost macrohomogeneous at a strain rate of 28×10^{-3} s^{-1}, while calculation gives $\dot{\varepsilon}_{ch} \approx 22\times10^{-3}$ s^{-1}. Thus, calculated results and experimental data for various materials are in good agreement.

Increasing the strain rate to $\dot{\varepsilon}_{ch}$ and higher qualitatively changes the deformation behavior: from localized discontinuous to homogeneous. This changes substantially the mechanical characteristics. As the strain rate is increased, the ultimate strength of steels 03Kh20N16AG6, 12Kh18N10T, and titanium alloy 3M decreases by 22, 19, and 18%, respectively; the elongation decreases by factors of 1.6, 1.2, and 1.9; and the contraction increases by 1.2, 1.2, and 1.1. The offset yield strength practically does not change. The ultimate strength of the AMg6 alloy is less sensitive to the strain rate: σ_{B} decreases by 4%, while the elongation does not decrease, but increases by a factor of 1.2. Such a small decrease in the ultimate strength of the alloy is due to weak thermal softening: the thermal-softening coefficient of this material $k_T = 1.5$ MPa/K, which is approximately one-third that for high-strength steels.

In [21], the discontinuous-yielding and mechanical characteristics of VT5-1 and OT4 titanium alloys were also substantially changed by increasing the strain rate from 0.28×10^{-3} to 1.4×10^{-2} s^{-1}. For example, the ultimate strength of the latter alloy decreased by 100 MPa.

Discontinuous yielding does not occur and the mechanical characteristics change because of the rapid heating of metals with low heat capacity at cryogenic temperatures, i.e., the adiabatic criterion fails. For example, during the deformation of 12Kh18N10T steel at a strain rate of 3×10^{-2} s^{-1}, its temperature increases by 100 K, according to (17).

This is much greater (approximately by the ratio of yield strengths) than $\Delta T_h = 25$ K achieved in AMg6 alloy at the same strain rate. These results are in agreement with the experimental data for type 304 chromium-nickel steel [22]: a Ø5.6 mm specimen is heated up to 120 K at a strain rate of 2.5×10^{-2} s^{-1}. Thus, the kinetic criterion

$$\dot{\varepsilon}/\dot{\varepsilon}_{ch} \geq 1 \tag{18}$$

limits the range of strain rates at which low-temperature discontinuous yielding does not occur in metals. Since a strain jump is rather quick, the effect of the testing machine on the specimen–machine system is insignificant at low strain rates. After the completion of the strain jump, however, the effect of external loading becomes decisive because further deformation is governed by the ratio between the rate of hardening of the cooled material and the rate of geometrical softening in the neck.

According to [21, 23], if the strain rate is increased to $\sim 1 \times 10^{-2}$ s^{-1}, then after the usual abrupt drop in the load, it does not increase, but continues to slowly decrease. This is because beginning from some critical strain rate $\dot{\varepsilon}_{cf}$, thermal relaxation after a strain jump has no time to be completed, and the external loading continues to increase the plastic deformation within the neck until rupture.

This scenario becomes possible when the rate of geometrical softening exceeds the rate of strain hardening and hardening due to cooling. For the corresponding strain rate $\dot{\varepsilon} = \dot{\varepsilon}_{cf} = \dot{\varepsilon}_{c_3}$, we have the following kinetic failure criterion:

$$\dot{\varepsilon} / \dot{\varepsilon}_{cf} \geq 1 \quad (19)$$

Such a limiting state is described by the following equation [12]:

$$\sigma_{ld} dF / Fdt = \theta d\varepsilon / dt + k_{\text{T}} dT / dt. \quad (20)$$

Since $dF / F = d\varepsilon$, Eq. (20) yields

$$\dot{\varepsilon}_{cf} = k_{\text{T}} \nu / (\sigma_{ld} - \theta), \quad (21)$$

where ν is the cooling rate. This equation is valid only if $\sigma_{ld} > \theta$, which is similar to the necking condition in the absence of discontinuous yielding. The inequality

$$\sigma_{ld} / \theta > 1 \quad (22)$$

is an additional failure criterion. As shown in [20, 23] for 03Kh20N16AG6 steel and titanium alloy, criterion (22) works when the specimen is prestrained.

The critical strain rate depends on the strain hardenability of the material. In typical cases, $\dot{\varepsilon}_{cf}$ is on the order of $10^{-2} - 10^{-1}$ s^{-1} [21, 23].

Effect of Prestrain

One of the factors that strongly affect the discontinuous yielding of metals is plastic prestrain. The effect of this factor on austenitic steels 03Kh20N16AG6, 03Kh13AG19, 12Kh18N10T and titanium alloy 3M was studied in [20, 23, 24]. Standard fivefold specimens with a gage diameter of 3 mm prestretched to a residual strain of 5, 10, or 20% were tested. The 03Kh20N16AG6 steel was also prestrained to 30%, while the 12Kh18N10T steel and 3M titanium alloy to 2.5% (the maximum value of ε_p for the latter was 10% due to relatively low ductility). The specimens were prestrained at a temperature of 293 K and tested in tension at temperatures of 77 and 4 K. The rigidity of the testing machine was 14.5 MN/m; the strain rate was 1.6×10^{-3} s^{-1}. Additional tests were also conducted for other values of these quantities.

In prestrained specimens, discontinuous yielding immediately follows the elastic region, and even the first serration on the tensile stress–strain curve has a high amplitude. In undeformed specimens, however, regular discontinuous yielding sets in after the small stage of homogeneous plastic deformation. It is obvious that the effect of prestrain on the beginning of discontinuous yielding is associated with the decreased initial value of the strain-hardening modulus. The magnitudes of the first elongation jump Δl_d and load drop ΔP_d monotonically increase with the prestrain ε_p.

Also, the higher the prestrain, the greater the ultimate and yield strengths and the lower the elongation and contraction. As the prestrain of 03Kh20N16AG6 steel is increased to $\varepsilon_p =$ 30%, these parameters change as follows: by 19 and 55% and by factors of 2.6 and 1.4. The number of discontinuities decreases by 48%. Prestraining to $\varepsilon_p \geq 10$ % causes an increase in the total strain $\delta_\Sigma = \varepsilon_p + \delta$ to 10%, irrespective of the level of ε_p.

The tensile stress–strain curves of 12Kh18N10T steel, which undergoes martensitic transformations during deformation and cooling, shows a considerably different behavior: as temperature is decreased, there appears a well-defined yield plateau; at a temperature of 4 K, the stress and strain discontinuities within this region do not change in amplitude. The yield strength increases substantially only if $\varepsilon_p \geq 10\%$, the elongation decreasing insignificantly.

The 03Kh13AG19 steel is unstable too, but its stress–strain curve does not have a yield plateau and its mechanical and discontinuous-yielding characteristics vary with ε_p in a way similar to the 03Kh20N16AG6 steel.

Usual changes in the mechanical characteristics with an increase in prestrain, i.e., increase in strength (insignificant for σ_B) and decrease in ductility, are also observed in the titanium alloy 3M.

The influence of various types of mechanical treatment on the properties of alloys was studied in [24]. Tests on 03Kh20N16AG6 steel specimens prestrained by twisting through an angle of 90 rad/m and by surface cold working revealed the following. In the former case, discontinuous yielding sets in immediately after the elastic stage and the changes in strength and ductility are similar to those in the specimens with $\varepsilon_p = 10\%$, but the decrease in the number of discontinuities is much greater (by 47%). In the latter case, the yield strength increases by 13%, δ and ψ decrease by 17% both, and the absolute and relative number of discontinuities decrease by 26% and 9%. As in the original specimens, there is the stage of homogeneous deformation before regular discontinuous yielding. This makes fast failure at high strain rates impossible.

As indicated above, the number of discontinuities also depends on the rigidity of the testing machine. To study the combined effect of the rigidity of the testing machine and the prestrain of the specimen on its mechanical and discontinuous-yielding characteristics, steel 03Kh20N16AG6 ($\varepsilon_p = 10\%$) was tested against two different values of rigidity: 12 and 5.7 MN/m. The reduction in rigidity did not change the yield strength, slightly decreased the ultimate strength, and increased the true stress at rupture by 16%. The ductility characteristics changed considerably: the elongation decreased by 22%, while the contraction increased by a factor of 1.15. The number of strain jumps halved, while the ratio N / δ increased by a factor of 1.5. These results are in qualitative agreement with those obtained earlier for undeformed specimens tested in a machine with a rigidity of 12 and 8.5 MN/m (Table 2). At lower rigidity of the testing machine, however, the prestrain caused much stronger changes in the mechanical characteristics of steel than those made by the rigidity alone, which is due to a change in strain hardening, the number of discontinuities remaining proportional to the rigidity.

Experiments show that prestraining in combination with cooling to a temperature of 4 K greatly promotes hardening, so that the offset yield strength reaches 3.4–3.7 or 240–270 % (03Kh20N16AG6), depending on the type of mechanical treatment and material. The other tested metals were hardened to a lesser degree; the maximum ratio $k_g = \sigma_{0.2}^{\prime T} / \sigma_{0.2}^{293}$ at $T = 4$ K (the prime refers to a specimen prestrained at given temperature T) was reached after prestretching (Table 3).

This coefficient allows assessing the degree of hardening of a material after either mechanical treatment or cooling. The ratio $\sigma_{0.2}^{\prime T} / \sigma_{0.2}^{\prime 293K}$ represents the contribution of low-temperature hardening to the total hardening k_g. The maximum temperature-dependent fraction of $\sigma_{0.2}^{\prime T} / \sigma_{0.2}^{\prime 293K}$ is reached in all the materials, except for the steel 12Kh18N10T, after precompression. After precompression, the specimens are hardened differently than after other types of mechanical treatment because of the Bauschinger effect. The maximum decrease in the yield strength was observed at a temperature of 77 K. For example, this decrease was greater than 300 MPa in 12Kh18N10T steel. However, the difference $\left(\sigma_{0.2}^{\prime T} - \sigma_{0.2}^{T}\right)$ becomes positive at T = 4 K in all the materials, except for the steel 12Kh18N10T, i.e., the Bauschinger effect is not manifested at this temperature and the material is hardened instead of being softened. This increase in resistance to initial plastic deformation is obviously due to the onset of discontinuous yielding.

Besides yield strength, the mechanical treatment also changes other mechanical characteristics of the specimens. The ultimate strength is affected only by prestretching. At ε_p = 10%, for instance, σ_B is increased by 14% in the steel 03Kh20N16AG6 and by 17% in the steel 03Kh13AG19. In this latter steel, σ_B is also increased by 20% after pretwisting. In the steel 12Kh18N10T, σ_B starts increasing at $\varepsilon_p >$ 10%, unlike the alloy 3M where the change of this characteristic is insignificant.

Mechanical treatment, especially prestretching reduces the ductility of metals. At T =4 K and ε_p = 10%, for instance, the elongation δ of the 3M titanium alloy and 03Kh13AG19, 03Kh20H16AG6, and 12Kh18N10T steels is decreased by factors of 1.80, 1.44, 1.20, and 1.06, respectively.

It should be noted that the deformability of the 03Kh20N16AG6 and 12Kh18N10T steels is reduced insignificantly. The elongation δ of the latter material starts decreasing only at $\varepsilon_p \geq 10$ %. This is accompanied by a decrease in the number of strain jumps in all the materials, whereas the relative number N / δ of such discontinuities either increases (03Kh20N16AG6 and 12Kh18N10T) or slightly decreases (the other two metals).

Thus, the 03Kh20N16AG6 and 12Kh18N10T steels are structural materials that possess a large resource of ductility, either in the initial state or after various types of prestraining.

Table 3. Relative increase in offset yield strength, $\sigma_{0.2}^{\prime T} / \sigma_{0.2}^{\prime 293K}$, after prestraining and cooling to 4 K

Prestraining method	Steel 03Kh13AG19			Steel 12Kh18N10T			Alloy 3M		
	293 K	77 K	4.2 K	293 K	77 K	4.2 K	293 K	77 K	4.2 K
Twisting	1.56	2.62	3.08	1.05	1.24	1.56	1.03	1.56	1.82
Surface cold working	1.28	2.26	2.70	1.00	1.15	1.44	1.01	1.51	1.80
Compression	0.77	1.40	2.41	0.87	0.92	1.16	0.95	1.27	1.75
Tension	1.64	2.50	3.21	1.21	1.49	1.60	1.18	1.73	2.04
Initial state	1	1.89	2.20	1	1.15	1.29	1	1.51	1.68

Effect of the Scale Factor

To study the scale effect in liquid helium [25], we selected the following materials: austenitic chromium–nickel steel 12Kh18N10T, aluminum alloy AMg6, titanium α-alloys VT5-1 and OT4. The cylindrical specimens of 12Kh18N10T steel and AMg6 alloy had diameters of 2, 4, and 6 mm and gage length $L_0 = 30$ mm. The titanium alloy specimens were cut from a 2-mm sheet in the rolling direction and had a gage width of 2, 5, 10, and 17 mm and gage length $L_0 = 30$ mm. The specimens were tested in tension at a strain rate of 2.8×10^{-3} s^{-1}. The test data show that increasing the dimensions of the specimens causes a significant change in the discontinuous-yielding and mechanical characteristics of the majority of materials.

As the diameter of the 12Kh18N10T steel specimens is increased, the stress–strain curve spreads widely, the elongation increases by a factor of 1.5 (Table 4), the number of strain jumps decreases substantially, and the magnitudes of elongation (Δl_d) and strain (ε_d) jumps increase.

In low-strength aluminum alloy AMg6, however, the scale effect is practically absent. Titanium alloys are similar in strength, but the OT4 (Table 5) is more ductile. With increase in the cross-sectional area to 35 mm^2, the yield strengths of both materials abruptly decrease and so do the maximum stress drops $\Delta\sigma_{d_{\max}}$, serrations alternating with regions of homogeneous deformation.

The specimens of less ductile alloy VT5-1 with $F_0 = 20\ \text{mm}^2$ fail during the first discontinuity preceded by monotonic plastic deformation.

Titanium alloys differ from steel and aluminum alloy by lattice, their specimens differ by geometry, and there is no geometrical similarity.

As follows from (15), the energy of a strain jump is determined by the strength of a material (starting stress), the rigidity of the specimen–machine system, and the cross-sectional area of the specimen. Low-temperature experiments use specimens of various sizes, including microsize. Testing machines may differ in rigidity by an order of magnitude. To make the test conditions energy equivalent, the following dimensionless criterion is used [26]:

Table 4. Discontinuous-yielding and mechanical characteristics of specimens of various sizes at a temperature of 4 K

Material	$2R_0$, mm	σ_B, MPa	$\sigma_{0.2}$, MPa	δ, %	ψ, %	σ_d, MPa	$\Delta\sigma_{d_{max}}$, MPa	N	N/δ
12Kh18N10T	2	1792	567	31.0	49.6	779	492	113	365
	4	1732	714	32.1	41.7	790	510	59	184
	6	1756	723	46.1	41.9	748	677	47	102
AMg6	2	562	192	24.7	32.0	191	37	267	1080
	4	530	198	20.6	15.4	193	42	214	1040
	6	556	195	28.5	16.4	193	47	228	800

Table 5. Discontinuous-yielding and mechanical characteristics of titanium-alloy specimens of various sizes at a temperature of 4 K

Material	F_0, mm^2	σ_B, MPa	$\sigma_{0.2}$, MPa	δ, %	ψ, %	σ_d, MPa	$\Delta\sigma_{d_{max}}$, MPa	N	N/δ
OT4	4	1455	1341	5.8	22.9	1341	312	11	190
	10	1443	1353	11.0	17.8	1353	304	11	100
	20	1429	1346	8.0	13.4	1346	72	5	63
	35	1426	996	4.1	11.5	1087	35	12	293
VT5-1	4	1460	1405	6.2	4.0	1920	320	7	113
	10	1420	1395	4.9	4.1	1910	210	5	102
	20	1400	1390	4.2	0.7	1910	0	0	0
	35	1420	960	4.0	3.2	1720	20	8	250

$$\chi = \sigma_0^2 R_0 / C\theta$$

The energy similarity of discontinuous yielding of specimens of various materials and sizes is provided at $\chi = \text{const}$ by varying the rigidity of the testing machine

$$C_m = R\left(\frac{\chi\theta_m}{\sigma_m^2} - \frac{1}{k_{01}}\right)^{-1}$$

where σ_m and θ_m are the average starting stress and the average strain-hardening modulus; $k_{01} = 0.1\pi E$ is the rigidity coefficient of a proportional cylindrical fivefold specimen. The value of χ is determined by specifying some parameters characterizing machine rigidity, material, and specimen.

Effect of Stress Concentration

Stress concentration is one of the major factors defining the strength and lifetime of cryogenic equipment. In [27], 03X20Kh16AG6 steel specimens were tested in tension at temperatures of 293, 77, and 4 K. The specimens were 5 mm in diameter and had an annular groove with a depth of 1 mm and corner radius ρ = 2.0, 1.5, 0.5, 0.25, 0.1 mm. The corresponding theoretical stress concentration factor K_t is equal to 1.45, 1.54, 2.10, 2.75, and 3.60. The specimen–machine system had a rigidity of 12 MN/m. The notched specimens were tested to determine their ultimate strength $\sigma_В^n$, true stress at rupture S^n, and contraction ψ^n. The discontinuous yield strength σ_d^n was determined at a temperature of 4 K. It can be measured because the material jumps into plastic state at the stage of elastic deformation. The tests show that as K_t is increased to 2.1 at temperatures of 293 and 77 K, $\sigma_В^n$ increases and then remains almost constant, ψ^n monotonically decreases, and S^n decreases from 2030 to 1650 MPa at 293 K and from 2920 to 2200 MPa at 77 K. At 4 K, however, the curve of variation in these quantities with K_t is nonmonotonic and has sharp bends. These changes occur when K_t changes from 1.54 to 1.45 and from 2.75 to 3.60 (ρ changes from 2.0 to 1.5 mm and from 0.25 to 0.1 mm, respectively), which is due to discontinuous yielding. Also, the number of strain jumps changes from 2 to 1 and from 1 to 0.

Whether the effect is manifested apparently depends not only on K_t, but also on ρ. The higher the stress rigidity, the lower the amplitude of discontinuities, and at ρ=0.1 mm discontinuous yielding does not occur, leading to quasibrittle fracture. Besides strongly constrained deformation, the absence of discontinuous yielding can also be attributed to non-adiabaticity, which is because the deformed length of the specimen is small, commensurable with ρ. In this case, volumetric heat transfer becomes highly intensive, thus suppressing local temperature jumps in the minimum cross-section.

It is obvious that strength increases with N at the expense of considerable strain hardening near the stress concentrator during nonisothermal discontinuous yielding. It is the sharp rise in temperature during a discontinuity that has a softening effect when deformation is constrained, which is manifested as a substantial contraction with increase in ρ. Such hardening results from a strain jump, when $\sigma_d^n < S^n$, unless the specimen is ruptured during it. Rupture occurs during the first discontinuity that immediately follows the elastic stage if there is a supercritical store of elastic energy, and rupture occurs after the discontinuity if the strain (or loading) rate is supercritical. Premature failure results in a decrease of $\sigma_В^n$ to σ_d^n. Since the elastic energy W at a given load is determined by the rigidity C, the inverse effect of increase in the number of strain jumps and, hence, increase in $\sigma_В^n$ and S^n can be achieved by increasing C. This effect of rigidity should be taken into account in standard mechanical tests at given temperature.

Also, specimens with a crack as a very sharp stress concentrator were tested. An annular fatigue crack with a depth of ~0.7 mm was made in specimens with a diameter of 6 mm. At a temperature of 4 K, the specimens failed in the linear region of the tensile stress–strain curve. In this case, the rated critical stress σ_{C0}, corresponding to σ_B^n and S^n, was 2510 MPa; i.e., further increase in the stress concentration no longer changes these characteristics.

The critical stress intensity factor K_c for a specimen of given diameter was 92.3 $\text{MPa}\cdot\sqrt{\text{m}}$. Unlike an axisymmetric crack, the fracture of a compact specimen with an edge crack subject to eccentric tension is preceded by the abrupt formation of a plastic zone by the discontinuous-yielding mechanism. This circumstance suggests that only one type of specimens should be used to measure the crack resistance of materials.

Qualitatively similar results on the influence of the inhomogeneous stress state caused by a stress concentrator on strength were also obtained in [28] for titanium alloy 3M and aluminum alloy AMg5.

The effect of a set of notches in the form of a metric screw thread on the discontinuous yielding, strength, and ductility of steel 03Kh20N16AG6 was studied in [27]. To this end, M3 and M3×0.35 (normal and fine) threads were cut on the entire gage length (15 mm) of the specimens. The strength of such specimens determined using the effective thread diameter is insignificantly different from the strength of unnotched specimens, and the characteristic corresponding to σ_B^n exceeds σ_B by 18% for both threads at a temperature of 4 K. The elongation δ' at this temperature is much shorter than that in unnotched specimens. The ratios of the number of strain jumps N'/N in the threaded and smooth specimens are 0.60 and 0.83 for the normal and fine threads, respectively.

Effect of the Shape of Specimens

It is of interest how the strength and deformability of metals depend on the shape of specimens in connection with the effect of discontinuous yielding.

Conical and tubular specimens of austenitic steels 03Kh20N16AG6, 12Kh18N10T, and titanium alloy 3M were studied in [29]. These were standard fivefold specimens with a gage diameter of 3 mm. The conical specimens had the angle $\alpha = 1°$ between the generatrix and the longitudinal *z*-axis, a major diameter of 3 mm, and a gage length of 15 mm. The tubular specimens had an outside gage diameter of 3, 5.5, and 6 mm and the ratio of wall thickness to radius Δ/R ranging from 0.09 to 0.5. The uniaxial-tension tests were conducted at temperatures of 293, 77, and 4 K and a strain rate of 1.6×10^{-3} s^{-1}. The tests data for standard and nonstandard specimens were then compared. Since the stress state of the latter was linear (or nearly linear) during loading, characteristics formally corresponding to the standard yield strength (only for the tubular specimens), ultimate strength, and elongation were determined.

The tests on the conical specimens show that with decrease in temperature, the ultimate strength does not change, while the elongation decreases in the 03Kh20N16AG6 steel, remains constant in the titanium alloy, and nonmonotonically increases the 12Kh18N10T steel. This pattern is qualitatively similar to the variation in the elongation of the original (cylindrical) specimens of the same materials. The ratio δ'/δ decreases in the

03Kh20N16AG6 steel and 3M alloy and increases in the 12Kh18N10T steel. At cryogenic temperatures, the stress–strain curve of the latter has a yield plateau, which becomes serrated in liquid helium and sloping in the conical specimens.

It is longer by 80% than the yield plateau of the cylindrical specimens. Note that pretwisted cylindrical specimens at a temperature of 4 K have a similar tensile stress–strain curve [24].

If the stress–strain curve can be described by Eq. (2), then the elongation before necking can be expressed as

$$\delta' = \frac{1}{L} \cdot \frac{m[(1+kL)^{1-1/m} - 1]}{k(1-1/m)} \tag{23}$$

where $k = 2\text{tg}\alpha / R_0$.

As is seen from formula (23), the elongation of conical specimens depends on the hardening index m, the initial cross-sectional radius, and the gage length. If there is no hardening ($m = 0$ in (23)), then $\delta' = 0$. Also, $\delta' \to m$ as $R_0 \to \infty$ or $L \to 0$; $\delta' \to 0$ as $R_0 \to 0$ or $L \to \infty$; $\delta' \to 0$ as $\text{tg}\alpha \to \infty$ and $\delta' = \varepsilon_b = m$ as $\alpha \to 0$ (cylindrical specimen). Thus, δ' may vary from 0 to m.

The power law (2) can only be used to describe the stress–strain curve of 03Kh20N16AG6 steel. For this material, the difference between the calculated and experimental values of δ' is insignificant at $T \geq 77$ K and is 40% at 4 K, which is due to change of the deformation mechanism. Additional tests on conical specimens of 03Kh20N16AG6 steel show that a slight increase in the cone angle (to α =2°) leads to a substantial decrease in δ' and to an increase in the stress at which jerky flow sets in.

The elongation of the tubular specimens decreased substantially at all temperatures; the less the ratio Δ/R, the greater the decrease.

The steel 03Kh20N16AG6 and titanium alloy 3M demonstrated the sharpest reduction in deformability at a temperature of 4 K. Contrastingly, the 12Kh18N10T steel became more ductile upon cooling.

The steel 12Kh18N10T and alloy 3M showed only minor alterations of the yield strength, whereas the 03Kh20N16AG6 steel specimen became much stronger with a decrease in the wall thickness and in temperature: the difference between the offset yield strengths of the tubular ($\sigma'_{0.2}$) and solid ($\sigma_{0.2}$) specimens reached 440 MPa. The ultimate strength of all the materials remained almost constant, unlike the solid specimens.

It is obvious that the change in deformation and some strength characteristics of the specimens of this shape is due to the change in the stress–strain state after necking, especially at $\Delta/R << 1$. While the solid cylindrical or plate specimens showed the early stage of quasihomogeneous deformation, the thin-walled tubular specimens underwent quite large strain jumps beginning immediately after the elastic stage, as in prestrained specimens.

As mentioned above, each strain jump in the solid cylindrical specimens was accompanied by the occurrence of a neck. The same was observed in the tubular specimens with $\Delta/R = 0.33$, 0.5. The thin-walled tubular specimen displayed spiral shear bands (necks) that did not reached the free lateral surface because it was absent. Tubular 03Kh20N16AG6 steel specimens with $\Delta/R = 0.17$ and a thin longitudinal notch were tested as well.

Due to the occurrence of several skew necks, they deformed easier and their elongation increased by 20%. The number of strain jumps increased by more than 50% and the value of the difference $\sigma_{0.2} - \sigma'_{0.2}$ (hardening) decreased by 15%. Thus, the deformability and load-carrying capacity of even simple rods become much more sensitive to geometrical factors during discontinuous yielding. The above discussion raises the problem of determining the mechanical characteristics of metallic materials at temperatures below 30 K. The multiple necking of cylindrical and plate metal specimens in liquid helium was studied in [30, 31] to select the optimal type of specimens for standard tests. In this case, the following ratio (stress "rigidity" within the neck) may be used as a criterion: $k_r = \sigma_m/\sigma_i$, where $\sigma_m = (\sigma_1 + \sigma_2 + \sigma_3)/3$; σ_i is the stress intensity; σ_1, σ_2, σ_3 are the principal stresses. It was shown that the stress state of plates, unlike cylinders, deviates from being linear by more than 50%. This changes substantially the mechanical characteristics. Tests on specimens of steels 03Kh20N16AG6, 12Kh18N10T and titanium alloys 3M and VT5-1 at $T = 4$ K have revealed a 10% difference in ultimate strength and a 60% difference in elongation. However, this difference is insignificant at temperatures of 293 and 77 K at which discontinuous yielding does not occur. It is thus recommended to give preference to cylindrical specimens to obtain comparable test data.

MODELING DISCONTINUOUS DEFORMATION

At set initial temperature, discontinuous deformation is manifested as a property of the specimen–machine (part–structure) system depending on some of its characteristics, which can be divided into three groups: (i) material characteristics, (ii) characteristics of the system and the loading conditions, and (iii) design and technology factors, including the scale factor. Up to now, the dependence of the effect on many factors has been studied only fragmentarily, in narrow ranges. The mathematical simulation of discontinuous deformation may be an effective way to considerably reduce sophisticated and expensive multiple-factor experiments.

In [32], a method of impulsive additional loading of a statically preloaded specimen at temperatures of 293 and 77 K was proposed and used to establish, theoretically and experimentally, the possibility of generating and studying effects that are similar to low-temperature discontinuous yielding, associated with design/technology factors, and observed only at cryogenic temperatures. It was shown for 03Kh20N16AG6 steel, as an example, that with increase in the rate of loading, the prestrain of the specimen, or the stored elastic energy of the testing machine, discontinuous deformation causes a sharp reduction in resistance to deformation.

The discontinuous yielding of metals can be studied in detail over the widest possible ranges of these factors by way of mathematical simulation. We use a local one-dimensional nonlinear multiparameter model formulated as a differential equation describing the dynamic equilibrium of the specimen–machine system during a strain jump [33]:

$$\ddot{\varepsilon}_d(1+\varepsilon_d)+a\dot{\varepsilon}_d+b\varepsilon_d(1+\varepsilon_d)+c\varepsilon_d^m=p\sin\omega t \tag{24}$$

where ε_d is the longitudinal discontinuous strain; t is time; a,b,c,p,m,ω are constants,

$$\varepsilon_d \geq 0;\ \dot{\varepsilon}_d \geq 0;\ \varepsilon_d(0)=\dot{\varepsilon}_d(0)=0;\ \varepsilon_d(\tau_D)=\varepsilon_D;\ \dot{\varepsilon}_d(\tau_D)=0, \tag{25}$$

where τ_D is the duration of a discontinuity; the factor $(1+\varepsilon_d)$ corrects for the decrease in the cross-sectional area of the specimen in the yield region. This equation has been derived for an unloaded system in which deformation is due only to the stored elastic energy. The discontinuity sets in once the critical stress σ_0 has been reached and the load has been switched off. Its driving power is the excessive stress in the region of catastrophic flow, resulting from abrupt and intensive adiabatic heating and thermal softening of the material. It is hypothesized that the temperature jump ΔT_m is proportional to the starting stress σ_0 with inelastic heat fraction as a coefficient and that the length l of the yield region is equal to the cross-sectional radius of the cylindrical specimen and is taken to be its gage length. Strain hardening is determined by the envelope of the peaks on the tensile stress–strain curve, and the hardening characteristics for a given discontinuity are generally functions of the starting stress.

Formula (24) includes a material model. It is an equation of mechanical state that provides a simplified description of deformation resistance: thermal softening is balanced by strain hardening and strain-rate hardening:

$$\sigma=\sigma_y(T)+A\varepsilon_d^m+\mu\dot{\varepsilon}_d \tag{26}$$

A and m are the hardening characteristics; μ is the viscosity coefficient; $\sigma_y(T)=\sigma_0-k_T\Delta T_m\sin\omega t$, ω is the circular frequency, k_T is the softening coefficient, and $\sin\omega t$ is a function increasing during the first quarter of the period and describing the process of heating with time $0\leq t\leq\tau_D$. The first two terms in (26) represent the original equation that fits the envelope of the peaks of the tensile stress–strain curve for the chosen materials:

$$\sigma=\sigma_{0.2}+A_i\varepsilon_t^{m_i} \tag{27}$$

where A_i and m_i are the initial values of the hardening characteristics; ε_t is the total strain.

The coefficients of Eq. (24) are given by $a=\mu/u$, $b=K/u=C/M$, $c=A/u$, $p=k_T\Delta T_m/u$, $K=Cl/F$, $u=Ml/F$ (F is the cross-sectional area of the specimen). The parameters of the model are the quantities appearing in these coefficients, which are the mechanical characteristics of the metal and the dynamic characteristics of the testing machine. The former group includes the parameters σ_0, A, m, μ. The latter group includes the movable added mass M and the rigidity C of the specimen–machine system, which represent, respectively, the inertial and positional forces in Eq. (24). The parameter F can be found using the equation $R=R_0/\sqrt{1+\varepsilon_t}$.

We study 03Kh20N16AG6 stable austenitic steel and AMg5 aluminum alloy. Such materials are widely used in the cryogenic technology. The considerable difference of their strength allows identifying the role of this important factor. The initial temperature T_0 for the process of interest is 4 K, which corresponds to the boiling point of liquid helium used to cool superconducting magnetic systems.

To identify the model, it is necessary to determine the mechanical characteristics of the materials and some parameters of discontinuous deformation. To this end, we conducted tension tests on cylindrical specimens with gage diameter $2R_0=3$ mm and gage length $L_0=15$ mm in liquid helium.

The rated strain rate was $1{,}6\times10^{-3}$ s^{-1}, the added mass was 300 kg, the rigidity of the force transmission chain of the machine was 14.5 MN/m, and the rigidity of the specimen–machine system was 12.0 MN/m for steel and 8.14 MN/m for aluminum alloy. The test data are summarized in Tables 1 and 2, and the values of the parameters and other constants in Table 6.

The mechanical parameters are determined from $m=\sigma_b\varepsilon_b/(\sigma_b-\sigma_0)$ and $A=(\sigma_b-\sigma_0)/\varepsilon_b^m$, where σ_b and ε_b are the true stress and the true strain corresponding to the maximum load, and $\varepsilon_b=\varepsilon_{b_i}-\varepsilon_t$, $\sigma_b=\sigma_{b_i}=const$. The quantity σ_0 can be found from (27) for given ε_t. To calculate m_i and A_i, it is necessary to substitute $\sigma_{0.2}$ for σ_0, and ε_{b_i} for ε_b (Table 6).

The value of T_m needed to determine the coefficient p in Eq. (1) is calculated from the formula $\Delta T_m=k_\sigma\sigma_0\beta(\sigma_0)$, where $k_\sigma=\Delta T_{\max}/\sigma_b$ is the coefficient of proportionality (Table 6), and the energy conversion coefficient can be found from formula (12), where $m=m_i$, $\sigma_y=\sigma_{0.2}$.

The thermal-softening coefficient can be determined from the formula $k_T=\left(\sigma_b^{77\text{K}}-\sigma_b^{293\text{K}}\right)/216$ after additional tests on specimens at temperatures of 293 and 77 K (Table 6). The value of ω is chosen so as to maximize the value of ε_d. This parameter can be represented as $\omega=k_\omega\sqrt{C/M}$, where $\sqrt{C/M}$ is the natural frequency, k_ω is the coefficient of proportionality (Table 6).

Table 6. Mechanical parameters of the model and material characteristic

Material	σ_{b_i}	A_i	ε_{b_i}	m_i	μ	k_T	ΔT_{max}	k_σ	ω	k_ω
	MPa				MPa·s	MPa/K	K	K/MPa	s^{-1}	
AMg5	706	859	0.304	0.404	4	1.75	61.5	0.087	133	0.665
03Kh20N16AG6	2214	2240	0.300	0.650	10	5.0	185	0.084	145	0.725

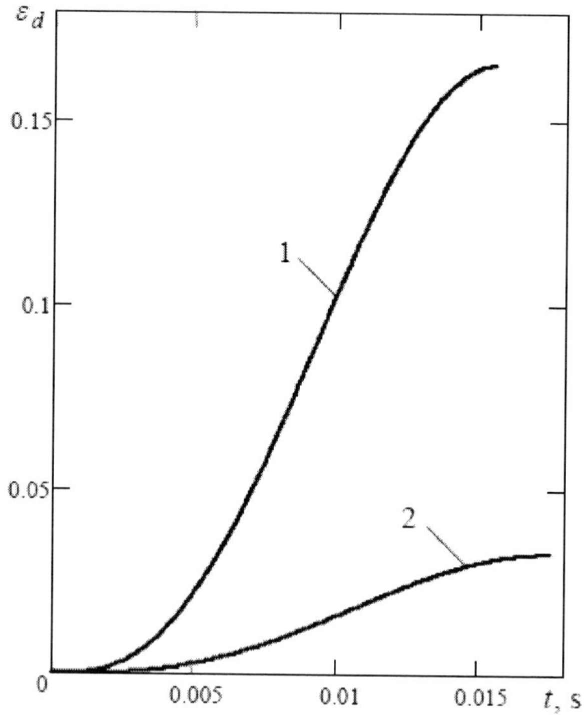

Figure 3. Development of a strain jump for 03Kh20N16AG6 steel (1) and AMg5 aluminum alloy (2) at $\sigma_0/\sigma_b = 0.81$.

The differential equation of the model was solved using a finite-difference program. The solution is the curve of ε_d versus t (Figure 3), which was used to determine ε_D and τ_D. The kinetics of a strain jump are similar in both materials, but the values of ε_D are as different as the strength characteristics, which is indicative of special importance of the parameter σ_0 characterizing the strength of the materials.

The calculated and experimental (determined from stress–strain curves) values of elongation jump $\Delta l_d = \varepsilon_D l$ are in good agreement.

The influence of the parameters [34] was analyzed as follows. The value ε_{D_0} obtained for the parameter values used in the tests was taken as a reference. The reference calculation was conducted using the values of ε_t for steel and aluminum alloy as medium ones for a highly serrated stress–strain curve.

In the figures below, they are represented by open circles.

Using these values of ε_t in (27), we determined σ_0 (1819 and 623 MPa, respectively) and the coefficients of Eq. (24). Next we performed numerical experiments by varying one of the parameters in Eq. (24) over a wide range. The values of ε_D were found for every parameter value.

The strain values were plotted against the values of each parameter, the other parameters being kept constant. The effect of σ_0 was analyzed by finding its value from Eq. (27) for a certain value of ε_t. The values *A*, *m*, and β were calculated as well.

The most important mechanical parameter is the starting stress which determines ΔT_{m} and the coefficient *p* in (24), i.e., the driving force. As already mentioned, it is related to the total strain of the specimen, and $\bar{\sigma}_d \le \sigma_0 \le \sigma_b$. The curves in Figure 4 clearly demonstrate the role of strength and ductility in the effect. Typical points corresponding to σ_d, σ_b, ε_d, ε_b can be seen in the figure.

As the starting stress increases, the strain jump increases intensively reaching 0.05 and 0.23 for the aluminum alloy and steel, respectively. In the latter case, it tends to δ, a measure of the strain at rupture (Table 2). The effect is not structurally critical, however, for low-strength materials such as AMg5. In high-strength steels and alloys, the stress σ_0 ranges from 1000 to 2500 MPa.

Figure 5 shows how the characteristics of the testing machine affect the discontinuous deformation of the 03Kh20N16AG6 steel. For the aluminum alloy, the curves are similar. All the coefficients of Eq. (24) depend on the added mass. The magnitude of the strain jump decreases with *M*, and as $M \to \infty$, the process becomes static because $\tau_D \to \infty$ and the damping factor is zero. Then ε_D asymptotically tends to 0.23 in the steel and to 0.043 in the alloy.

The effect of rigidity is determined by the stored elastic energy whose fraction, according to (15), is converted into work of deformation.

Then $\Delta W \to 0$ and $\varepsilon_D \to 0$ as $C \to \infty$ and, on the contrary, $\Delta W \to \infty$ and $\varepsilon_D \to \infty$ as $C \to 0$, which is confirmed by Figure 5, and it is clear how ε_D behaves with variation in *C*.

This factor, as an external effect on the strain jump, should be taken into account if the rated strain rate is high. For the preloaded system, the basic equation of the model is as follows [34]:

$$\ddot{\varepsilon}_d(1+\varepsilon_d) + a\dot{\varepsilon}_d + b\varepsilon_d(1+\varepsilon_d) + c\varepsilon_d^m = p\sin\omega t + q(1+\varepsilon_d)t \quad (28)$$

where $q = Q/u$, $Q = KV/l$, V is the cross-head speed, the other notation being the same as in (24). The solution of the equation must satisfy conditions (25), as in the case of the non-preloaded system. This model is applicable to strain-rate controlled loading conditions used in standard tension tests on metals.

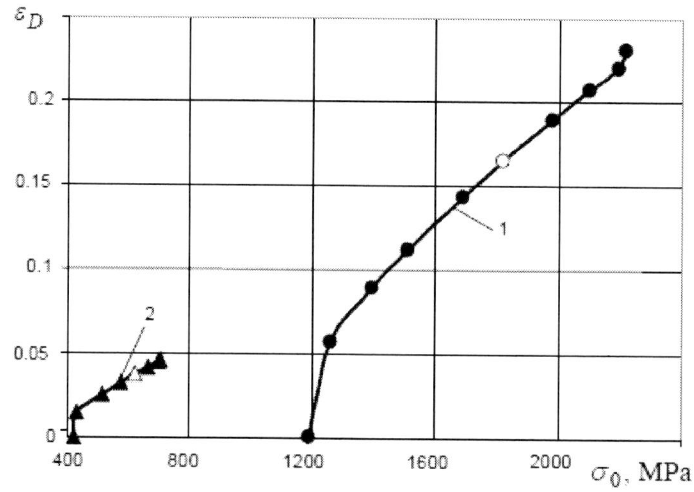

Figure 4. Dependence of the strain jump amplitude on the starting stress for steel 03Kh20N16AG6 (curve 1) and aluminum alloy AMg5 (curve 2).

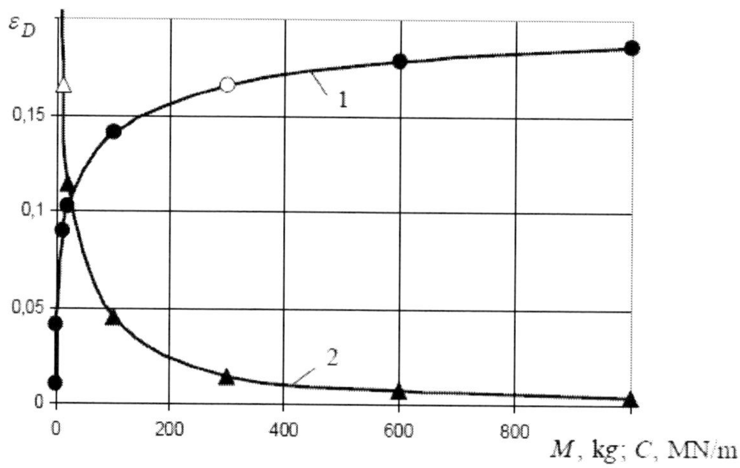

Figure 5. Strain jump versus parameters of the loading machine: mass M (1) and rigidity C (2) for steel 03Kh20N16AG6.

The analysis of the effect of additional loading using model (28) shows that $\dot{\varepsilon}_D$ varies linearly with V: $\varepsilon_D = \varepsilon_{D_0} + k_{\text{v}} V$, where k_{v} is the coefficient of proportionality, or

$$\varepsilon_D = \varepsilon_{D_0} + k_{\dot{\varepsilon}} \dot{\varepsilon}_n \tag{29}$$

Since $V = \dot{\varepsilon}_n L_0$, where $\dot{\varepsilon}_n$ is the rated strain rate and L_0 is the initial gage length of the specimen, the coefficient of proportionality $k_{\dot{\varepsilon}} = k_{\text{v}} L_0$. We have $k_{\text{v}} = 5.4 \cdot 10^{-3}$ s/mm,

$k_{\dot{\varepsilon}} = 8.1 \cdot 10^{-2}$ s for the steel and $k_{\mathrm{v}} = 9.59 \cdot 10^{-3}$ s/mm, $k_{\dot{\varepsilon}} = 14.4 \cdot 10^{-2}$ s for the aluminum alloy. In static tests, $\dot{\varepsilon}_n < 1$ s^{-1}.

It can be seen that the effect of additional loading on the aluminum alloy is considerably stronger than on the steel, namely, by a factor of 1.8 (the ratio of the coefficients of proportionality in (29)). In this case, the jump duration is noticeably longer. In [34], these mathematical models were used to analyze the dependence of low-temperature discontinuous yielding on eight decisive factors, including the scale effect. Each factor was varied from zero to infinity.

PRACTICAL APPLICATIONS

Studies show that the discontinuous deformation of metals at given ambient temperature is strongly dependent on the parameters $\dot{\varepsilon}_n$, C, M of the testing machine and on the scale factor. These parameters also have some effect on the mechanical characteristics measured in standard tests. Constraints for some test parameters based on experimental data were proposed in [16, 17]. An important application of these results is the theoretical justification of standards on tension tests of metals at very low temperatures.

The rated strain rate is a parameter that, unlike others, can easily be changed in experiment, and its effect on discontinuous deformation and mechanical characteristics is the strongest [7, 20, 21]. This factor is also responsible for the instability of deformation. Generally, there are three critical strain rates [16]. The first critical strain rate $\dot{\varepsilon}_{c_1}$ keeps the heat generation low during the subcritical stage of stable plastic deformation, and the effect occurs. The second strain rate $\dot{\varepsilon}_{c_2}$ completely damps it. If $\dot{\varepsilon}_n \geq \dot{\varepsilon}_{c_3}$, then thermal relaxation continues after the completion of the strain jump, and local adiabatic plastic deformation sets in, followed by unloading and fracture of the specimen. Finally, the third critical strain rate $\dot{\varepsilon}_{c_4}$ that leads to fracture during the strain jump can be found from model (28). Using formula (29) and the equality $\varepsilon_D = \delta$ as a criterion, we obtain 2.3 s^{-1} for the steel and 2.2 s^{-1} for the aluminum alloy, which fall outside the static range.

Let us examine the possibility of standardizing the strain rate for static-tension tests of metals in liquid helium. Tests in this medium are the most popular and are of the most practical importance in the temperature range $T < 30$ K. At $T_0 = 4$ K, all structural metallic materials, including austenitic steels, titanium, aluminum, and copper alloys, exhibit discontinuous yielding. It is necessary to set some maximum rise ΔT_h in the temperature above its rating caused by the deformation of the specimen. Let $\Delta T_{h_{\mathrm{m}}} \leq 1$ K [20], which is a quite severe constraint. Then formula (17) can be used to calculate the lower (or first) critical strain rate $\dot{\varepsilon}_{cl} = \dot{\varepsilon}_{c_1}$ for a certain material and its specimen of certain standard size. In liquid helium, heat transfer occurs most intensively, in the so-called nucleate boiling regime. In this case, $q = 5000$ W/m^2·K. Calculation gives $\dot{\varepsilon}_{c_1} \leq 2 \times 10^{-3}$ s^{-1}.

Since the critical strain rates are related as $\dot{\varepsilon}_{c_1} < \dot{\varepsilon}_{c_2} << \dot{\varepsilon}_{c_3}$, it is ε_{c_1} that should be standardized. Experiments give $\dot{\varepsilon}_{c_1} \leq 2\times 10^{-3}$ s^{-1} [7, 20, 21].

Let ε_D in the additionally loaded system do not exceed $\varepsilon_{D_0} = 0.165$ (for 03Kh20N16AG6 steel) by more than 5%. From (29) we obtain $\dot{\varepsilon}_n \leq 0{,}103$ s^{-1} greatly exceeding the standard values $\dot{\varepsilon}_n \leq 1.6\times 10^{-3}$ s^{-1} [35] and $\dot{\varepsilon}_n \leq 1.0\times 10^{-3}$ s^{-1} [36], which may be considered quite justified. Note that when $\dot{\varepsilon}_n = 1.6\times 10^{-3}$ s^{-1}, the strain increases above ε_{D_0} by a negligible amount (0.3%). Even if the strain rate is increased by an order of magnitude, the strain will increase by merely 0.7%. This leads us to the important conclusion that the system is practically nonloaded at standard strain rates.

When the strain rate is very small, the parameter C is the most significant. After the critical stress is reached, local discontinuous deformation occurs spontaneously at the expense of the stored elastic energy, which is equal to $W = \sigma_0^2 F^2 / 2C$ at the beginning of the strain jump. Some of it, usually a greater fraction, is stored in the testing machine, while the other fraction in the specimen.

At relatively high temperatures, the rigidity may be manifested only on the decreasing portion of the stress–strain curve (decrease in the load, $dP/dt < 0$, indicates that the stored elastic energy is converted into work of deformation).

When plastic deformation periodically loses stability, there are multiple load drops. As a result, the effect of rigidity is accumulated. A decrease in the rigidity C_m of the testing machine and, hence, the rigidity C of the specimen–machine system leads to an increase in ε_D, a decrease in the number of discontinuities, and earlier fracture of the specimen, i.e., shortening of the stress–strain curve up to fracture during the first discontinuity [18]. The characteristics δ and σ_{B} decrease too.

Therefore, this parameter should be standardized to obtain reliable repeated values in standard tests on metals.

The rigidity C_m is of the order of 10 MN/m, which corresponds to the initial portion of the $\varepsilon_D - C$ curve, where the dependence is the strongest (Figure 5). Linearizing it on the interval 1.0–20 MN/m including the reference point (for 03Kh20N16AG6 steel) yields the equation

$$\varepsilon_D = 0{,}295 - k_c C, \tag{30}$$

where $k_c = 0.011$ m/MN. Let again ε_D do not exceed $\varepsilon_{D_i} = 0.165$ by more than 5%. Then $C \geq 11.0$ MN/m follows from (30). Since $C = C_s C_m / (C_s + C_m)$, where $C_s = 70$ MN/m is the rigidity of a fivefold steel specimen, we have $C_m \geq 13.05$ MN/m.

This result confirms the recommendations [17] and justifies a machine compliance of no higher than 0.7×10^{-4} mm/N (corresponding to $C_m \geq 14.3$ MN/m) established by the Ukrainian Standard "Metals: A method for tension testing of metals and alloys at low and cryogenic temperatures."

The dependence of ε_D on M (Figure 5), however, is not so strong in the range typical of testing machines. For ε_D to increase by 5%, it is necessary that M be increased to 465 kg. Such an effect of the added mass may be considered insignificant, and there is no need to standardize the parameter M. However, it is well to bear in mind that "lighter" machines are preferable.

An important issue is choosing standard specimen sizes for discontinuous yielding conditions. Because multiple necking accompanies strain drops and the stress state of plate specimens is far from being linear, it is recommended to give preference to cylindrical specimens. Since the scale effect is strong [25, 26], it is necessary to narrow down the range of standard specimen sizes as much as possible to obtain comparable values of mechanical characteristics.

It may be recommended to use fivefold cylindrical specimens with an initial diameter of 3 or 4 mm for testing at temperatures below 20 K. These are the closest possible values that allow using machines with a low maximum load. The points corresponding to these diameters, as in the specimen with a diameter of 7 mm (F=38 mm^2) established by the standard [36], lie on the shallow branch of the curve of strain jump versus F. As the cross-sectional area of the specimen is increased by a factor of 3 (from 12.6 to 38 mm^2), the strain jump increases only by 14% [34].

This study reveals that the strain jumps in high-strength materials can be very large; i.e., this phenomenon is critical to the strength and load-carrying capacity of highly stressed structural members at very low temperatures. The catastrophic-deformation or failure criterion is given by

$$\varepsilon_D / \varepsilon_{cr} \geq 1$$

where ε_{cr} is the strain at rupture or ultimate strain.

Here the practically important question arises of what a measure of strength of a material could be in such unusual conditions. It was shown that the ultimate strength strongly depends on the loading conditions and can be reached if deformation is nonisothermal and inhomogeneous.

Therefore, it is reasonable that the set of standard mechanical characteristics include the starting stress as the upper discontinuous yield point σ_{hd} (the stress corresponding to the first upper peak on the initial elastic portion of the stress–strain curve) and the lower discontinuous yield point σ_{ld} (the stress corresponding to the lower load peak). The former allows evaluating the strain jump and associated risk for the material, while the latter defines the level at which the premature jump caused by an external impulsive (electromagnetic, thermal, or mechanical) load is impossible and is a limit for unconditionally stable states of the material [16]. The ultimate stresses can be determined in this case from the formula

$$[\sigma] = \sigma_{ld}/n_f$$

where n_f is the factor of safety.

Conclusion

The program of relevant experimental and theoretical studies of the mechanical behavior of metallic materials in liquid helium has been implemented. The main subject of these studies is the low-temperature discontinuous yielding of metals and their strength in such conditions. Its mechanisms and accompanying thermal and energy conversion effects have been considered.

Phenomenological criteria defining the conditions in which this effect occurs have been formulated and related. A two-parameter (temperature and stress) criterion of instability of plastic deformation of metals has been derived. It employs an experimentally determined material constant (discontinuous yield strength σ_d) to calculate both the maximum critical temperature and the critical stress at given temperature.

The effect of the most important factors related to the loading conditions (specimen–machine rigidity and strain rate) on the discontinuous yielding of austenitic steels and aluminum and titanium alloys has been studied experimentally. The rigidity of the specimen–machine system is a factor that determines whether discontinuous yielding occurs or not. Even after a relatively small increase in the strain rate, this effect is no longer manifested, which leads to a considerable decrease in such material characteristics as ultimate strength and elongation.

If a strain drop occurs immediately after the stage of elastic deformation, an increase in the strain rate leads to rupture of the specimen. Appropriate kinetic criteria have been proposed.

The rigidity of the system depends on the stored elastic energy which goes to meet the energy demands of the discontinuous yielding process. With decrease in rigidity and, hence, increase in energy, the elongation jump increases and the stress–strain curves become shorter. This results in reduced ductility and strength.

Discontinuous yielding affects the manner in which stress concentration influences the mechanical characteristics, deformation, and failure of metals. The higher the stress concentration, the fewer there are strain jumps. Discontinuous yielding does not occur at all in specimens with a sharp notch.

The effect of various types of prestrain (pretension, precompression, pretwisting, surface cold working) on the discontinuous yielding of specimens has been examined. The magnitude and type of prestrain have a strong influence on the discontinuous-yielding and mechanical characteristics. A change of the deformation mechanism in alloys is responsible for the Bauschinger effect, i.e., increase in the yield strength of a metal when the direction of the load is reversed.

Prestrain as well as stress concentration (technology factor) combined with low rigidity of the structure or supertcritical strain or loading rates (operational factor) may lead to rupture at cryogenic temperatures.

On the other hand, a combination of strain hardening and low-temperature hardening can effectively be used to increase the ultimate stresses of highly ductile metals for structural elements of cryogenic equipment operating under maximum static loads at constant low temperature.

The study of the scale effect in and the stress state of various specimens during unstable plastic deformation has made it possible to formulate recommendations on sizes of standard specimens for static tension tests at temperatures below 30 K. To study the scale effect, a dimensionless criterion for the energy similarity of discontinuous yielding has been proposed and justified.

A method for the experimental simulation of a strain jump at room temperature has been developed and shown to be capable of reproducing and studying effects similar to those observed in liquid helium.

To theoretically formalize the dependence of the magnitude of strain jump in metals on eight decisive factors, a nonlinear multi-parameter mathematical model has been set up. It employs the concept of adiabatic deformation of metals and allows for the effect of the loading system. To identify the model, specimens of 03Kh20N16AG6 steel and AMg5 aluminum alloy were tested in liquid helium and nitrogen and at room temperature. The dependence of strain jumps on the mechanical characteristics of the material, the dynamic characteristics of the testing machine, rated strain rate, and scale effect has been established numerically for the widest possible ranges of these factors. It has been shown that the starting stress, the strain rate, and the rigidity of the testing machine have the strongest effect on discontinuous yielding. The model has been validated by comparing its results with experimental data.

An important practical result is the scientific justification of the advanced method for tension tests of metals at temperatures below 20 K. Recommendations on the standardization of the strain rate, the rigidity of the testing machine, and the type and sizes of specimens have been formulated.

A method for standardizing the strength characteristics of steels and alloys at cryogenic temperatures has been proposed and justified. A distinctive feature of the proposed approach to determining the ultimate stresses is that the lower discontinuous yield stress (above which the deformation process can lose stability) rather than the ultimate strength is used as a measure of strength.

Thus, the studies performed made it possible to describe the discontinuous yielding of metals—an important scientific problem in the field of low-temperature strength.

REFERENCES

[1] Pustovalov, V. V. *Low Temp. Phys.* 2008, V.34, P. 683-724.

[2] Bobrov, V. S., Zaycev, S. I., Lebyodkin, M. A. *Fizika Tverdogo Tela.* 1990, V. 32, № 10, P. 3060-3065 (in Russian).

[3] Vorob'ev, E. V. *Strength of Materials.* 1990, V. 22, № 6, P. 835-840.

[4] Basinski, Z. S. *Proc. Roy. Soc.* 1957, A-240, P. 229-242.

[5] Seeger, A. *Dislocations and Mechanical Properties of Crystals* / J. Wiley and Sons, New York, 1957, 243 p.

[6] Wessel, E. T. *Trans. ASM.* 1957, V. 49, № 1, P. 149-172.

[7] Strizhalo, V. A., Bugaev, V. Yu., Medved', I. I. *Strength of Materials.* 1990, V. 22, № 1, P. 1-7.

[8] Macdougall, D. *Exp. Mech.* 2000 V. 40, № 3, P. 298-306.

[9] Erdman, J. C., Jahoda, J. A. *Appl. Phys. Letters*. 1964, V. 4, P. 204-205.
[10] Stryzhalo, V. A., Vorob'ev, E. V. *Materials Science*. 2002, V. 38, № 5, P. 692-697.
[11] Smirnov, V. G., Solntsev, Yu. P. *Strength of Materials*. 1992, V. 24, № 12, P. 765-771.
[12] Stryzhalo, V. A., Vorob'ev, E. V. *Strength of Materials*. 1993, V. 25, № 8, P. 576-583.
[13] Johnson, V. J. *Properties of Materials at Low Temperatures*. - N.Y.: A Compedium, Pergamon Press, 1961. - 270 p.
[14] Shpeizman, V. V., Nikolaev, V. I., Smirnov, B. I., Lebedev, A. B. *Physics of the Solid State*. 2000, V. 42, № 6, P. 1066-1069.
[15] Stryzhalo, V. A., Vorob'ev, E. V. *Strength of Materials*. 1994, V. 26, № 10, P. 713-717.
[16] Vorob'ev, E. V. *Mechanica*. 2006, № 1(57), P. 17-21.
[17] Stryzhalo, V. A., Vorob'ev, E. V. *Strength of Materials*. 2011, V. 43, № 3, P. 268-275.
[18] Vorob'ev, E. V. *Strength of Materials*. 2001, V. 33, № 1, P. 62-65.
[19] Stryzhalo, V. A., Vorob'ev, E. V. *Strength of Materials*. 1997, V.29, № 3, P. 269-274.
[20] Vorob'ev, E. V. *Strength of Materials*. 2000, V.32, № 6, P. 549-554.
[21] Strizhalo, V. A., Bugaev, V. Yu., Medved', I. I. *Strength of Materials*. 1990, V.22, №9, P. 1279-1283.
[22] Iino, Y. and Suzuki, K. *6th Int. Congr. Exp. Mech*. (Portland, Ore, 5–10 June, 1988). – London; Bethel, 1988, V. 1, P. 106–111.
[23] Strizhalo, V. A., Vorob'ev, E. V., Novogrudskii, L. S. *Strength of Materials*. 1995, V.27, №8, P. 434-440.
[24] Vorob'ev, E. V. *Strength of Materials*. 2001, V.33, № 6, P. 573-577.
[25] Strizhalo, V. A., Bugaev, V. Yu., Medved', I. I. *Strength of Materials*. 1990, V.22, № 5, P. 698-704.
[26] Vorob'ev, E. V. *Strength of Materials*. 2002, V.34, № 4, P. 375-380.
[27] Vorob'ev, E. V., Strizhalo, V. A. *Strength of Materials*. 2003, V.35, № 2, P. 136-139.
[28] Vorob'ev, E. V., Anpilogova, T. V. *Strength of Materials*. 2007, V.39, № 2, P. 219-221.
[29] Vorob'ev, E. V. *Strength of Materials*. 2002, V.34, № 5, P. 451-455.
[30] Vorob'ev, E. V. *Strength of Materials*. 2008, V.40, № 3, P. 350-355.
[31] Vorob'ev, E. V. *Strength of Materials*. 2008, V.40, № 4, P. 439-444.
[32] Strizhalo, V. A., Vorob'ev, E. V. *Strength of Materials*. 1997, V.29, № 3, P. 269-274.
[33] Vorob'ev, E. V., Anpilogova, T. V. *Strength of Materials*. 2011, V.43, № 1, P.77-86.
[34] Vorob'ev, E. V., Anpilogova, T. V. *Computational Materials Science*. 2013, V.68, P. 66–72.
[35] GOST 22706-77 Metals. Method for tension tests at the temperature -100 up to -269^{0}C.
[36] ISO 19819:2004(E) Metallic Materials – Tensile Testing in Liquid Helium.

In: Helium: Occurrence, Applications and Biological Effects ISBN: 978-1-62808-612-6
Editor: Simon B. Davis

Chapter 2

THE USE OF HELIUM WITH OXYGEN AT PRESSURE AS A BREATHING GAS MIXTURE

Jean-Claude Rostain*
Aix-Marseille University, UMR-MD2, Faculté de Médecine Nord, Marseille, France

ABSTRACT

The use of compressed air or nitrogen-oxygen mixture in diving induce from 30 metres of sea water several neurological and psychomotor disturbances due to the increase of partial pressure of nitrogen and to its narcotic potency called nitrogen narcosis. To avoid these disturbances, nitrogen has been replaced by helium, an inert gas with a low narcotic potency in relationship with its very low lipid solubility. With helium-oxygen synthetic gas mixture, depths higher than 50 metres or pressure higher of 0.5 MPa have been reached up to 6.1MPa. However, from 1 MPa physiological and neurophysiological changes different from the symptoms of nitrogen narcosis have been reported. In man, these changes consist of electroencephalographic modifications from 1MPa, occurrence of tremor and sleep disturbances from around 2MPa, myoclonia and dysmetria from 4MPa, and in experimental animals, epileptic seizures for pressure higher than 8MPa. Theses disturbances have been regrouped in a high pressure nervous syndrome (HPNS), as helium did not seem implicated in these disruptions. Indeed, if helium has narcotic properties, the signs and symptoms will be similar to or worse than 0.9MPa compressed air at about 4.1MPa. However owing to antagonism of the weak narcotic effect of helium by the increased pressure (called the pressure reversal effect) there is no helium narcosis. However, in the light of results obtained with narcotic gases added to helium oxygen mixture, such as nitrogen or hydrogen, mood changes and some somesthetic hallucinations reported at pressure higher than 4-5 MPa could be in relationship with a narcotic effect of helium rather than a pressure effect. Moreover hallucinatory behaviours have been reported in monkeys in helium oxygen mixture for pressure of 8MPa and above.

Neuropharmacological and neurochemical studies (microdialysis and voltammetry) performed with helium-oxygen mixture at pressure have demonstrated changes in amino-

* Tel. 33 4 9169 89 06; Fax. 33 4 91 65 38 51; jean-claude.rostain@univ-amu.fr.

acid (gama-amino-butyric acid and glutamate) and monoamine (dopamine and serotonin) neurotransmission regulations at the level of basal ganglia, brain structures implicated in the regulation of the motor, loco-motor and cognitive processes.

INTRODUCTION

All mammals including man exposed to increasing pressure of breathing gas mixture show disturbances at the level of the nervous system.

At sea level, atmospheric pressure is 0.1 MPa (0.1 megapascal ~ 1Kg/cm^2 ~ 1 atmosphere or 1 bar). Underwater, the weight of water increases with depth. At 10 metres of sea water (msw) for example, each cm^2 of the body is under an increase of pressure that corresponds to a weight of a column of water of 1000 cm of high on 1 cm^2 of area i.e. 1000 cm^3 of water weighting 1 kg. Thus, the pressure due to water at 10 metres is 1 kg/cm^2 or 1 bar. When the depth increases progressively from 10 m to 20 m, the pressure gradually increases from 1 bar to 2 bars. As a result, each time a man goes 10 metres under water he is exposed to an increase of 1 bar of pressure. Thus at a depth of 10 metres the absolute pressure equals 2 atmospheres (1 of atmospheric pressure and 1 for the 10 metres of water). At 20 metres, the absolute pressure increases to 3 bars (1 of atmospheric pressure and 2 for the 20 metres of water), at 100 m.s.w., it will be 11 bars (1 of atmospheric pressure and 10 bars for the 100 metres of water). The relationship between pressure and volume (Boyle - Mariotte's law) requires that the man must breathe a gaseous mixture the pressure of which is equal to the ambient pressure.

The quantity of gas dissolved within the body tissues is a function of the partial pressure of each gas making up the mixture (Henry's law), the partial pressure of each gas being equal to the product of the ambient pressure and the percentage of this gas in the respiratory mixture. For instance at atmospheric pressure, air is composed of 21% of oxygen and 79 % of nitrogen. Consequently the partial pressure of oxygen is 0.210 bar and that of nitrogen is 0.790 bar. At 30 m.s.w. , the absolute pressure is 4 bars and the partial pressure of oxygen is 4 x 0.21 = 0.84 bar and that of nitrogen is 4 x 0.79 = 3.16 bars.

Both the pressure itself (pressure per se) and the partial pressures of the gases in the respiratory mixture, will affect the organism. In other way, oxygen at pressure is itself toxic: pressure of 2-3 bars affect the nervous system, and over a long period of time even pressure higher than 0.5 bars will affect the respiratory system.

Thus, increasing the pressure of air produces disturbances in the central nervous system characterized by behavioural troubles and changes in mental processes (table 1) from an absolute pressure of 4 bars (Bennett and Rostain, 2003a). These disturbances were attributed from the works of Behnke et al. (1935) to nitrogen increasing partial pressure and called nitrogen narcosis (Bennett and Rostain, 2003a). This is due to nitrogen which becomes narcotic when its partial pressure is increased. Its narcotic potency is in relationship with its solubility in oil or lipid. The studies performed on the oil solubility of the inert gases (i.e. gases which are not metabolized) have shown that three inert gases have a narcotic potency lower than that of nitrogen: hydrogen, neon and helium. The studies on the narcotic potency of these three gases agree that helium has almost no narcotic activity (Behnke et al., 1935, Case and Haldane, 1941, Marshall, 1951, Bennett, 1966, Brauer et al., 1968).

Table 1. Signs and symptoms of nitrogen narcosis in man which occur in mean from 30 msw (3 bars of relative pressure or 4 bars of absolute pressure or 4 ATA)

NITROGEN NARCOSIS IN MAN From 4ATA
Temporo - spatial disorientation Euphoria Hallucinations Memory troubles Mood changes Impaired neuromuscular coordination Psychomotor and intellectual decrements

Consequently to avoid the physiological problems which occurred with nitrogen at pressure, synthetic gas mixtures were used which include oxygen and diluent gas other than nitrogen. Such mixtures are made up in part of oxygen, the partial pressure of which must remain within acceptable limits, thus requiring a decrease in its percentage with the increase in pressure. In addition, the diluent gas should be as inert as possible for the organism and should have a narcotic potency lower than nitrogen. The use of synthetic helium-oxygen mixture in which the oxygen partial pressure is maintained at 0.04 MPa, has allowed man to reach pressure higher than 1.1 MPa of absolute pressure. Nevertheless, helium-oxygen which one breathes at high pressure causes a group of disturbances different from those of nitrogen narcosis, which are nervous in origin, are attributed to a pressure effect and are grouped together under the heading High Pressure Nervous Syndrome (H.P.N.S.) (Brauer et al., 1969, Fructus et al., 1969, Rostain and Naquet, 1974)

THE USE OF HELIUM AS A DIVING GAS

As said before, the choice has thus turned to helium which seems to have almost no narcotic potency. The use of a synthetic helium-oxygen respiratory mixture where the oxygen partial pressure is maintained at values close to those of the atmospheric pressure has allowed divers to reach great depths. The first deep dives (equal or greater than 100 m.s.w.) were carried out in the sixties. After the Precontinent operations (Fructus and Chouteau, 1963) numerous dives in hyperbaric chambers and at sea were carried out between 100 and 300 m.s.w. each one going a bit deeper (Cabarrou et al., 1966, Aquadro and Chouteau, 1967, Bennett, 1967, Fructus and Delauze, 1968, Waldvogel and Bulhmann, 1968). Dives reached and went beyond 300 m.s.w. between 1968 and 1970 (Fructus et al., 1969, Overfield et al., 1969, Overrath et al., 1969, Salzano et al., 1970).

Nonetheless during these dives in a helium-oxygen atmosphere a certain number of new facts appeared. During a rapid compression to 180 m.s.w., Bennett (1966) described the existence of nausea, vertigo and most markedly tremors which the author attributed to helium (helium tremor). Later, during dives to between 300 and 365 m.s.w. Brauer et al. (1969) and Fructus et al. (1969) showed a whole series of disturbances, different from those of nitrogen narcosis, which those authors attributed to pressure and named the High Pressure Nervous

Syndrome (HPNS), events later confirmed by Bennett and Towse (1971) during a dive to 457 m.s.w.

HPNS in Helium-Oxygen Mixture

HPNS is composed of neurological disturbances and of electrophysiological changes, the most important of which are shown in Table 2. The first signs which consist of electroencephalographic changes generally appear between 10 and 20 bars.

Table 2. Signs and symptoms of the high pressure nervous syndrome in helium-oxygen mixture. The first one occurred between 11 ATA. and 20 ATA

HELIUM-OXYGEN PRESSURE EFFECTS
IN MAN

Up to 62 ATA

BEHAVIORAL SYMPTOMS	ELECTROPHYSIOLOGICAL CHANGES
> 16 ATA TREMOR (8 – 12 Hz)	> 11 ATA EEG: INCREASE IN SLOW WAVES, DECREASE IN FAST ACTIVITIES
> 40 ATA FASCICULATIONS, MYOCLONIA	> 30 ATA CHANGES IN EVOKED POTENTIALS IN CORTICAL EXCITABILITY CYCLES
> 40 ATA DYSMETRIA	> 20 ATA SLEEP: INCREASE IN STAGES 1 and 2 DECREASE IN STAGES 3 and 4, REM
> 35 ATA SOMNOLENCE	> 25 ATA HYPER-REFLEXIAA
> 15 ATA COGNITIVE IMPAIREMENTS Up to 20 %	

Neurological Disturbances

Tremor is first observed at the extremities of the limbs during periods of activity. The measurements were carried out with an accelerometer placed on the middle finger of the dominant hand during the "serment" test. Tremor begins to be recorded between 15 and 20 bars. It increases with the depth and progressively affects all the limbs and the trunk. The power spectra carried out on the signals recorded from the accelerometer give a frequency of 8 to 12 Hz (Rostain and Lemaire, 1973, Rostain and Naquet, 1974, 1978, Fructus and Rostain, 1978) (Figure 1).

Fasciculation and myoclonus (muscular jerks) first appear at the extremities of the limbs (at about 400 to 500 meters). They then reach the limbs and finally the muscles of the neck, the face and the body trunk (Fructus and Rostain, 1978).

Dysmetria is seen during classical neurological tests ("fleuret" and finger to nose tests) and appears at depths greater than 400-500 m. s. w. (Fructus and Rostain, 1978).

Under some conditions drowsiness occurs when the subject is left at rest, but it is easily overcome by activity.

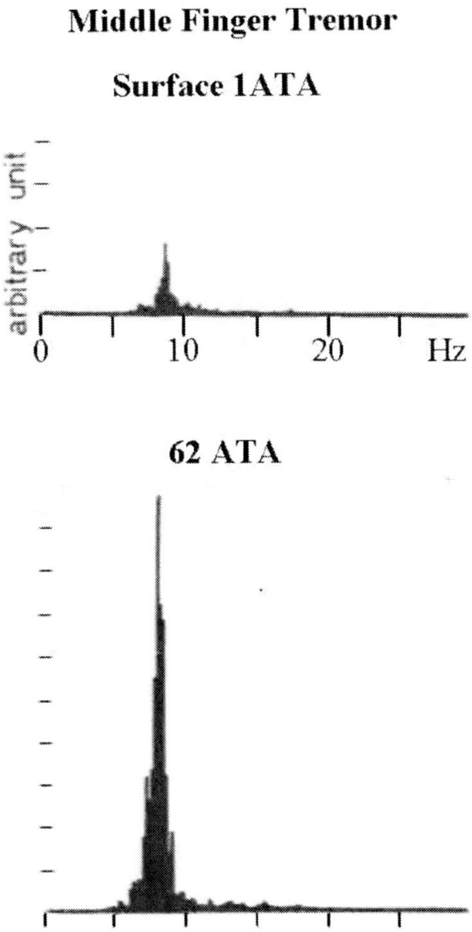

Figure 1. Power spectra of the middle finger tremor at 1 ATA and at 62 ATA in helium-oxygen mixture.

Beside these neurological symptoms, decreases in performance have been recorded. A decrement up to 20% was recorded in psychomotor and intellectual tests.

Electrophysiological Disturbances

Electroencephalographic (EEG) changes have been recorded from 10 bars. These changes (Fig 2) consist at first by a decrease of alpha waves (8-14 Hz) which occurred in the posterior region of the scalp from 10 bars (Rostain, 1993, Bennett and Rostain, 2003a) and of an increase of slow wave activities particularly that of the theta frequency (4-7 Hz) that occurs around 20 bars sometimes accompanied by that of the delta frequency (1-4 Hz).These changes predominate in the anterior region of the scalp (Figure 3). Often there is also a decrease of the faster activities (beta 14-40 Hz) (Rostain and Naquet, 1974, 1978, Rostain, 1993, Bennett and Rostain, 2003b).

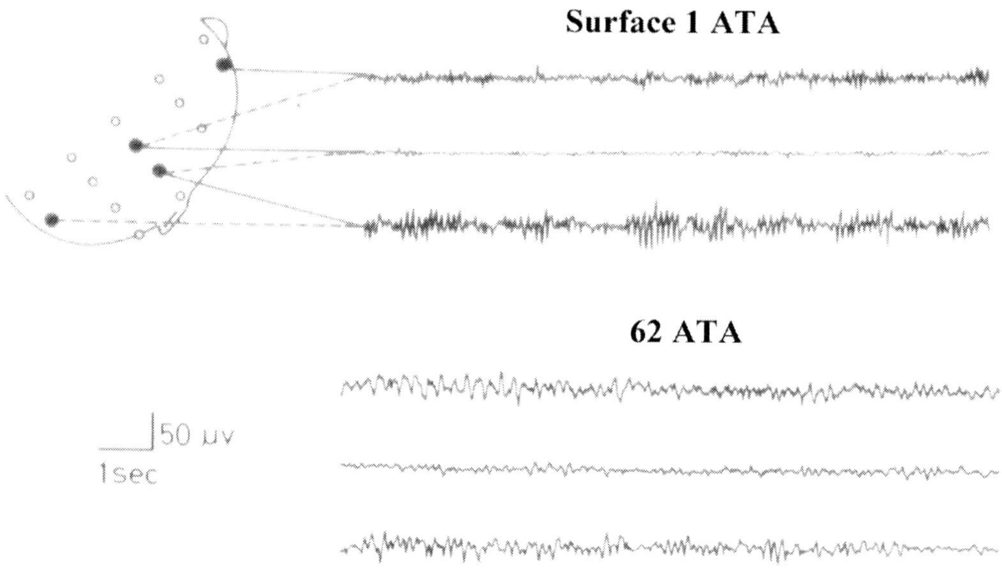

Figure 2. EEG recording of one subject during a control period at 1 ATA and at 62 ATA. In each condition, three leads are represented: fronto polar-central, central-mid temporal, mid temporal-occipital. One can see the increase of slow waves on the fronto central lead at 62 ATA and the decrease of the alpha activity in the posterior lead.

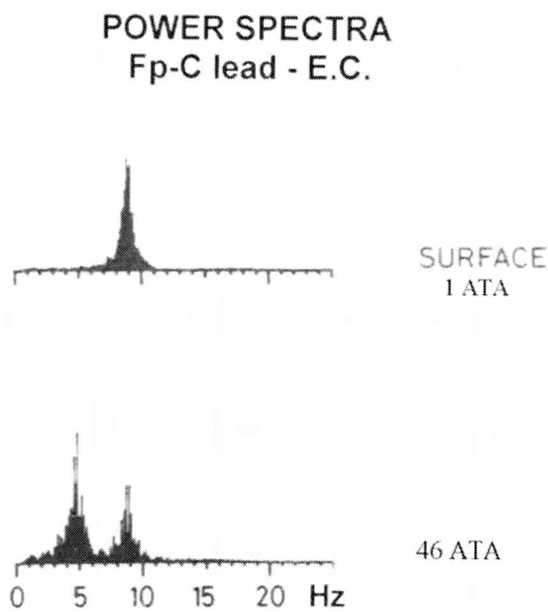

Figure 3. Power spectra of EEG activities in fronto-polar-central lead at 1 ATA (top) and 46 ATA (Bottom). At 1 ATA, the EEG is characterized by an activity predominant in the alpha frequency (8-12 Hz). At 46 ATA this activity is decreasing and we recorded an increase in theta EEG frequency (4 to 6 Hz).

There are also changes of the cortical evoked potentials brought about by different sensory modalities (visual, somatic, auditory) (Rostain and Dimov, 1976, Bennett and Rostain, 2003b). The cortical excitability cycles are also modified and indicated a cortical hyperexcitability (Rostain and Dimov, 1976). Some reflexes are modified and shown generally hyper-reflexia (Roll et al., 1978, Harris, 1979a, b, Harris and Bennett, 1983 Parmentier et al., 1985, Bennett and Rostain, 2003b).

Studies of sleep (electroencephalographic) have also shown changes: awake periods, stages 1 and II of sleep (light sleep) increase; stages III et IV (deep sleep) lessen ; periods of paradoxical sleep are unstable and decrease after 40 bars. (Rostain and Lyagoubi, 1971, Rostain et al., 1973, 1988, 1991, 1997).

Other symptoms such as nausea and vertigo have also been described under certain conditions (Bennett and Rostain, 2003b).

Narcosis in Helium-Oxygen Mixture

Based on the lipid solubility hypothesis, the narcotic effect of helium would occur around 400 msw (41 ATA) (Bennett and Rostain, 2003a). However, pressure counteracts this weak narcotic potency according to the pressure reversal effect and the critical volume hypothesis (Lever et al., 1971, Miller et al., 1973, Bennett and Rostain, 2003a, b). The symptoms that occur are different from those observed in narcosis and they are called the High Pressure Nervous Syndrome (HPNS) (Table 2). It is generally considered that helium is not narcotic. However, from recent data obtained during experimental dives with narcotic gases added to helium at great pressure (Stoudemire et al., 1984, Fructus et al., 1984), there were mood changes or sensory hallucinations reported in some cases in helium-oxygen dives to pressure greater than 4 MPa (400m), which could be due to a narcotic effect of helium rather than a pressure effect (Rostain, 1994, Rostain et al., 1999). (Table 3). Moreover, hallucinatory behaviour has been also reported in monkeys breathing a helium-oxygen mixture at pressures of 8 MPa and above (Rostain, 1987, Gardette et al., 1994, Rostain et al., 1994), which could be due to the narcotic effect of helium at high pressure (Rostain et al., 1994).

Table 3. Symptoms attributed to narcosis from 41 ATA in man

HELIUM NARCOSIS
Weak narcotic potency and Pressure reversal effect induce High Pressure Nervous Syndrome Narcotic symptoms From 41 ATA mood changes hallucinations

Characteristics of HPNS in Experimental Animal Models

Most of these symptoms have been described in animal models exposed to high pressures of helium (Brauer, 1975, Halsey, 1982) (Table 4). In the rat and the monkey neurological disturbances have also been noted. Tremor at 10 to 14 Hz and myoclonus were reported. In free moving animals, motor and hyperlocomotor activities have been recorded. For pressures greater than those reached by man (higher than 70 bars), convulsions have been recorded (Rostain et al. 1984a, 1986a).

Table 4. Signs and symptoms of the high pressure nervous syndrome in rats and monkeys in helium-oxygen mixture

HPNS in animal models (Rats and Monkeys) in helium-oxygen mixture
Up to 101 ATA

Behavioral changes	Electrophysiological disruptions
> 30 ATA Tremor (10-14 Hz) > 20 ATA Motor and locomotor hyperactivities > 40 ATA Myoclonus > 81 ATA Muscular spasms > 90 ATA Convulsions	> 15 ATA EEG changes: Increase of slow waves. > 90 ATA paroxystic activity, epileptic seizure. > 40 ATA Changes in evoked potentials, cortical > 40 ATA excitability and reflexes (monkeys)

Electrophysiological changes have also been recorded. An increase of the slow wave EEG activities (theta and delta) has been found together with a decrease of the activities of more rapid activity frequencies (alpha and beta). EEG activities of a paroxysmal type with a pointed character or a spike and wave have also been recorded for pressure never reached by man (greater than 70 bars.). Such paroxysmal activities are usually followed by a generalized epileptic seizure of the tonic-clonic type (Rostain et al., 1984a, 1986a). Somatic evoked potentials, cortical excitability cycles and reflexes are also modified in monkeys (Hugon et al., 1981, 1983). The electrical activity of diverse brain structures has been studied in vivo in the rat. These include the hippocampus, thalamus and caudate nucleus and show an increase in slow waves (Fagni et al., 1985, Rostain et al., 1986a). However, epileptic seizures do not appear first in any of these three structures (Rostain et al., 1986a); when they occur they are secondary to the electrographic seizure recorded at the cortical level.

The Role of Hyperbaric Variables in the Occurrence and Development of HPNS in Helium-Oxygen Mixture

Rate of Increased Pressure

The compression and the type of compression (speed as a function of depth, stops at specific levels) play a large part in the occurrence of HPNS in the intensity of its diverse symptoms.

The first very deep dives by man with a helium-oxygen mixture were performed with very fast rates of compression. The COMEX Physalie 1 to IV dive series in 1968 were performed with a compression rate of 180 m.s.w. per 1 hour; they induced disturbances in the central nervous system beyond 200 rn.s.w., and the depth of 365 m.s.w. was not exceeded because these were too severe (Brauer et al., 1969, Fructus et al., 1969). Subsequent experirnents were carried out with slower speeds of compression, and depths of 457 m.s.w., 500 m.s.w. or 610 m.s.w. were reached (Bennett and Towse, 1971, Rostain and Naquet, 1974, Fructus et al., 1976). These experiments have shown that the depth at which the symptoms of HPNS occur tends to increase while their intensity tends to decrease if compression follows a slow exponential time course interrupted by holds at intermediate depths. HPNS is enhanced by continuous and/or rapid compression.

Influence of Depth (or Pressure) in Helium-Oxygen Breathing Mixture

The effects of pressure itself have been described during hydrostatic compression of aquatic animals. The works of Regnard (1891), Ebbecke (1936), Cattell (1936), for example, have shown that the principal effect of pressure is a stimulation of the central nervous system from 50 bars. With pressure of 200 to 300 bars spontaneous muscular contractions appear, followed by paralysis of the animal. Pressure of 400 to 500 bars is lethal. Miller (1972) compressed tritons with increasing hydrostatic pressure or with pressures of mixed gases. He observed in the first case behavioural changes at about 100 bars and paralysis of the animal at 200 bars. In the second case, with helium, the effects were similar but appeared at a slightly greater depth; protection by the gas being linked to its narcotic potency. Fish under compression (Barthelemy and Belaud, 1972, Barthelemy et al., 1981) show pressure induced changes from 20 bars.

ln mammals the effects of pressure per se are difficult to define because the effects of compression are so closely associated with the pressurizing respiratory mixture. Nevertheless, the phenomena which one observes during a prolonged stay at constant pressure might be in part linked to the pressure effect itself.

Origins and Mechanisms of HPNS in Helium-Oxygen Mixture

The origins and the mechanisms of HPNS begin to be understood from the neurochemical studies performed during the last 30 years.

In order to try to explain the events which occur at pressure, various authors have proposed several years ago, the hypothesis of hypoxia (Bennett, 1966; Chouteau and Imbert, 1971; Hyacinthe et al., 1973) or have postulated that there are disturbances of the membrane structure (Bennett et al., 1975 , Miller, 1975) or of the energetic process (Rostain and Charpy, 1976).

Electrophysiological studies have shown that the cerebellum is probably not the origin of hyperbaric tremor (Kaufmann et al., 1978, Fagni, 1987) but that the spinal cord is affected by the hyperbaric environment and could produce, in the spinal animal, rhythmic muscular activities, but which differ from those recorded in the intact animal (Kaufmann et al., 1979, Fagni, 1987, Fagni et al., 1982). This suggested a role for the higher nervous centres in the aetiology of hyperbaric tremor.

The muscular spasms and myoclonus without electroencephalographic changes of a paroxystic type (spikes, spike and wave) seen in the rat and in the baboon Papio papio (Rostain et al., 1984a, 1986a) suggest that they have a non-cortical origin. The muscular disturbances could originate in the peripheral nervous system or in the lower part of the brain (spinal cord, brain stem) and could be similar to the non epileptic myoclonus described by Hallett et al. (1977). In support of this hypothesis, the results of Kaufmann et al. (1979) in the spinal rat and those of Fagni et al. (1982) on the spinal cat show that the spinal cord plays a role in HPNS. In addition, Brauer et al. (1981) have shown with 5-deoxyglucose that Type I seizures in the mouse (which are similar to muscular seizures in the baboon Papio papio, Rostain, 1980) would have a sub-cortical origin found in the ventral region of the thalamus, the posterior hypothalamus and the pontine and medullar reticular formation.

All these studies have demonstrated the complexity of HPNS. The variables of the hyperbaric environment have differential actions on the different nervous structures which react according to their own sensitivity (Rostain, 1980). In general, the results obtained during experiments carried out both in animals and men, suggest that there is some dysfunction of numerous biochemical mechanisms.

Thus in the baboon Papio papio the epileptic seizure localized in the posterior region of the cortex in a helium-nitrogen-oxygen mixture, while generalised and of tonic-clonic type in the absence of nitrogen, could be linked to hyperexcitability of this region in relation to a change in the level of gamma amino butyric acid (GABA); an inhibitory neurotransmitter. Similar phenomena have been observed in photosensitive baboons when the level of this neurotransmitter is modified by pharmacological manipulations (Meldrum and Horton, 1974; Meldrum et al., 1979).

Again, in the baboon Papio papio changes in cortical excitability recorded during the study of somatic evoked potentials could also be related to disturbances of the GABAergic system, as could cycles of excitability. The vigilance decreased and sleep disturbances (which in the baboon can become total sleep deprivation at depths greater than 1000 m.s.w.) suggest the possibility of disruption of the noradrenaline-serotonin system, insofar as it can be implicated in the regulation of vigilance.

Likewise, the increase of tremor could be related to an increase in catecholamines (increase in physiological tremor, Marsden, 1978). The disturbances of the monoamines could be related to behavioural changes and hallucinatory type episodes.

In other respects, the study of reflexes carried out on the baboon Papio papio and on man (Fagni et al., 1980; Hugon et al., 1980, 1981, Harris, 1979a, b, Harris and Bennett, 1983) demonstrate a certain number of variations which suggests that there are disturbances of the cholinergic system at least in the peripheral nervous system.

Therefore, the study of HPNS mechanisms and neurophysiological effects of compression proceeds via a neuropharmacological and neurochemical approach associated with an electrophysiological study of the nervous structures likely to be implicated in the appearance of HPNS. Such studies have been undertaken by various laboratories to try to both alleviate diverse symptoms of HPNS and to identify the neurotransmitters which could be implicated. The studies were particularly focalized on amino acid neurotransmitters and on monoamines (Meldrum et al., 1983, Forni and Rostain, 1984, 1989, Rostain et al., 1986, Wardley-Smith et al., 1986, Rostain and Forni, 1995, Abraini and Rostain, 1991a, Darbin et al., 1997b; 1999; Kriem et al., 1995, 1996a, b, 1998b, Balon et al., 2002 a, b.).

1. Amino Acid Neurotransmitters

The neurophysiological results obtained from several experiments have suggested that there are disturbances of GABA mediated neurotransmission. Experiments performed with drugs which enhance neurotransmission mediated by GABA by a variety of mechanisms were used. The implication of this neurotransmitter (Rostain et al., 1984e, 1986a) has been demonstrated in some EEG changes. Moreover the use of sodium valproate known to have complex pharmacological actions beside the reduction of EEG changes, also increased the threshold pressure for tremor onset (from 400 to 800 m.s.w.) and myoclonus (from 700 to 900 m.s.w.)

The effect of this drug on several HPNS symptoms might be the result of these various actions. Indeed, sodium valproate produces increased cortical content of GABA (Chapman et al., 1982). The mechanism includes some inhibition of GABA transaminase (Anlezark et al., 1976, Simler et al., 1973) and some inhibition of succinic-semialdehyde dehydrogenase (Van der Laan et al., 1979) in addition to a possible small stimulation of glutamic acid decarboxylase, an enzyme involved in the synthesis of GABA (Loscher, 1981). Sodium valproate leads to a decrease in the concentration of aspartate in the cortex and hippocampus of the rat brain (Chapman et al., 1982). Aspartic acid is another putative amino acid neurotransmitter which has an excitatory action. Consequently the action of sodium valproate on tremor and myoclonus would be related to its action on aspartate. To test this hypothesis, the injection of 2-amino-7-phosphonoheptanoic acid (2APH), a compound which is an antagonist of excitation produced by dicarboxylic amino acids with preferential activity at N-methyl-D-aspartate preferring receptors (Perkins et al., 1981), ncreased the threshold pressure for tremor in rats from 400 m.s.w. to 800 m.s.w. (Meldrum et al., 1983, Wardley-Smith et al., 1986). The results obtained from these studies on GABA and aspartate seem to show that GABA does not play an essential role in the appearance of HPNS. The injection of those drugs which increase the GABA level reduces the general excitability of the nervous system and could have a beneficial effect on certain symptoms. The works of Bichard and Little (1985) on the isolated spinal cord and of Little and Thomas (1986) on the superior cervical ganglion in vitro confirm this conclusion by showing that the enhancement of GABA by a pharmacological agent can reduce the hyperexcitability brought on by compression, but it is insufficient to suppress HPNS symptoms.

However, changes in neurotransmission of the aspartic acid type brought about by the increase in pressure of the helium-oxygen mixture could be implicated in the occurrence of hyperbaric tremor. Focal injection of NMDA into the ventrolateral thalamic nucleus and more particularly into the substantia nigra significantly lowered the threshold pressure for tremor by about 30-40% and increased its intensity. In contrast, the injection of 2 APH significantly increased the threshold pressure by about 40-50% and decreased its intensity (Millan et al., 1989, 1990).

Such a change produced by a hyperbaric environment has been confirmed by Fagni et al. (1987) and Zinebi et al. (1988) who have recorded an increase in the sensitivity of the receptors to N-methyl-D-aspartate in hippocampal slices exposed to high pressures of helium. Indeed, this *in vitro* preparation includes excitatory neurotransmission (Schaffer-commissural afferents from hippocampic pyramidal cells CA3), which activates two types of amino acid receptors of pyramidal cells CA1: receptors sensitive both to quisqualate and kainate, and receptors sensitive to NMDA. It also includes inhibitory neurotransmission with receptors sensitive to GABA (pyramidal cells CA1).

These studies have shown that hyperbaric pressure produced a decrease of excitatory and inhibitory amino acid neurotransmissions which reached 50% at 81-101 ATA (8.1-10.1 Mpa) (Fagni et al., 1987a, Zinebi et al., 1988a). These observations are in agreement with those obtained by other authors in neuromuscular junction in crustacea (Kendig and Grossman; 1987). However, an intrinsic hyperexcitability was recorded in the target cell (pyramidal cells CA1) which may induce epileptiform activities (Fagni et al., 1987a). Neuropharmacological studies performed *in vitro* in hippocampal slices show that excitatory responses obtained with application of quisqualate or kainate were not modified under pressure. In the same way, the inhibitory responses obtained by application of GABA or muscinol ($GABA_A$ receptor agonist), were not modified with hyperbaric pressure (Fagni et al., 1987b, Zinebi et al. 1988b). Similar results have been obtained from the rat superior cervical ganglion *in vitro* (Little and Thomas, 1986). These results suggest that the decreases in excitatory and inhibitory neurotransmission are related to a decrease in the release of neurotransmitter implicated glutamate and GABA and not to a deficit of receptors.

Moreover, the neuropharmacological studies have also shown that the excitatory effects of NMDA on pyramidal cells CA1 were enhanced under pressure (Fagni, 1987). This hypersensitivity of NMDA receptor and the decrease of GABA activity recorded under pressure may explain the intrinsic hyperexcitability of CA1 cells (Zinebi et al., 1988b) and the epileptiform activities recorded in rat hippocampal slices *in vitro* (Fagni et al., 1985).

This hypothesis may be supported by the fact that the use of specific antagonists of NMDA receptors (5-aminophosphonovalerate, 5APV or 2APH) and the use of substances which enhanced the GABA inhibition (nipecotic acid, muscinol or diazepam) when administered together suppressed the excitatory effects of pressure in hippocampal slices *in vitro* (Fagni et al., 1989, Zinebi et al., 1990). Pressure produced an increase in the sensitivity of NMDA receptors and a decrease of GABAergic neurotransmission which led to hyperexcitability. Similar results have been reported by Grossman and Kendig (1988), in excitatory and inhibitory synapses of isolated nerve-muscle preparation of 2nd and 4th segments of lobster. These results and the explanation of the effects of pressure on hippocampal cortex could be extended to other structures (neocortex, spinal cord), which have the same excitatory and inhibitory amino acid controls. These results suggest that drugs which enhance GABA neurotransmission and antagonize NMDA activities reduce HPNS in whole animals.

Recent experiments performed on GABA neurotransmission with the agonist of $GABA_A$ and $GABA_B$ receptors (muscimol and baclofen) or GABA reuptake inhibitor (nipecotic acid), or GABA transaminase inhibitor, Gamma Vinyl GABA (GVG) injected in the substantia nigra, suggested that changes in GABA neurotransmission in the substantia nigra reticulata (SNr) but not in the substantia nigra compacta (SNc), play a crucial role in the control of motor and locomotor hyperactivity (Kriem et al., 1998a, Balon et al., 1999, 2002a, b). Moreover, the administration of NMDA receptor antagonists (2-APH) into the SNr reduced the LMA (Kriem et al., 1999) and the myoclonia while the NMDA receptor agonist (PDA) increase the myoclonia. The administration of the glutamate uptake inhibitor L-trans-PDC in the substantia nigra attenuated the locomotor hyperactivity. These results, obtained at the level of SNr, suggest that high helium pressure induced a decrease in GABAergic inputs to the SNr (Kriem et al., 1998a) with a concomittant more severe reduction in activity of the glutamatergic inputs in the SNr, so that the main effect would be a reduction in activity of the GABAergic output from th SNr to the thalamus. If such, it is likely local infusion into the SNr of L-trans-PDC, which increases selectively glutamate level by preventing its uptake could have restored the neurochemical balance between GABA and glutamate neurotransmissions and thus attenuated the development of LMA (Abraini et al., 1999). This agrees with data obtained by Millan et al. (1989) that has shown that injection into the SNr of NMDA, a glutamate receptor agonist, had a sedative effect. Furthermore, the works of Abraini et al. (1999) confirm that LMA and myoclonia have different neural origins and further suggest that glutamate in the SNr would be involved in the development of myoclonia. Others have shown implication of other neural structures such as pedunculopontine nucleus, ventrolateral thalamic nucleus, red nucleus, entopeduncular nucleus (Millan et al., 1989, 1990, 1991) and the globus pallidus (Kriem et al., 1999, Abraini et al. 1999, Darbin et al., 2000) in the development of the different symptoms of HPNS. Furthermore, experiments using non-NMDA receptor antagonists indicate that they are sometimes effective against tremor but generally ineffective against seizures (Wardley -Smith and Meldrum, 1984, Wardley-Smith et al., 1987, Pearce et al., 1994) and that non-NMDA receptor agonists administered in substantia nigra or the globus pallidus reduce LMA (Kriem et al., 1999).

Recent microdyalisis studies in the striatum of free moving rats have shown an increase of glutamate concentrations under helium pressure. (Darbin et al., 2001, Vallée et al., 2009a) that indicated at least glutamate is not reduced in this structure by helium pressure.

However, all these studies did not exclude that disruptions of other neurotransmissions by pressure could be implicated in the occurrence of some symptoms of HPNS in helium-oxygen breathing mixture.

2. Monoamine Neurotransmitters

Numerous authors have suggested a role for cerebral monoamines in the occurrence of some HPNS symptoms in helium-oxygen mixture. Thus Brauer et al. (1978) found that the hyperbaric seizure threshold is decreased by the depletion of monoamines resulting from administration of reserpine. Bowser-Riley et al. (1982) affected HPNS by administering L-DOPA (precursor of dopamine) systemically or by intracerebroventricularly injecting neurotoxins of catecholaminergic neurons such as 6-hydroxydopamine (6-0HDA) or of indolaminergic neurons such as 5-6 and 5-7 dihydroxytryptamine (DHT). Measurements of monoamines have been carried out on mice brain homogenates at the end of a pressure experiment (Koblin et al., 1980, Daniels et al., 1981). However, it is difficult to draw conclusions from these measurements as they were performed in whole brain homogenates and included several variables resulting from the effects of compression and decompression. In addition, these methods cannot take into account disorders of neurotransmission either pre or post-synaptically.

The use of electrochemical techniques developed in vivo for animals (Adams, 1978, Gonon et al., 1978, Marsden, 1979) enables continuous measurement of monoamines at a predetermined point in the nervous system. A multifiber carbon electrode developed by Forni (1982) which is sensitive to dopamine and allows in vivo measurement for several weeks as described by Forni and Nieoullon (1984) and El Ganouni et al. (1987) has been implanted in the striatum of hamsters and rats. This structure is rich in dopaminergic endings and is implicated in the control of motor, locomotor and some cognitive processes that are disrupted by helium pressure. The differential pulse voltammetry (DPV) studies demonstrated an increase in extracellular dopamine level which occurred around 30 bars (Forni and Rostain, 1989, Abraini and Rostain, 1991a, Rostain and Forni, 1995). The increase of dopamine at 80 or 90 ATA (8 or 9 MPa) was around 25-30% with a return to basal value during the decompression. This increase is independent of changes of temperature or partial pressure of oxygen, which occur in the hyperbaric experiments (Forni and Rostain, 1989, Rostain and Forni, 1995). Consequently, the increase of extracellular DA in the caudate nucleus is pressure-dependent; it was also found in the nucleus accumbens (Abraini and Rostain, 1991d). This increase is correlated with the locomotor and motor hyperactivity (LMA) described in free-moving animals (Tomei et al., 1991, Abraini and Rostain, 1991a). Similar changes have been reported in monkeys exposed to high pressure of a helium-oxygen mixture (Forni and Rostain, 1984, Cosson et al., 1989).

Various drugs have been used in rats to research the origin of this increase. Alpha-para-methyl-para-tyrosine (inhibitor of DA synthesis), apomorphine (agonist of D1/D2 receptors), SKF 38393 (D1 agonist) and quinpirole (D2 agonist), all of which decrease the extracellular DA level at normal pressure (Abraini and Rostain, 1991b,c), showed that pressure suppressed this effect and dopamine returned to reference values or to slightly higher values. These results suggest that the increase in extracellular DA might be related to increase release by mobilization of stored DA. Paul and Philp (1989), using HPLC measurements of endogenous DA and 3-4 dihydroxyphenylacetic acid (DOPAC), a metabolite of DA, concluded that pressure increased the release of DA without any action on re-uptake or degradation. Studies with reserpine, a depletor of DA vesicle stores have indicated that the striatal DA increase in helium-pressure was the consequence of an increase of DA synthesis and DA release (Fechtali et al., 1994). Moreover, studies with agonists and antagonists of D1 and D2 receptors (Abraini and Rostain, 1991c, Abraini et al., 1992) have indicated a different action of pressure on D1 and D2 receptors. According to the hypothesis of Abraini and Fechtali (1992), high pressure would produce interactive disorders in the link mechanisms of regulation between the dopaminergic and amino-acidergic systems.

Other studies in free-moving rats using microdialysis have confirmed the pressure induced DA striatal increase (Requin and Risso, 1992, Darbin et al., 1997a, b) and the implication of both, D1, D2 and NMDA receptor in this change (Darbin et al., 1999). Moreover the development of LMA requires both striatal D1, D2 and NMDA receptor activities (Darbin et al., 1999, 2000) and pressure enhanced NMDA activity in the striatum and also in the globus pallidus which appeared to be implicated in the occurrence of these motor disturbances and to be involved in myoclonia (Kriem et al., 1999, Darbin et al., 2000). Recent works performed on GABAa or GABAb receptors by the use of agonists (muscimol or baclofen) injected focally in SNr or SNc have shown the implication of these receptors in the striatal dopamine release and that helium pressure inhibits the activity of GABA neurotransmission at the level of substantia nigra which induces a DA increase (Balon et al., 1999, 2002b).

Other monoamines such as serotonin (5-HT) could be implicated in HPNS in helium-oxygen breathing mixture. Neuropharmacological studies have indicated that some behavioral symptoms in rats seem to be related to activation of the 5-HT1a receptor subtype (Wardley-Smith et al., 1990). These behavioral changes are reduced in rats pretreated by para-chloro-phenyl-alanine (PCPA), which depleted 5-HT levels in brain, and are increased by pretreatment with a 5-HT uptake inhibitor, Wy 27587. David et al., (1988) suggest the involvement of serotonin in the enhancement of rat spinal excitability under hyperbaric pressure. The hyperbaric alteration in spinal excitability in rats pretreated with the classical 5-HT receptor antagonist drug metergoline was suppressed. Studies performed on the role of serotoninergic regulation on nigro-striatal and meso-limbic pathways have demonstrated that the 5-HT1B and 5-HT3 receptors are activated by helium pressure and are implicated in part in the DA increase and LMA development (Kriem et al., 1995, 1996a). Moreover, the antagonists of these receptors blocked the occurrence of myoclonia. This last result suggests that serotonin could be implicated in this symptom.

Microdialysis studies have indicated that helium pressure induces an increase of serotonin in the striatum which could explain the activation of these receptors (Darbin et al., 1997b, 2010, Vallée et al., 2009b). At the level of nucleus accumbens, the use of 5-HT2-C receptors agonist has indicated that this receptor plays a crucial role in the development of LMA and in the pressure-induced increase in DA accumbens release. In contrast, the administration of 5-HT2A receptors antagonist had no effect on LMA and DA increase but reduce myoclonia (Kriem et al., 1996b, 1998b). These results suggest that helium pressure may simultaneously induce an increase in 5-HT transmission at the level of 5-HT2A receptors and a decrease in 5-HT transmission at the level of 5-HT2C receptors (Kriem et al., 1998b). Consequently these 5-HT receptors would be located on different neurons, so that one of which enhances while the other attenuates DA release. The 5-HT2A receptor would be located at the post synaptic level on GABA neurons and the 5-HT2C receptor would be an autoreceptor; the activation of this receptor by an agonist reduced the increase of 5-HT release produced by pressure and consequently a reduction of DA release (Kriem et al., 1998b).

These basic researches have increased the data on the origin and mechanisms of HPNS. Helium pressure disrupts the neurochemical interactions between the systems and the structures of the central nervous system, changes specifically some neurotransmission by acting on synthesis or receptor activities. Daniels et al. (1998) have shown in Xenopus oocytes, that NMDA receptors are sensitive to pressure and Mor and Grossman (2006, 2007) have reported that isolated NMDAR response was augmented at high pressure. Helium pressure could have an action on ion channels (Etzion and Grossman, 1999).

CONCLUSION

More studies are needed to understand the mechanisms implicated in the action of inert gas at pressure. The reduction or even the suppression of some clinical symptoms obtained by narcotic gases added to a helium-oxygen mixture supports the lipid theories. However there were also other effects recorded with some symptoms of HPNS which indicated that the lipid theory is insufficient to explain alone the effects of pressure and inert gases at pressure.

Recently, protein theories have been proposed due to results obtained from some experiments with inhalational anesthetics which have been interpreted as evidence for a direct anesthetic-protein interaction (Franks and Lieb, 1982, 1984, 1991, 1994). The question is whether inert gases do exert binding processes on proteins at raised pressure. Data obtained by Abraini et al. (1998) with two inert gases (nitrogen and argon) and an anesthetic gas nitrous oxide (N2O) seem to indicate that inert gases bind directly to a modulatory site of a protein receptor and act as allosteric modulators. The results clearly showed whatever the gas used, the pressure required to produce a 100% loss of righting reflex increased significantly as the compression rate increased. The rate at which compression was applied influenced the anesthetic potencies of these 3 gases in a sigmoidal fashion rather than a linear fashion as the lipid theory would suggest. The sigmoidal curve indicates a gas-protein interaction. The gas could bind to modulatory sites of protein receptors, producing conformational changes and thereby make channel opening more or less favourable.

Colloc'h et al. [2007] have examined using X-ray crystallography the binding characteristics of the Xenon (Xe) and nitrous oxide (N2O) that act by blocking the NMDA receptor. Xe and N2O would first bind to brain intracellular proteins possessing large hydrophobic cavities, which constitute easy targets for inhalational anesthetics, thereby disrupting the activity of these proteins in a manner sufficient to induce moderate neuronal dysfunctions and leading to the early stages of anesthesia (amnesia and hypnosis). If gas concentration rises, the smaller hydrophobic gas-binding cavities within the NMDA receptor would then begin to be filled, thereby disrupting the receptor function and leading to surgical anesthesia. Similar step-by-step mechanisms of general anesthesia, which assume a causal link between the behavioural effects of anesthesia, from amnesia and hypnosis to ''surgical anesthesia'' and the progressive occupation of the anesthetic binding sites from globular proteins to ion channel receptors, may occur for other types of inhaled anesthetics or narcotic gases and/or ion channel receptors, such as the GABAa receptor, which is thought to be the molecular target of most volatile anesthetics and NMDA receptor, that are together the target of inert gases at pressure.

As the binding sites between molecules of the gas and protein are hydrophobic cavities, a link could be made between the membrane theory and the protein theory; and such mechanisms may also explain a number of critical exceptions to the Meyer-Overton rule.

References

Abraini, J.H. and Rostain, J.C. 1991a. Pressure-induced striatal dopamine release correlates hyperlocomotor activity in rats exposed to high pressure. *J. Appl. Physiol. 71*: 638-643.

Abraini, J.H. and Rostain, J.C. 1991b. Effects of the administration of a-methyl-p-tyrosine on the striatal dopamine increase and the behavioral motor disturbances in rats exposed to high pressure. Pharmacol. *Biochem. Behav. 40*: 305-310.

Abraini, J.H. and Rostain, J.C. 1991c. Effects of helium-oxygen pressure on dopaminergic receptors. *Undersea Biomed. Res .18* (Suppl), 55.

Abraini, J.H. and Rostain J.C. 1991d. Dopamine increase in the nucleus accumbens of rats exposed to high pressure. *Neuroreport 2*: 233-235.

Abraini, J.H. and Fechtali, T. 1992. A hypothesis regarding possible interactions between the pressure-induced disorders in dopaminergic and amino-acidergic transmission. *Neurosci. and Biobehav. Rev. 16:* 597-602.

Abraini, J.H., Fechtali, T. and Rostain, J.C. 1992. Pressure reversed extracellular striatal dopamine decrease produced by D1 receptor agonist SKF 38393, and D2 receptor agonist LY 171555, but failed to change the effect of the activation of both Dl and D2 receptors. *Neuroscience 50*: 395-402.

Abraini, J.H., Rostain, J.C. and Kriem, B. 1998. Sigmoidal compression rate-dependence of the narcotic potency of inert gases in rats: implication for lipid vs protein theories of inert gas action in the central nervous system. *Brain Res. 808:* 300-304.

Abraini J.H., Kriem, B. and Rostain, J.C. 1999. Administration of the glutamate uptake inhibitor L-trans-PDC in the globus pallidus and the substantia nigra reticulata, but not in the striatum attenuates the psychostimulant effect of high helium pressure on locomotor activity in the rat. *Neurosci. Res.* 35: 273-279.

Adams, R.N. 1978. *In vivo* electrochemical recording: A new neurophysiological approach. Trends *Neurosci. 1*: 160-163.

Aquadro, C.F. and Chouteau, J. 1967. Problems of extreme duration in open sea saturation exposure. In: C.J. Lambertsen [ed.] Underwater Physiology III. pp. 98-108. Williams and Wilkins, Baltimore USA.

Anlezark, G., Horton, R.W., Meldrumn B.S. and Samaya, M.C.B. 1976. Anticonvulsivant action of ethanolamine-a-sulphate and di-n-propylacetate and the metabolism of y-aminobutyric acid (GABA) in mice with audiogenic seizures. *Biochem. Pharmacol. 25*: 413-417.

Balon N, Kriem, B. and. Rostain, J.C. 1999. Implication of GABA receptors in regulation of the nigro striatal pathway of rats exposed to helium-oxygen pressure. *Undersea and Hyperbaric Med. 26* (suppl): 82.

Balon, N., Kriem, B., Dousset, E., Weiss, M. and Rostain, J.C. 2002a. Opposing effects of narcotic gases and pressure on the striatal dopamine release in rats. *Brain Res. 947:* 373-379.

Balon, N., Kriem, B., Weiss, M., and Rostain, J.C. 2002b. GABAergic modulation in the substantia nigra of the striatal dopamine release and of the locomotor activity in rats exposed to helium pressure. *Brain Res. 948*: 82-92.

Balon, N., Kriem, B., Weiss, M., and Rostain, J.C. 2002c. Indirect presynaptic modulation of striatal dopamine release by GABA B receptors in the rat substantia nigra. *Neuroscience Lett. 325*: 33-36.

Barthélémy, L. et Belaud, A. 1972. Constatations physiologiques et physiopathologiaques faites sur un poisson (Anguilla anguilla L) en conditions hyperbares. *Bull. Med. Sub. Hyp. 8*: 33.

Barthelemy, L., Belaud, A. and Saliou, A. 1981. A study of the specific action of "per se" hydrostatic pressure on fish considered as a physiological model.. In: A.J. Bachrach and M.M. Matzen [eds.]. Underwater Physiology VII. pp. 641-649. Undersea Medical Society, Bethesda Md, USA.

Behnke, A.R.,. Thomas, R.M. and Motley, E.P. 1935. The psychologic effects from breathing air at 4 atmospheres pressure. *Am. J. Physiol. 112*: 554-558.

Bennett, P.B. 1966. The Aetiology of Compressed Air Intoxication and Inert Gas Narcosis. International Series of Monographs in Pure and Applied Biology; Zoology Division, Vol. 31, Pergamon Press, Oxford, U.K.

Bennett, P.B. 1967. Performance impairment in deep diving due to nitrogen, helium, neon and oxygen. In: C.J. Lambertsen [ed.] Underwater Physiology III. pp 327-340. Williams and Wilkins, Baltimore, USA.

Bennett P.B. and Rostain, J.C. 2003a. Inert gas narcosis. In: A.O. Brubakk, T.S. Neuman (eds). Bennett and Elliott's Physiology and Medicine of Diving. (5th ed). , pp300-322. London: Saunders

Bennett, P.B. and. Rostain, J.C. 2003b. The High Pressure Nervous Syndrome. In: A.O. Brubakk and T.S. Neuman [eds.] Bennett and Elliott's Physiology and Medicine of Diving. (5th ed). pp. 323-357. London: Saunders.

Bennett, P.B., Simon, S. and Katz, Y. 1975. High pressure of inert gases and anesthesia mechanisms. In : B.R. Fink [ed.] Molecular Mechanism of Anesthesia. Progress in Anesthesiology. pp. 367-403. Raven Press, New-York, USA.

Bennett, P.B. and Towse, E.J. 1971. The High Pressure Nervous Syndrome during a simulated oxygen-helium dive to 1500 ft. *Electroenceph. clin. Neurophysiol. 31*: 383-393.

Bichard, A.R. and Little, H.J. 1982. Drugs that increase y-aminobutyric acid transmission protect against the High Pressure Neurological Syndrome. *Br. J. Pharmacol. 76*: 447-452.

Bichard, A.R. and Little, H.J. 1985. Effects of high pressure helium on gamma amino butyric acid release from isolated spinal cord. *J. Neurochemistry. 44*: 999-1005.

Bowser-Riley, F., Dobbie, J.A., Paton, W.D.M. and Smith, E.B. 1982. A possible role for monoaminergic inhibition in the high pressure nervous syndrome. *Undersea Biomed. Res. 9 (Suppl)*: 32-33.

Brauer, R.W. 1975. The high pressure nervous syndrome: animals. In: P.B. Bennett and D.H. Elliott [eds.]. Physiology and medicine of diving and compressed air work. pp. 231-247. Bailliere Tindall, London., UK.

Brauer, R.W., Beaver, R.W. and Sheehan, M.E. 1978.The role of monoamine neurotransmitters in the compression rate dependence of HPNS convulsions.. In: C.W. Shilling and M.W. Beckett [eds] Underwater Physiology VI. Pp. 49-59. Fed Am Socs Exp Biol, Bethesda, MD, USA

Brauer, R.W., Dimov, S., Fructus, X., Gosset, A. and Naquet, R. 1969. Syndrome neurologique et électrographique des hautes pressions. *Rev. Neurol. 121*: 264-265.

Brauer, R.W., Mansfield, W.M., Beaver, R.W. and Gillen, H.W. 1981.The HPNS as a composite entity-consequences of an analysis of the convulsion stage. In: A.J. Bachrach and M.M. Matzen [eds.] Underwater Physiology VII. . pp. 391-399. Undersea Med. Soc., Bethesda, Md, USA.

Brauer R.W., Way, R.O. and. Perry, T.A. 1968. Narcotic effects of helium and hydrogen and hyperexcitability phenomenon at simulated depths of 1500 to 4000 ft of sea water.. In: B.R. Fink [ed] Toxicity of Anesthetics. pp 241-255. Williams and Wilkins, Baltimore, USA.

Brauer, R.W. and Way, R.O. 1970. Relative narcotic potencies of hydrogen, helium, nitrogen and their mixtures. *J. Appl. Physiol. 29*: 23-31.

Cabarrou, P., Harmann, H., Weiner, K.H., Alinat, P. and Fust, H.D. 1966. Introduction à la physiologie de "Homo aquaticus". *Presse Med. 74* : 2771-2773.

Case, E. M. and Haldane, J.B.S. 1941. Human physiology under high pressure. *J. Hyg. Camb. 41*: 225-249.

Cattel, McK. 1936. The physiological effects of pressure. *Biol. Rev. 11*: 441-475.

Chapman, A., Keane, P.E., Meldrum, B.S. Simiand, J. and Vernieres, J.C. 1982. Mechanisms of anticonvulsant action of valproate. *Prog. Neurobiol. 19*: 315-359.

Chouteau, J. et Imbert, G. 1971. La limitation hypoxique de la plongée profonde de longue durée. *Maroc Med. 52* : 229-236.

Colloc'h, N., Sopkova-de Oliveira Santos, J., Retailleau, P., Vivarès, D., Bonneté, F., Langlois d'Estainto, B., Gallois, B., Brisson, A., Risso, J.J., Lemaire, M., Prangé, T., Abraini, J.H. 2007. Protein crystallography under xenon and nitrous oxide pressure: comparison with in vivo pharmacology studies and implications for the mechanism of inhaled anesthetic action. *Biophys J. 92*: 217-224.

Cosson, P., Etienne, J. and Gillard, J. 1989. Voltammetry in the striatum of chronic implanted monkey during exposure to hydrogen-oxygen mixtures at 51 ATA: Detection of dopamine.. In: J.C. Rostain, E. Martinez and C. Lemaire [eds.]. High Pressure Nervous Syndrome, 20 Years Later. pp 239-249. ARAS-SNHP Publications, Marseille, France.

Daniels, S., Green, A.R., Koblin, D.D., Lister, R.G., Little, H.J., Paton, W.D.M., Bowser Riley, F., Sha, W.S.G and Smith E.B. 1981. Pharmacological investigation of the High Pressure Neurological Syndrome: Brain Monoamine concentrations.. In: A.J. Bachrach and M.M. Matzen [eds.]. Underwater Physiology VII. p. 329-336. UMS, Bethesda, Md, USA.

Daniels S., Roberts, R. and William, N. 1998. Effects of high pressure on post synaptic ionotrophic receptors. In: P.B. Bennett, I. Demchenko and R.E. Marquis [eds.]. High Pressure Biology and Medicine.: pp 22-31. University of Rochester Press, New York, USA.

Darbin, O., Risso, J.J. and Rostain, J.C. 1997a. A new system analysis of motor and locomotor activities associated with a midrodialysis study of pressure-induced dopamine increase in rats. *Physiol. Behav. 62:* 367-371.

Darbin, O., Risso, J.J. and Rostain, J.C. 1997b. Pressure induces striatal serotonin and dopamine increases: a simultaneous analysis in free moving microdialysed rats. *Neuroscience Lett 238*: 69-72.

Darbin, O., Risso, J.J. and Rostain, J.C. 1999. The full expression of locomotor and motor hyperactivities induced by pressure requires both striatal dopaminergic and NMDA receptors activities in rat. *Neuroscience Lett. 267*: 149-152.

Darbin, O., Risso, J.J. and Rostain, J.C. 2000 High pressure enhanced NMDA activity in the striatum and the globus pallidus: relationships with myoclonia and locomotor hyperactivity in rat. *Brain Res 852:* 62-67.

Darbin, O., Risso, J.J. and Rostain, J.C. 2001. Helium-oxygen pressure induces striatal glutamate increase: a microdialysis study in freely-moving rats. *Neuroscience Lett. 297*: 37-40.

Darbin, O., Risso, J.J. and Rostain, J.C. 2010. Dopaminergic control of striatal 5-HT level at normobaric condition and at pressure. *Undersea Hyperb Med., 37(3)*: 159-166.

David, H.N., Balon, N., Rostain, J.C. and Abraini, J.H. 2001. Nitrogen at raised pressure interacts with the GABAA receptor to produce its narcotic pharmacological effects in the rat. *Anesthesiology. 95:* 921-927.

Ebbecke, U. 1936. Uber das verhalten des Zentralnervensystems (Ruckenmarksfrosh) unter Einwirkung hoher Drucke. *Pflug. Arch. 237*: 785-789.

El Ganouni, S., Forni, C. and Nieoullon, A. 1987. In vitro and in vivo characterization of the properties of a multifiber carbon electrode allowing long-term electrochemical detection of dopamine in freely moving animals. *Brain Res. 404*: 239-256.

Etzion, Y. and Grossman, Y. 1999. Spontaneous NA+ and CA2+ spike firing of cerebellar Purkinje neurons at high pressure. *Pflugers Arch. 437*: 276-284.

Fagni, L. 1987. Effets des hautes pressions d'hélium sur l'excitabilité cellulaire et la transmission synaptique dans le système nerveux central. DSc Thesis, University of Mediterranée, Marseille, France.

Fagni, L., Hugon, M. and Rostain, J.C. 1980. Facilitation des potentiels évoqués somesthésiques (PES) en plongée pfofonde a saturation. *J. Physiol. (Paris) 76* : 17A.

Fagni, L., Soumireu-Mourat, B., Carlier, E. and Hugon, M. 1985. A study of spontaneous and evoked activity in the rat hippocampus under helium-oxygen high pressure. *Electroenceph. clin. Neurophysiol. 60 :* 267-275.

Fagni, L., Weiss, M., Pellet J. and Hugon M. 1982. The possible mechanism of the Pressure-Induced motor disturbances in the cat. *Electroenceph. clin. Neurophysiol. 53*: 590-601.

Fagni, L., Zinebi, F. and Hugon, M. 1987a. Evoked potential changes in rat hippocampal slices under helium pressure. *Exp Brain Res 65*: 513-519.

Fagni, L., Zinebi, F. and. Hugon, M. 1987b. Helium pressure potentiates the N-methyl-D-aspartate and D, L-homocysteate induced decreases of field potentials in the rat hippocampal slice preparation. *Neurosciences letters. 81* : 285-290.

Fagni, L., Zinebi, F., Hugon, M. and Rostain, J.C. 1989. Hypothèses sur les mécanismes du syndrome nerveux des hautes pressions. *Rev. Scient. Tech. Defence 2*: 115-128.

Fechtali, T., Abraini, J.H., Kriem, B. and Rostain, J.C. 1994. Pressure increases de novo synthesized striatal dopamine release in free moving rats. *Neuroreport 5*: 725-728.

Forni, C. 1982. Realization of a new multifiber electrochemical device allowing continuous *in vivo* measurements of neuromediators. *J. Neurosci. Meth. 5*: 167-171.

Forni, C. and Nieoullon, A. 1984. Electrochemical detection of dopamine release in the striatum of freely moving hamsters. *Brain Res. 297*: 11-20.

Forni, C. and Rostain, J.C. 1984. Influence des hautes pressions de mélange hélium-oxygène sur la libération de dopamine.. In: Nome, T., Susbielle, G., Comet, J., Jacquin, M. and Sciarli, R. [eds.] Proc Xth Congress of European Undersea Biomedical Society, pp, 31-36. Marseille: France. EUBS.

Forni, C. and Rostain; J.C. 1989. Effect of helium-oxygen pressure on dopamine detected *in vivo* in the striatum of hamsters. *J. Appl. Physiol. 67*: 1617-1622.

Franks, N.P. and Lieb, W.R. 1991. Stereospecific effects of inhalational general anaesthetic optical isomers on ion nerve channels. *Science; 254:* 427-430

Franks, N.P. and Lieb, W.R. 1982. Molecular mechanisms of general anaesthesia. *Nature*; *300*: 487-493.

Franks, N.P. and Lieb, W.R. 1984. Do general anaesthetics act by competitive binding to specific receptors? *Nature 310*: 599-601.

Franks, N.P. and Lieb, W.R. 1994. Molecular and cellular mechanisms of general anaesthesia. *Nature 367*: 607-614

Fructus, X.R. 1987. Hydrogen, pressure and HPNS. In: Brauer RW [ed].. Hydrogen as a Diving Gas. . pp. 125-138. Undersea and Hyperbaric Medical Society, Bethesda, MD,

USA.

Fructus, X., Agarate, C., Naquet, R. and Rostain, J.C. 1976. Postponning the "High Pressure Nervous Syndrome" (HPNS) to 1640 feet and beyond.. In: C.J. Lambertsen [ed.]. pp. 21-33. Underwater Physiology V. FASEB, Bethesda, Maryland, USA.

Fructus, X. and Chouteau, J. 1963. Aspects physiologiques de la vie sous pression. *Médecine de la plongée.* 1-17

Fructus, X. and Delauze, H.G. 1968. Aspects opérationnels de la plongée profonde. *Rev. Pysiol. Sub. Med. Hyp. 1:* 50-58.

Fructus, X., Gardette, B., Carlioz, M. and Giran, Y. 1984. Hydrogen narcosis, In: Nome, T., Susbielle, G., Comet, M., Jacquin, M. and Sciarli, R. [eds.]. Proceeding of Xth Congress EUBS, pp87-96 Marseille, France. EUBS.

Fructus, X., Naquet, R., Gosset, A., Fructus, P. and Brauer, R.W. 1969. Syndrome Nerveux des Hautes Pressions. *Marseille Med.* ***6***: 509-512.

Fructus, X. and Rostain, J.C. 1978. HPNS: A clinical study of 30 cases.. In: C.W. Shilling and M.W. Beckett [eds.] Underwater Physiology VI. pp 3-8. FASEB, Bethesda, Md. USA

Gardette, B. 1987. Hydra IV and Hydra V: human deep hydrogen dives 1983-1985. In: R.W. Brauer [ed.]. Hydrogen as a diving gas. pp. 109-117. UHMS publication 69 (WS-HYD) Bethesda, Md, USA.

Gardette, B. and Gortan, C. 1994. Mice and monkeys deep dives in heliox, hydrox and hydreliox gas mixtures - synthesis of COMEX "Hydra" programme. In: P.B. Bennett and R.E. Marquis [eds]. Basic and applied high pressure biology. pp 173-184. University Press of Rochester, New-York, USA.

Gonon, F., Cespuglio R., Ponchon, J.L., Buda, J., Jouvet, M., Adams, R.N. and Pujol, J.F. 1978. Mesure électrochimique continue de la libération de dopamine réalisée *in vivo* dans le neostratum du rat. *CR Acad. Sci. (Paris) 286*: 1203-1206.

Grossman, Y. and Kendig, J.J. 1988. Synaptic integrative properties at hyperbaric pressure. *J. Neurol. 60:* 1497-1511.

Hallet, M., Chadwick, D., Adam, J. and Marsden, C.D. 1977. Reticular reflexe myoclonus. A physiological type of human post hypoxic myoclonus. *J. Neurol. Neurosurg. Psychiat. 40*: 253-264.

Halsey, M.J. 1982. Effects of High Pressure on the Central Nervous System. *Physiological Rev. 62*: 1341-1377.

Harris, D.J. 1979a. Observations on the knee-jerk in Oxy-Helium at 31 and 43 bar. *Undersea Biomed Res. 6 :* 55-74.

Harris, D.J. 1979b. Hyperbaric hyperreflexia : tendon-jerk and Hoffmann reflexes in man at 43 bars. *Electroenceph. Clin. Neurophysiol. 47* : 680-692.

Harris, D.J. and. Bennett, P.B. 1983. Force and duration of muscle twitch contractions in humans at pressures up to 70 bar. *J. Appl. Physiol.: Respirat. Environ. Exercise Physiol. 54* : 1209-1215.

Hugon, M., Fagni, L.et Rostain, J.C. 1980. Cycle d'excitabilité des reflexes monosynaptiques et mécanismes cholinergiques. Observations de physiologie hyperbare chez les primates. *J. Physiol. (Paris),76*: 19a.

Hugon, M., Fagni, L., Rostain, J.C. and Seki, K. 1981. .Somatic evoked potentials and reflexes in monkey during saturation dives in dry chamber. In: A.J. Bachrach and M.M.

Matzen [eds.] Underwater Physiology VII. pp. 381-390. Undersea Medical Society, Bethesda, Md. USA.

Hugon, M., Fagni, L. and Seki, K. 1983. Deep sea diving: human performance and motor control under hyperbaric conditions with inert gas. In: J.L. Desmedt [ed.]. Motor control Mechanisms in Health and Disease. pp. 829-849. Raven Press. New-York, USA.

Hyacinthe, R., Broussolle, B. et Desbruyeres, D. 1973. Affinité de l'hémoglobine pour l'oxygène chez l'homme en milieu hyperbare. *Bull. Med. Sub. Hyp. 10* : 44-46.

Kaufmann, P.G., Bennett, P.B. and Farmer, J.C. 1978. Effect of cerebellar ablation on the high pressure nervous syndrome in rats. *Undersea Biomed. Res. 5*: 63-70.

Kaufmann, P.G., Finley, C.C., Bennett, P.B. and Farmer, J.C. 1979. Spinal cord seizures elicited by high pressure of helium. *Electroenceph. clin. Neurophysiol. 47*: 31-40.

Kendig J.J. and. Grossman, Y. 1987. How can hyperbaric pressure increase central nervous system excitability? In: H.W. Jannash, R.E. Marquis and A.M. Zimmerman [eds.]. Current Perspectives in High Pressure Biology. pp 159-169. Academic Press. New York. USA.

Koblin, D.D., Little, H.J., Green, A.R., Daniels, S., Smith E.G. and Paton, W.B.M. 1980. Brain monoamines and the high pressure neurological syndrome. *Neuropharmacol. 19*: 1031-1038.

Kriem, B., Abraini, J.H. and Rostain, J.C. 1996a. Role of 5-HT1B receptor in the pressure-induced behavioral and neurochemical disorders in rats. *Pharmacol. Biochem. Behav. 53*: 257-264.

Kriem, B., Cagniard, B., Rostain, J.C. and Abraini, J.H. 1998a. Modulation by GABA transmission in the substantia nigra compacta and reticulata of locomotor activity in rat exposed to high pressure. *Neuroreport 9*: 1343-1347.

Kriem, B., Rostain, J.C. and. Abraini, J.H. 1995. Involvement of 5HT3 receptor in the pressure-induced increase in striatal and accumbens dopamine release and the occurrence of behavioral disorders in free-moving rats. *Neurosciences Lett. 197*: 57-60.

Kriem, B., Rostain, J.C. and. Abraini, J.H. 1996b. Contribution of central 5-HT2 receptors in the occurrence of locomotor activity and myoclonia in freely-moving rats exposed to high pressure. *Neuroreport 7*: 2687-2690.

Kriem, B., Rostain, J.C. and. Abraini, J.H. 1998b. Crucial role of the 5-HT2C receptor, but not of the 5-HT2A receptor, in the down regulation of simulated dopamine release produced by pressure exposure in freely moving rats. *Brain Res. 796*: 143-149.

Kriem, B., Rostain, J.C. and. Abraini, J.H. 1999. Administration of either non-NMDA receptor agonists or NMDA receptor antagonists into the substantia nigra reticulata or the globus pallidus reduces the psychostimulant effect of high helium pressure on locomotor activity in rats. *Neuroreport 10*: 3777-3783.

Lever, M.J., Miller, K.W., Paton, W.D.M and Smith, E.B. 1971. Pressure reversal of anaesthesia. *Nature. 231*: 368-371.

Little, H.J. and Thomas, D.L. 1986. The effects of anaesthetics and high pressure on the responses of the rat superior cervical ganglion in vitro. *J. Physiol. (London). 374*: 387-399.

Loscher, W. 1981. Valproate induced changes in GABA metabolism at the subcellular level. *Biochem. Pharmacol. 30*: 1364-1366.

McLeod, M., Bennett, P.B. and Cooper, R.L. 1988. Rat brain catecholamine release at 1, 10, 20 and 100 ATA heliox, nitrox and trimix. *Undersea Biomed. Res. 15*: 211-221.

Marsden, C.D. 1978. The mechanisms of physiological tremor and their signifiance for pathological tremors. In: J.E. Desmedt [ed]. Physiological tremor, pathological tremors and clonus. Prog. clin. Neurophysiol. 5 : 1-16.

Marsden, C.D. 1979. Functional aspects of 5-hydroxytryptamine neurones: Application of electrochemical monitoring *in vivo*. *Trends Neurosci.* *5*: 1-16.

Marshall, J.M. 1951. Nitrogen narcosis in frogs and mice. *Am. J. Physiol.* *166*: 699-711.

Meldrum, B.S. and. Horton, R.W. 1974. Neuronal inhibition mediated by GABA and patterns of convulsions in baboons with photosensitive epilepsy (Papio papio). In: P. Harris and C. Mandley [eds.] The natural History and Management of Epilepsy. pp. 55-64. Livingstone, Edinburgh, UK.

Meldrum, B.S, Menini, C., Naquet, R., Laurent, H. and Stutzmann, J.M. 1979. Proconvulsivant, convulsivant and other actions of the baboon Papio papio. *Electroenceph. clin. Neurophysiol.* *47*: 383-395.

Meldrum, B.S., Wardley-Smith, B., Halsey, M.J. and Rostain, J.C. 1983. 2-amino-phosphonoheptanoic acid protects against the high pressure neurological syndrome. *European J. Pharmacol,* *87*: 501-502.

Millan, M.H., Wardley-Smith, B., Halsey, M.J. and Meldrum, B.S. 1989. Studies on the role of the NMDA receptor in the substantia nigra pars reticulata and entopedeuncular nucleus in the development on the high pressure neurological syndrome in rats. *Exp. Brain Res.* *78*: 174-178.

Millan, M.H., Wardley-Smith, B., Halsey, M.J. and Meldrum, B.S. 1990. Effect of NMDA and 2-amino-7-phosphonoheptanoate focal injection into the ventrolateral thalamic nucleus on the high pressure neurological syndrome in the rat. *Brain Res.* *507*: 354-356.

Millan, M.H., Wardley-Smith, B., Halsey, M.J. and Meldrum, B.S. 1991. Brain nuclei and neurotransmitters involved in the regulation of the high pressure neurological syndrome in the rat. *Neuropharmacology* *30*: 1351-1355.

Miller, K.W. 1972. Inert gas narcosis and animals under high pressure. In: M.A. Sleigh and A.G. Macdonald [eds.]. The effects of pressure on organisms. pp. 363-378. Symposia of the society for experimental biology. Cambridge university press. UK.

Miller, K.W. 1975. The pressure reversal of anesthesia and the critical volume hypothesis. In: B.R. Fink [ed.]. Molecular Mechanisms of Anesthesia. pp.341-351. Raven Press, New York, USA.

Miller, K.W., Paton, W.D, Smith, R.A. and. Smith, E.B. 1973. The pressure reversal of general anesthesia and the critical volume hypothesis. *Molecular Pharmarcol.* *9:* 131-143.

Mor, A. and Grossman, Y. 2006. Modulation of isolated N-methyl-d-aspartate receptor response under hyperbaric conditions. *Eur. J. Neurosci.* *24*: 3453-3462.

Mor, A. and Grossman, Y. 2007. High pressure modulation of NMDA receptor dependent excitability. *Eur. J. Neurosci.* *25*: 2045-2052.

Overfield, E.M., Saltzman, H.A., Kylstra, J.A. and Salzano, J.V. 1969. Respiratory gas exchange in normal men breathing 0.9 0J0 oxygen in helium at 31.3 ATA. *J. Appl. Physiol.* *27*: 471-475.

Overrath, G., Matthys, H., and. Buhlmann, A. A. 1969. Saturation experiment at 31 ATA in an oxygen-helium atmosphere. *Helv. Med. Acta,* *35*: 180-200.

Parmentier, J.L., Harris, D.J. and. Bennett, P.B. 1985. Central and peripheral causes of hyperreflexia in humans breathing 5 % trimix at 650 m. *J. Appl. Physiol.* *58*: 1239-1245.

Pearce, P.C., Maclean, C.J., Shergil, H.K, Ward, E.M., Halsey, M.J., Tindley, G., Pearson, J. and Meldrum, B.S. 1994. Protection from high pressure induced hyperexcitability by the AMPA/Kainate receptor antagonists GYKI 52466 and LY 293558. *Neuropharmacology 33*: 605-612.

Perkins, M.N., Stone, T.W., Collins, J.F. and Curry, K. 1981. Phosphonate analogues of carboxylic acids as amino acid antagonists on rat cortical neurones. *Neurosci. letters. 23*: 333-336.

Regnard, P. 1891. Recherches expérimentales sur les conditions physiques de la vie sous les eaux. Masson, Paris, France.

Requin, M. and Risso, J.J. 1992. Effects of high pressure on dopamine release in freely moving rats: a microdialysis study. *Neuroscience Lett. 146*: 211-214.

Roll, J.P., Lacour, M., Hugon, M. and Bonnet, M. 1978. Spinal reflex activity in man under hyperbaric heliox conditions (31 and 62 ATA). In: C. W. Shilling and M. W. Beckett [eds.]. Underwater Physiology VI. pp. 21-28. FASEB, Bethesda Md, USA.

Rostain, J.C. 1980. Le syndrome nerveux des hautes pressions chez l'homme et le singe Papio papio. D.Sc. Thesis. Université d' Aix-Marseille I, France.

Rostain, J.C. 1987. The high pressure nervous syndrome at the central nervous system level. In: Jannasch HW, Marquis RE, Zimmerman AM (eds). Current perspectives in high pressure biology. pp 137-148. London: Academic Press.

Rostain, J.C. 1993.The nervous system: man and laboratory mammals. In: A.G. Macdonald ed). Advances in comparative and environmental physiology: effects of high pressure on biological systems: pp. 198-238. Springer Verlag, Berlin, Germany.

Rostain, J.C. 1994. Nervous system at pressure. In: Bennett PB, Marquis RE (eds). Basic and applied high pressure biology. pp 157-172. Rochester: University press of Rochester.

Rostain, J.C. and Charpy, J.P. 1976. Effects upon the EEG of psychometric performance during deep dives in helium-oxygen atmosphere. *Electroenceph. clin. Neurophysiol. 40*: 571-584.

Rostain, J.C. and Dimov, S. 1976. Potentiels évoqués visuels et cycle d'excitabilité au cours d'une plongée simulée a - 610 m en atmosphère hélium-oxygène (physalie VI). Electroenceph. clin. *Neurophysiol. 41:* 287-300.

Rostain, J.C., Dumas, J.C., Gardette, B., Imbert, J.P and Lemaire, C. 1984a. Effects of addition of nitrogen during rapid compression of baboon. *J. Appl. Physiol. Respirat. Environ. Exercise. Physiol. 57:* 332-340.

Rostain, J.C. and Forni, C. 1995. The effects of high pressures of various gas mixtures on rat striatal dopamine detected in vivo by voltammetry. *J. Appl. Physiol. 78*: 1179-1187.

Rostain, J.C., Gardette, B., Gardette-Chauffour, M.C. and Forni, C. 1984b. HPNS of the baboon during helium-nitrogen-oxygen slow exponential compressions. *J. Appl. Physiol.: Respirat. Environ. Exercise Physiol. 57:* 341-350.

Rostain, J.C., Gardette-Chauffour, M.C., Gourret, J.P and Naquet, R. 1988. Sleep disturbances in man during different compression profiles up to 62 bars in helium-oxygen mixture. *Electroenceph. clin. Neurophysiol. 69:* 127-135.

Rostain, J.C., Gardette-Chauffour, M.C. and Gardette, B. 1994. HPNS during a deep hydrogen-helium-oxygen dive up to 701 meters. *Undersea and Hyperbaric Med 21 (suppl)* : 40.

Rostain, J.C., Gardette-Chauffour, M.C. and. Naquet, R. 1997. EEG and sleep disturbances during dives at 450 msw in helium-nitrogen-oxygen mixture. *J. Appl. Physiol. 83*: 575-

582.

Rostain, J.C., Gardette-Chauffour, M.C. and Gardette, B. 1999. Hydrogen, a gas for diving: a mini review. *Undersea and Hyperbaric Med*; *26 (suppl)* : 62.

Rostain, J.C. and Lemaire, C. 1973. Evolution du tremblement au repos et pendant l'effort au cours de plongées profondes en atmosphère hélium-oxygène. *Revue EEG. Neurophysiol. Paris. 3:* 203-206.

Rostain, J.C., Lemaire C., Gardette-Chauffour, M.C. and Naquet, R. 1984c Evolution of HPNS in 16 divers breathing He-N2-O2 during long stays at 45 bars. In: A.J. Bachrach and M.M. Matzen [eds.]. Underwater Physiology VIII. pp. 665-672. Undersea Medical Society, Bethesda Md, USA.

Rostain, J.C. et Lyagoubi, S. 1971. Etude électroencephalographique de la veille et du sommeil sous hyperbarie prolongee en atmosphere helium-oxygene (operation Janus 11). *Rev. EEG. Neurophysiol. 1:* 239-241.

Rostain, J.C. and Naquet, R. 1974. Le Syndrome Nerveux des Hautes Pressions : caractéristiques et évolution en fonction de divers modes de compression. *Rev. EEG. Neurophysiol. 4*: 107-124.

Rostain, J.C. and Naquet, R. 1978. Human neurophysiological data obtained from two simulated dives to a depth of 610 meters. In: C.W. Shilling and M.W. Beckett [eds.]. Underwater Physiology VI. pp 9-19. FASEB, Bethesda Md. USA.

Rostain, J.C., Regesta, G. and. Naquet, R.. 1973. Study of the organisation of sleep of divers staying at 300 meter depths in an oxy-helium atmosphere. *Electroenceph. clin. Neurophysiol. 34* : 788.

Rostain, J.C, Regesta, G., Gardette-Chauffour, M.C. and Naquet, R. 1991. Sleep organization in man during long stays in helium-oxygen mixture at 30 and 40 bars. *Undersea Biomed. Res. 18*: 21-36.

Rostain, J.C., Wardley-Smith, B., Forni, C. and Halsey, M.J. 1986. Gamma amino butyric acid and the high pressure neurological syndrome. *Neuropharmacol. 25*: 545-554.

Salzano, J.V., Rausch, D.C. and. Saltzman, H.A. 1970. Cardiorespiratory responses to exercise at a simulated seawater depth of 1000 feet. *J. Appl. Physiol. 28*: 34-41.

Simler, S., Ciesielski, L., Maitre, M., Randrianarisoa, H. and Mandel, P. 1973. Effect of sodium-n-dipropylacetate on audiogenic seizures and brain y-aminobutyric acid level. *Biochem. Pharmacol. 22*: 1701-1708.

Stoudemire, A., Miller, J., Schmitt, B.S.F., Logue, P., Shelton, D., Latson, P.A.G. and Bennett, P.B. 1984. Development of an organic affective syndrome during a hyperbaric diving experiment. Am *J Psychiatry 141*: 1251-1254.

Tomei, C., Abraini, J.H. and Rostain, J.C. 1991. A new device for behavioral analysis on rats exposed to high pressure. *Physiol. Behav. 49*: 393-396.

Vallée, N., Rostain, J.C., Boussuges, A. and Risso, J.J. 2009a. Comparison of Nitrogen Narcosis and Helium Pressure Effects on Striatal Amino Acids: A Microdialysis Study in Rats. *Neurochem Res. 34(5):* 835-844.

Vallee, N., Rostain, J.C., Risso, J.J. 2009b. How can an inert gas counterbalance a NMDA-induced glutamate release? *J Appl Physiol., 107(6)*: 1951-1958.

Van Der Laan, J.W., De Boer, Th. and Brulnvels, J. 1979. Di n-propylacetate and GABA degradation. Preferential inhibition of succinic semi aldehyde dehydrogenase and indirect inhibition of GABA-T. *J. Neurochem. 32:* 1769-1780.

Waldvogel, W. and Buhlmann, A.A. 1968. Man's reaction to long lasting overpressure exposure. Examination of the saturated organism at a helium-pressure of 21-22ATA. *Helv. Med. Acta. 34*: 130-150.

Wardley-Smith, B., Hudson, S., Dore, C.J., Charlett, A., Fletcher, A., Brammer, N.T., Minchin, M.C.W. and Wann, K.T. 1990. Exposure to high pressure may produce the 5-HT behavioral syndrome in rats. *Undersea Biomed. Res. 17*: 275-286.

Wardley-Smith, B. and Meldrum, B.S. 1984. Effect of excitatory amino-acid antagonists on the high pressure neurological syndrome in rats. *Eur. J. Pharmacol. 105*: 351-354.

Wardley-Smith, B., Meldrum, B.S. and Halsey, M.J. 1987. Effect of two novel dipeptide antagonists of excitatory amino acid neurotransmission on the high pressure neurological syndrome in the rat. *Eur. J. Pharmacol. 138*: 417-420.

Wardley-Smith, B., Rostain, J.C., Meldrum, B.S. and Halsey, M.J. 1986. Effect of 2-aminophosphonoheptanoic acid on the EEG of rats exposed to high pressure. *Undersea. Biomed. Res. 13:* 155-163.

Zinebi, F., Fagni, L. and Hugon, M. 1988a. The influence of helium pressure on the reduction induced in field potentials by various amino acids and on the GABA-mediated inhibition of the CA1 region of hippocampal slices in the rat. *Neuropharmacol. 27*: 57-65.

Zinebi, F., Fagni, L. and Hugon, M. 1988b. Decrease in recurrent and feed-forward inhibitions under high pressure of helium in rat hippocampal slices. *Eur. J. Pharmaco. 153:* 191-199.

Zinebi, F., Fagni, L. and Hugon, M. 1990. Excitatory and inhibitory amino-acidergic determinants of the pressure induced neuronal hyperexcitability in rat hippocampal slices. *Undersea Biomed. Res. 17*: 487-493.

In: Helium: Occurrence, Applications and Biological Effects ISBN: 978-1-62808-612-6
Editor: Simon B. Davis

Chapter 3

PHYSICS AND CHEMISTRY OF HELIUM CLUSTERS AND DROPLETS

J. Peter Toennies*

Max-Planck-Institut für Dynamik und Selbstorganisation, Göttingen, Germany

ABSTRACT

Many microscopic properties of the only natural superfluid element helium have only become available within the last 18 years through a number of novel experiments. The diffraction of matter-waves of small helium clusters have established the unusually large bond distance (5.2 nm) and extremely weak binding energy (10^{-7} eV) of the dimer and other exotic properties of small helium clusters. The high resolution spectroscopy of closed shell chromophore molecules attached to small clusters or embedded inside large superfluid droplets have revealed that as a result of superfluidity molecules rotate virtually without friction and that only a hand-full of He atoms are required. These and a number of other experiments demonstrate that helium droplets are the most gentle and coldest of all cryomatrices with enormous potential for a wide range of future applications.

Keywords: Superfluidity of helium, atomic clusters, infrared and visible spectroscopy
PACS: 36.40.-c, 67.40.-w, 36.20 Ng

I. INTRODUCTION

Helium is the second most abundant element in the solar system after the hydrogen atom and also the next simplest of all atoms. With only two electrons in a closed 1S shell all the physical properties of the atom are extremely well understood. The large energy of the first electronically excited state at 19.82 eV, enormous ionization potential of 24.5 eV and unusually small polarisability of $\alpha = 0.21$ Å, even compared to the H atom ($\alpha = 0.67$ Å^3),

* Email: jtoenni@gwdg.de.

make He the most stable and inert of all atoms. The stability of the He atom is also responsible for an exceedingly weak van der Waals interaction with all other atoms and molecules, consequently it is not involved in any chemistry, although a few compounds containing helium have been predicted [1] but none have been found. Up to about 15 years ago it was not even clear whether He could form a stable He_2 dimer. The discovery of the He dimer and its unusual properties will be discussed later in this review. The weak van der Waals potential between He atoms and the bosonic nature of ^4He have, however, a dramatic influence on its bulk behavior. For one it is the only substance which does not solidify down to the lowest temperatures. And it is the only liquid to exhibit superfluidity in its natural state.

Figure 1 shows the phase diagram of ^4He. For comparison the phase diagram of an ordinary substance, in this case argon, is shown in the inset. Helium is the only element that remains liquid down to the lowest temperatures. In place of a triple point as in all other elements helium has a λ-line connecting the points marked by λ ($P_\lambda = 0.05$ bar, $T_\lambda = 2.18$ K) and λ' in Figure 1 which separates the normal liquid (He I) from the superfluid (He II). The rare isotope ^3He, being a fermion, only becomes superfluid at about 10^{-3} K. The superfluidity of ^4He has been known for over 70 years when it was officially discovered in 1938 by Allen and Misener [2] and independently by Kapitza [3]. That something strange occurred at the λ-point had already been suspected ever since helium had been first liquefied by Heike Kamerling Onnes in 1908. Landau and Tisza proposed a two fluid model to explain the many well-known macroscopic manifestations of superfluidity such as (i) flow without resistance (ii) a vanishing viscosity, (iii) the ability to "creep" out of vessels against the force of gravity, and (iv) the fountain effect, which is facilitated by a type of Maxwell demon which allows only the superfluid fraction to pass through a porous plug while holding back the normal fluid fraction. Superfluidity and the closely related phenomenon of Bose Einstein Condensation (BEC), discovered experimentally in 1995, are the only laboratory manifestations of macroscopic coherent quantum matter with many surprising properties.

In his 1954 book entitled "Superfluids" Fritz London called superfluid helium the only representative of a particular 'fourth' state of aggregation besides the solid, liquid and gaseous states. Indeed, superfluidity is an ubiquitous process occurring throughout the universe in stars and in nuclei. Earlier in 1938 in the same year that superfluidity was discovered London called attention to the relationship between BEC and superfluidity [4]. BEC is now considered to be well understood as a result of the extensive theoretical activity stimulated by the laser experiments on Bose-Einstein condensates in highly diluted and ultra-cold ($\approx 10^{-9}$ K) alkali gases [5]. The phenomenon of superfluidity in comparatively dense strongly interacting liquid helium, however, still poses unsurmountable problems for a fundamental understanding. In their 1999 review of the theory of BEC Dalfovo, Giorgini, Pitaevski, and Stringari write: "The explicit connection between superfluidity and BEC is not trivial and has been the object of a longstanding and deep investigation in the last decades" [6].

Theory associates superfluidity with a macroscopic global wave function which encompasses the entire system. This macroscopic wave function is generally written as

$$\Psi(\mathbf{r}) = \Psi_0 \exp(\, iS(\mathbf{r})\,), \quad (1)$$

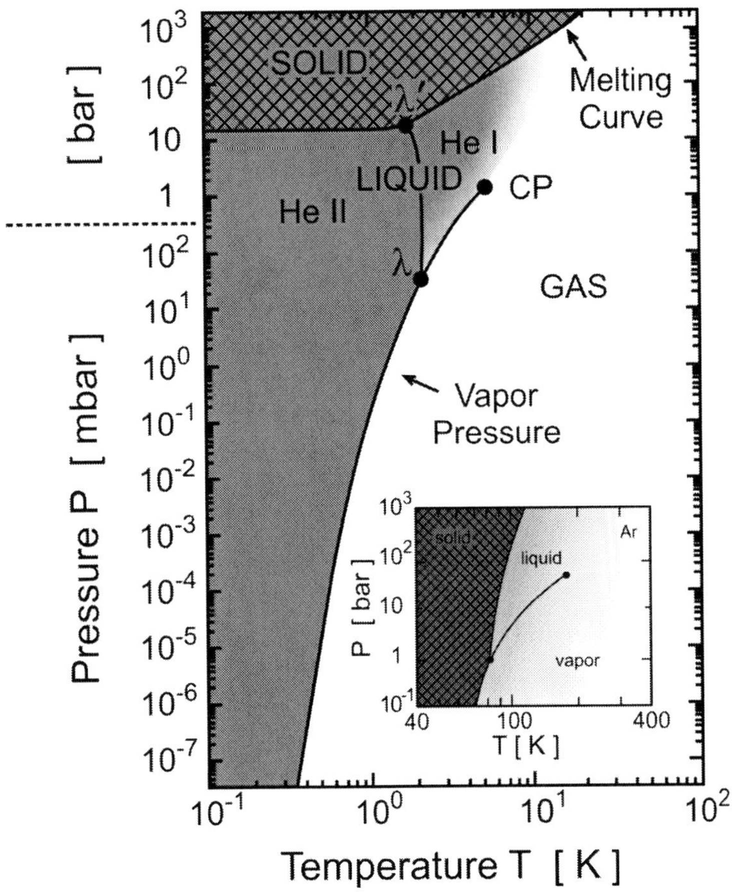

Figure 1. Pressure-temperature phase diagram of ^{4}He. The inset shows the phase diagram for argon which has a triple point typical of all other substances.

where $\Psi_0 = n_0^{1/2}$ (n_0 is the density of the condensate) and S(r) is the phase of the coherent condensate [7]. Since in the superfluid state all particles now have a common phase as opposed to the random phases of particles in normal liquid helium (He I) superfluidity is considered an example of spontaneously broken gauge symmetry and in this context Eq.(1) is referred to as the order parameter. The usual theories of superfluidity presuppose an infinitely extended system.

Up to rather recently it was not clear if finite sized systems could also exhibit superfluidity. Already in 1960 Migdal suggested that nuclei on the basis of their anomalously small moments of inertia were superfluid [8]. Then in 1989 Sindzingre, Klein and Ceperley demonstrated theoretically that small clusters of ^{4}He with only 64 atoms also had anomalously small moments of inertia and therefore should be superfluid [9]. Since about that time the study of superfluidity in clusters and droplets of ^{4}He has emerged as a major field of experimental and theoretical research with ramifications into many areas of physical and chemical research.

The present review attempts (1) to highlight how our present understanding of superfluidity has changed as a result of a variety of different types of molecular beam experiments on helium clusters and droplets and (2) to call attention to the remarkable opportunities which helium droplets provide for molecular spectroscopy and chemistry. The next section describes how beams consisting of isolated atoms, small clusters and large droplets are produced in the laboratory. The experimental techniques used to explore in detail the weak van der Waals interaction between two, three and more He atoms are discussed in the following section. These experiments may also be looked up as an attempt to explore superfluidity by a bottom-up approach starting from the dimer and extending up to clusters with about 50 atoms. The large size of the dimer and its simplicity have attracted new atomic physics investigations. The following section describes how high resolution molecular spectroscopy could be implemented to detect and explore superfluidity effects at the other extreme of large very cold droplets with thousands of atoms. Also the spectroscopy of systems consisting of only a few He atoms will be discussed. Next some additional demonstrations of superfluid effects are presented. The review closes by calling attention to new emerging areas of research involving superfluid He droplets. For recent more comprehensive reviews, the reader is referred to References [10],[11],[12], [13] [14] [15].

II. Beams of Helium Atoms, Cluser and Droplets

The experiments discussed in this review are made possible by using directed molecular beams consisting of isolated atoms, clusters or droplets. The beams are produced by a free jet expansion of either compressed helium gas or liquid through a (typically) 5 micron diameter thin walled orifice into a vacuum maintained by large vacuum pumps. At a source temperature greater than about 50 K the properties of the gas as it expands into vacuum are well described by an adiabatic (isentropic) transformation: $T^{\gamma} P^{1-\gamma}$ = const, where T and P are the local temperature and pressure and $\gamma = 1.67$ is the ratio of the specific heats of an ideal monoatomic gas. With the rapid drop in pressure the temperature drops accordingly until the pressure is so low that collisions cease to occur. In the case of helium very low temperatures are reached because the collision cross section for momentum transfer, σ, increases sharply at low temperatures approaching $\sigma = 259{,}000$ Å^2 at T = 0 [16]. This is, in fact, the largest known cross section for naturally occurring ground state atoms and molecules. The large cross section is related to the weak potential, which in turn reduces the amount of clustering and the corresponding amount of heat released in the expansion. These effects explain the very low ambient temperatures of only 10^{-3} K in the fully expanded gas and the related narrow velocity distributions with $\Delta v/v \approx 1$ %. Not only beams of atoms, but also beams containing clusters have narrow velocity distributions[17].

At lower source temperatures between about 10 and 30 K extensive cluster formation takes place in the expansion and droplets with sizes up to about 10^4 atoms are produced as shown in the bottom part of Figure 2 [12]. At even lower source temperatures between about 5 and 10 K depending on the source pressure a different type of expansion involving the normal liquid occurs [18]. Down to temperatures of about 6 K the liquid fragments after it emerges from the orifice producing large droplets with average sizes between 10^5 and 10^7 atoms. Below about 6 K the liquid fragments before leaving the orifice and even larger droplets with sizes up to 10^{11} atoms are formed [19]. At temperatures below about 4 K the liquid remains intact over longer distances before it breaks up into a regular sequence of huge droplets with diameters comparable with the orifice diameter ($\bar{N} \approx 10^{12}$-10^{13} atoms) [20].

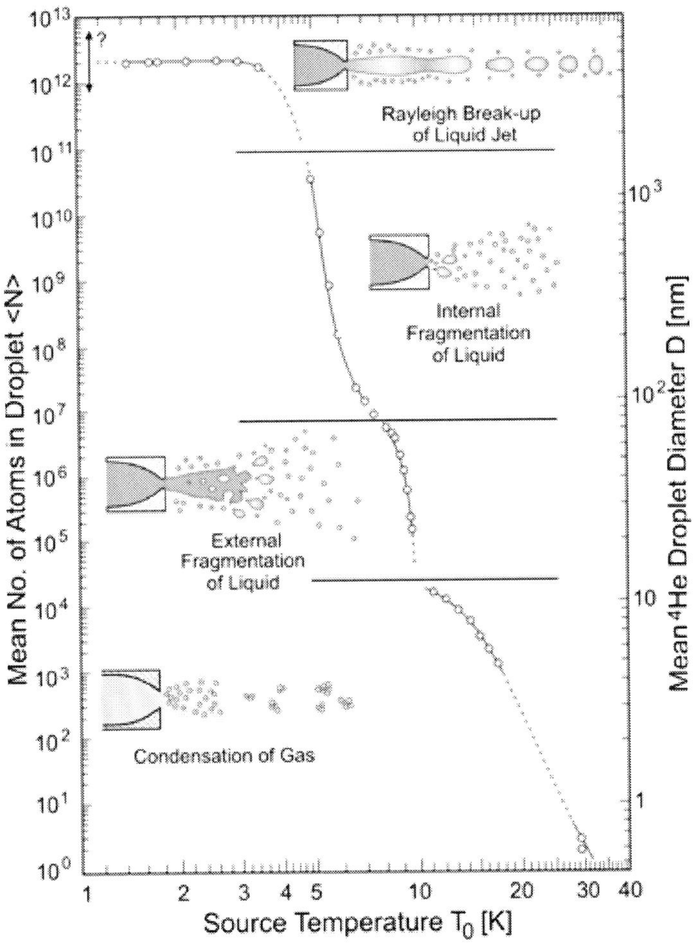

Figure 2. Overview of the mean number sizes (<N>) and diameters (D) of clusters and liquid droplets of ^{4}He in different types of operation of free jet expansions from a 5 micron nozzle. The data and curves are for a source pressure of 20 bar.

These different types of expansions, summarized in Figure 2, provide with a relatively simple apparatus clusters and droplets with sizes covering 13 orders of magnitude. Next we describe experiments which explore the physical properties of the small clusters with $\bar{N} \leq 50$. As mentioned in Section IV once formed the clusters and droplets rapidly cool by evaporation to temperatures of 0.37 K or less in the case of ^{3}He droplets

III. The Enormous Helium Dimer and Other Exotic Helium Clusters

The closed shell structure of the two electron atom means that the ^{4}He dimer is the simplest system for testing quantum chemical methods. For a long time since the first calculation by John Slater in 1928 it was not clear if the potential was sufficiently strong to support a bound state. Experiments based on the mass spectrometric detection of the ionization product He_2^+ were controvertial because of the possibility that it could also arise in the fragmentation of larger clusters [21, 22]. The ^4He trimer also is of special interest as the only candidate among the naturally occurring trimers and small nuclei to exhibit an excited state with remarkable properties called an Efimov state [23].

To detect these exceedingly fragile species a new experimental method was devised. Beams of small clusters were treated as matter-waves and diffracted from nano-structured transmission gratings [24]. The method takes advantage of the narrow 1 % relative velocity spread of helium cluster beams so that they constitute a highly coherent matter wave. A cluster consisting of N atoms has a distinct de Broglie wavelength given by $\lambda = h/(N \cdot m \cdot v)$, where m is the mass of one atom. The first order ($n=1$) diffraction angle ϑ for a given cluster is inversely proportional to its size

$$\vartheta(n=1) \cong \frac{\lambda}{d} = \frac{h}{Nm\,vd}, \tag{2}$$

where d is the slit width of the grating and v is the velocity. The diffraction technique has the unique advantage that it is entirely non-destructive since only wave packets which do not interact with the grating bars contribute coherently to the diffraction peaks. The experiments rely on the availability of silicon nitride transmission diffraction gratings with d = 100 nm [25, 26] so that e.g. the dimers ($\lambda(He_2) = 0.09$ nm) from a jet at a source temperature of $T_0 = 30$ K have a first order diffraction peak at ϑ (He) = 1.2 mrad = 0.07 deg. The excellent resolution achieved with this method is illustrated in Figure 3 which displays a series of diffraction patterns obtained with a source temperature of $T_0 = 24$ K (a) and three patterns measured at 6 K, one with increased angular resolution [27]. In the top-most panel (a) a large number of higher order diffraction peaks due to atoms, extending up to n=8, are seen. In measurements (not shown here) at $T_0 = 300$ K atom diffraction peaks up to the 23rd order could be resolved. Closer examination of the diffraction pattern in (a) reveals a small peak halfway between the large central peak and the first order atom peak, which because of its position must be due to the dimer. Third order diffraction of the dimer explains the smaller peak between the first and second order atom peak. At the lower source temperature of 6 K in (b) additional peaks appear between the central and first order atom peak which are assigned to the trimer and tetramer.

Before discussing parts (c) and (d) we dwell on some of the physical properties of these small quantum systems which have been derived using the diffraction technique. From the form factor (slit function), which is determined from the envelope of the intensities of the sequence of diffraction peaks with order n = 0, 1, 2 of a given species, the effective grating slit width could be determined. The slit width depends on the mean size of the cluster $\bar{R}$ according to the following simple relationship $s_{eff} = s_0 - \frac{\bar{R}}{2}$, where s_0 is the actual geometrical slit width. Since the accuracy of the diffraction method is greater than achievable with an electron microscope the average value of s_0 is more precisely obtained from the value of s_{eff} measured with the much smaller He atoms, for which $\bar{R} \approx 0$, extrapolated to the classical limit of infinite velocity. The diffraction method yields for the He dimer $\bar{R} = 5.2 \pm 0.4$ nm. Thus the helium dimer is the largest ground state diatomic molecule. Its bond distance is, for example, nearly two orders of magnitude larger than the 0.074 nm distance in the H_2 molecule [25, 26]. The binding energy is connected to the size by the simple expression $E_b = \frac{\hbar^2}{4m\bar{R}^2}$ .with the result that $E_b \cong 1.1 \pm 0.3 \cdot 10^{-3}$ K $\approx 10^{-7}$ eV. Since at this energy the classical outer reflection point in the dimer van der Waals potential is at about 1.5 nm. the dimer spends most of its time in the classically forbidden outer region of the long range potential.

This experimental result presently provides the most discriminating test of the most recent ab initio quantum chemical calculations which include relativistic electrodynamic effects. They predict values of $\bar{R} = 4.71 \pm 0.05$ nm and $E_b = 1.62 \pm 0.03 \bullet 10^{-3}$ [28], values which are nearly in agreement within the experimental errors. Recently the suggestion was made that the larger value found in the experiment was due to the contribution of about 2 Å from the long range Casimir Polder retardation, which was not included in the previous theories [29].

The effective slit widths were also determined from diffraction patterns of D_2, Ne, Ar, Kr and Xe over a range of velocities. The strong distinctly different decrease in the effective slit width with decreasing velocity for each of the above species was found to result from the long range van der Waals interaction $V(z) = -\frac{C_3}{z^3}$ of the atom (molecule) with the surfaces of the grating bars [30]. This is the first method which provides precise numbers for the important C_3 coefficient. The method has subsequently been applied to metastable He* and Ne* [31] and Li, Na K and Rb [32]. The sensitivity of the method can be improved by the implementation of an interferometer as first pointed out by Perreault and Cronin [33] and recently further investigated for Li atoms [34]. Recently the ratios of the C_3 coefficients for different atom pairs have also been found to provide a test of atomic structure calculations or, depending on the choice of the atoms, to provide evidence on the permittivity of the surface [35].

Ever since the 1970 prediction of a unique type of three body bound state by Efimov [23] there has been an intensive search to confirm experimentally its existence. According to Efimov if between 2 of 3 identical bosons the bond is critical ($E_b = 0$, scattering length = ∞) then there are an infinite number of 3 particle bound states. Previous to 2006, when an Efimov state was indirectly detected in Cs atoms laser cooled to 200 nK, [36] the helium trimer was considered to be the most likely system to have an Efimov excited state [37], [38]. In an attempt to detect such a state the diffraction technique described above was used to measure the size of the first excited Efimov state of the $^4He_3$ trimer. From the measured slit function the distance between any two of the atoms in the trimer was found to be $\bar{R} = 1.1 \pm 0.5$ nm. This value is compatible with an assumed mixture of 94 % ground state trimers (with $\bar{R} = 0.96$ nm) and 6 % of the huge excited state Efimov trimers ($\bar{R}$ = 7.97 nm) where the values of $\bar{R}$ were taken from theory [39]. These $\bar{R}$ values are in reasonable agreement with the experiment for the estimated fractional populations. Recently the same three-body recombination loss technique used with Cs atoms has been applied to laser cooled metastable He atoms, ^7Li, ^{39}K, ^{85}Rb, a three-spin component mixture of fermionic ^6Li and in a mixture ^{41}K and ^{87}Rb atoms [40]. However the direct detection of an Efimov molecule still remains elusive.

The diffraction experiments have also established the existence of 11 very tenuous boson-fermion ^4He_m ^3He_n clusters [41]. One of these $^4He_2$ $^3He_2$ is especially interesting. Of the six pair interactions, only that between the two ^4He atoms has a bound state since by virtue of their smaller masses the ^3He-^3He and ^4He-^3He pairs are not bound although, of course, the potential is the same as for $^4He_2$. Nevertheless, the $^4He_2$ $^3He_2$ cluster was found to be bound and since this is reminiscent of a Borromean state (a bound three body complex in which none of the pairs are bound) it was called a pseudo-Borromean system.

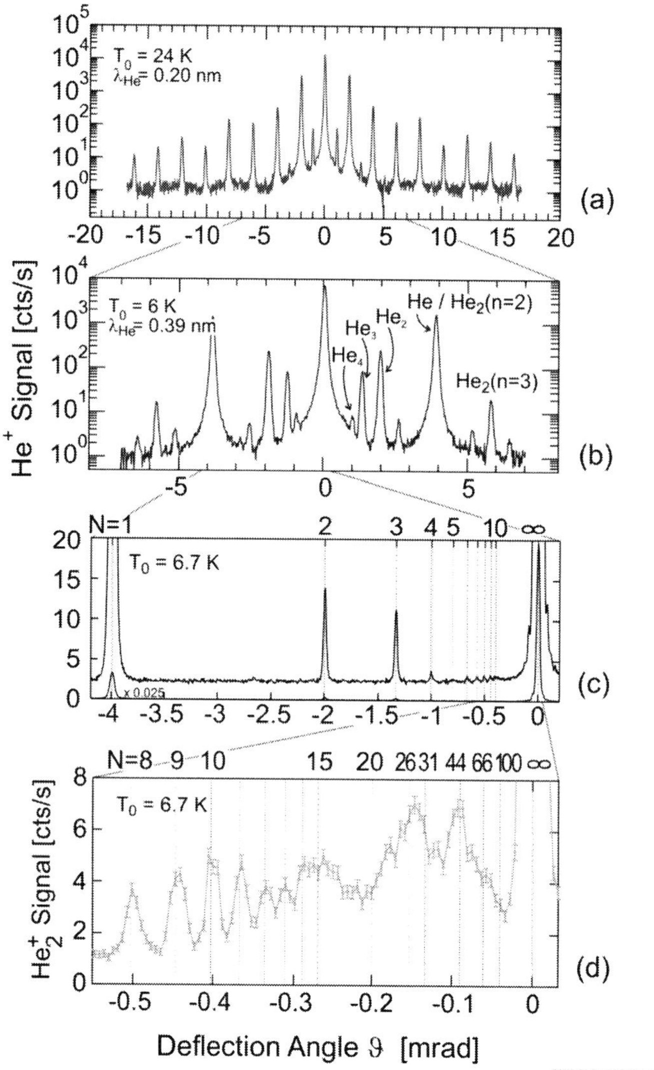

Figure 3. Diffraction patterns measured for different source temperatures and angular resolution. In (a) the diffraction pattern is dominated by higher order atom diffraction. In (b) first order diffraction peaks from the dimer, trimer, and tetramer are clearly resolved. In (c) the angular resolution is greatly improved. By detecting only He_2^+ fragments the background from n = 0 atom diffraction is eliminated and larger clusters up to about N = 13 are resolved.

The observation of the helium dimer has opened up new opportunities in atomic physics. Friedrich and coworkers have explored the possibility to laser dissociate the dimers by taking advantage of the large anisotropy to excite the rotations and thereby lead to their dissociation. They find that the critical angular momentum and the corresponding laser intensity depends on the internuclear distance and thus should provide a means to map out the potential and the dipole polarizability [42], [43]. Another theory has shown that strong laser fields with 5.6 x 10^{19} W/cm^2 can create very strongly bound helium molecules with unusual properties [44]. Recently the process of Interatomic Coulomb Decay (CDI), first predicted by Cederbaum and colleagues [45] [46], was observed in the helium dimer. Unlike other systems the kinetic energy release observed in a COLTRIMS coincidence experiment shows an oscillatory structure exposing the vibrational structure of an intermediate state of an excited $He_2{}^{*+}$ molecular ion [47]. In a related experiment the Coulomb explosion imaging of the helium dimer was found to provide a direct view of its nuclear wave function [48]. Very recently the Coulomb explosion technique has been used to measure the dimer bond distance. Surprisingly the result indicates that it is about 30% smaller than found in the diffraction experiments and in the theory.

The angular resolution of the apparatus used in the experiments Figure 3(a) and (b) discussed above was about 80 microrad. As illustrated in Figure 3(c) and (d) by improving the resolution to 20 microrad it was possible to extend the measuring range to much large clusters with sizes up to about $N \approx 100$ [49, 50]. By switching the detector to $He_2{}^+$, which is the most abundant fragment produced in the ionization of clusters, the signal-to-noise ratio for detecting clusters was greatly improved as seen in Figure 3(d). Although beyond about $N = 13$ the resolution was insufficient to distinguish clusters differing by a single atom, groups of abundant clusters with sizes centered at about N = 10-11, 14-15, 22, 25-27 and 44 were observed. Such magic numbers were totally unexpected since a number of theoretical studies [51, 52] had demonstrated that these could be ruled out for liquid ^{4}He clusters. The magic numbers were explained with the aid of sophisticated diffusion Monte Carlo calculations of the rotational-like and vibrational breathing low-lying excitation levels of the clusters [50]. These calculations revealed that the magic numbers coincided with the sizes of clusters which could just accommodate one more additional stable excitation level. It was argued that such clusters would have a significantly increased chance of being formed in the final stages of the growth process compared to the next smaller or larger clusters. Via this model the theoretical results could be confirmed and thus the unexpected magic numbers provide the first experimental insight into the elementary excitation levels of very cold bosonic clusters. These levels are consistent with those expected for a microscopic superfluid.

IV. Freely Rotating Molecules in Helium Droplets

Compared to the extensive experimental and theoretical studies of macroscopic superfluid phenomena experimental studies of microscopic superfluidity have up to recently been sparse. Attempts to use neutral molecules as weakly interacting probes by inserting them into bulk superfluid helium were largely unsuccessful since their high mobility enables the molecules to aggregate and eventually sink to the bottom of the container or float at the surface. It has however been possible to build up sufficient steady state concentrations of laser ablated metal atoms inside normal and superfluid liquid helium to carry out spectroscopic studies [53], but this technique so far has only been successful with metal atoms and molecules[54]. Generally because of the strong interaction of metal atoms with helium the spectral features have been broader than found in the gas phase [53].

Although molecular beams of helium droplets were produced already in 1961 by Becker and colleagues [55, 56] it was only in 1990 that it was realized that droplets - being liquid -

could pick up molecules either singly or in greater numbers which could then form clusters inside the droplets [57, 58]. In 1992 Scoles and coworkers succeeded in observing unusually sharp infra-red spectral features of the SF_6 molecule and SF_6 dimer attached to helium droplets [59]. Subsequent experiments from our group on the same system exhibited the complete set of rotational lines [60]. This result, which indicates that molecules rotate nearly freely inside a liquid, was totally unexpected. In previous matrix spectroscopic investigations in cryogenic liquids only broad unresolved peaks had been observed. Moreover, since the helium bath was known to enclose the embedded molecule with a compressed snowball-like layer of atoms [61] it was expected that the rotations would be greatly hindered.

Figure 4 shows the molecular beam apparatus used in these experiments [11]. The molecules of interest are admitted to a separate pick-up cell where they can be captured by the droplets on their way through the cell. Depending on the pressure in the cell either single or larger numbers of molecules are picked-up. When the frequency of an anti-parallel directed coaxial laser beam is in resonance with an optical transition the energy deposited in the molecule is rapidly transferred to the helium environment. Even weak infra-red excitations lead to the evaporation of several hundred helium atoms and a significant reduction in the size of the droplet. This decrease in turn causes a significant reduction in the cross section for ionization in the mass spectrometer detector. This depletion detection method is especially sensitive for helium droplets because of their very low heat of evaporation (7.2 K per atom for ^4He; 2.7 K per atom for ^3He).

Figure 5 compares the rotational spectrum of the linear carbonyl sulfide (OCS) molecule in ^4He droplets (Figure 5(b)) with the spectrum of the free molecules (Figure 5(a)) and with the spectrum inside non-superfluid ^3He droplets (Figure 5(c)) [62, 63]. Several differences and similarities between the ^4He droplet spectrum and that of the free molecule are noteworthy. For one the line widths in the droplets are only slightly broadened indicating only a very weak interaction of the freely rotating chromophore with the surrounding medium. Secondly the line spacing in the droplets is greatly reduced by about a factor three indicating that the embedded molecules have a larger effective moment of inertia (MOI) by the same amount.

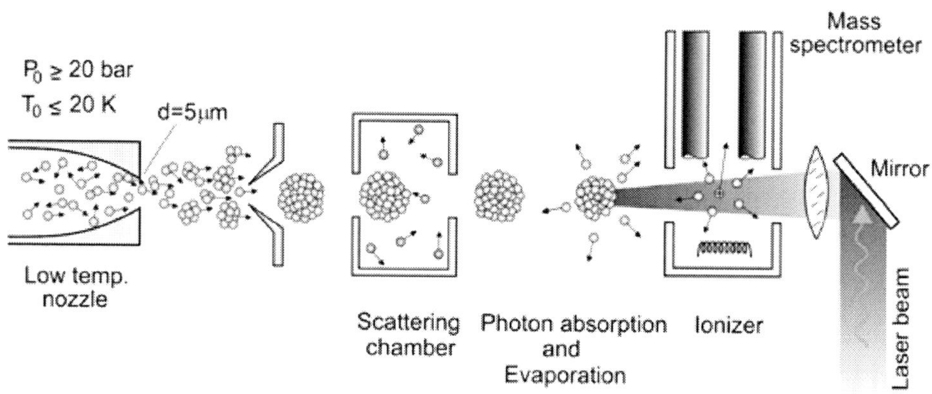

Figure 4. Schematic diagram of the laser-depletion apparatus used for the pick-up and depletion spectroscopy of molecules inside helium droplets. The resonant absorption of photons is detected via a decrease of the mass spectrometer signal resulting from the evaporative loss of helium atoms as the photon energy heats the droplet.

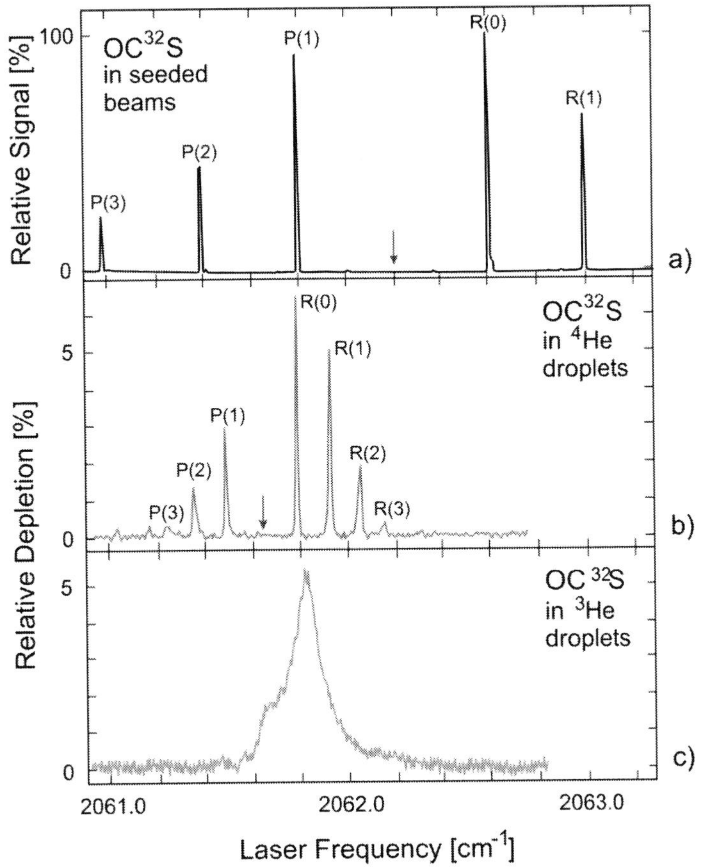

Figure. 5. Comparison of the rotational spectra of OCS (a) the free molecule cooled in an Ar seeded beam, (b) inside pure ^4He droplets with $\bar{N} = 6 \cdot 10^3$ atoms, (c) inside pure ^3He droplets with $\bar{N} = 1.2 \cdot 10^4$ atoms. The various lines correspond to different rotational transitions accompanying an OCS asymmetric stretch vibrational transition. The P-branch corresponds to $j \rightarrow j\text{-}1$and R-branch to $j \rightarrow j+1$ transitions. The lack of a $j \rightarrow j$ Q-branch (red arrow) indicates that the overall symmetry of the molecule is unchanged in the droplet.

The absence of a Q-branch both in the free molecule and in the ^4He droplet spectrum indicates that the helium environment has no effect on the symmetry. Since the line intensities follow a Boltzmann distribution the rotational temperature was determined to be 0.37 K much less than the temperature of 5 K in the seeded beam (Figure 5a)). Since bulk helium is superfluid below 2.12 K the droplets were expected to be sufficiently cold to be superfluid. The excellent agreement of the droplet temperature with earlier theoretical predictions of internal droplet temperatures indicates, moreover, that despite the weak coupling with the surrounding bath the rotational degrees of freedom are fully equilibrated.

The comparison with the ^3He droplet spectrum (Figure 5 (c)) is also very instructive. Since ^3He is a fermion, superfluidity in the bulk occurs at much lower temperatures of $3x10^{-3}$ K and can be excluded even at the predicted lower droplet temperatures of about 0.1 K [64]. The greatly broadened spectrum in the ^3He droplets, despite the much lower droplet temperature and weaker interaction due to the greater zero point energy of the lighter ^3He atoms , was initially interpreted as indicating that some new type of superfluidity of the ^4He

droplets explains the sharp rotational features [62]. This new form of superfluidity is often designated as *microscopic superfluidity*.

The first direct evidence that the ^{4}He droplets are indeed superfluid came in 1996 from the electronic excitation spectra of the $S_1 \rightarrow S_0$ transition of the glyoxal ($C_2H_2O_2$) molecule [65]. In addition to a sharp zero phonon line (ZPL) the spectrum exhibited an unusual sharply peaked phonon wing (PW) which was clearly separated from the zero phonon line. Similar PWs are well-known for species trapped in low temperature condensed matrices where they are found to be essentially proportional to the density of the available phonon states. Whereas usually the PW is broad and partly merged with the ZPL the glyoxal PW is sharply peaked and well separated from the ZPL. Since its energy coincides with the peculiar density of states of superfluid helium with a roton gap this experiment provided the first direct evidence that the ^{4}He droplets are indeed superfluid in the macroscopic sense. Confirmation comes from the spectrum of glyoxal in pure ^{3}He droplets [66] in which the sharply peaked phonon wing is replaced by a gradually falling intensity starting at the ZPL and extending to higher energies. In ^{3}He the roton gap in the density of states is smeared out and partly concealed by particle-hole pair excitations at the Fermi level [67].

To establish the smallest number of helium atoms needed for microscopic superfluidity the spectra in mixed ^{4}He/^{3}He droplets with different numbers of added ^{4}He atoms were investigated. In the mixed droplets the lighter ^{3}He atoms form an outer shell enclosing an inner ^{4}He cluster surrounding the OCS molecule in its interior. With 40 ^{4}He atoms the P- and R-lines became barely visible and with 60 ^{4}He atoms were fully resolved. On the basis of these experiments it was concluded that about 60 ^{4}He atoms, the same number predicted by theory [9] are required for microscopic superfluidity [62]. Since the outer ^{3}He atoms have a much smaller heat of evaporation (2.7 K versus 7.2 K for ^{4}He) the mixed droplets are much colder with a temperature of 0.15 K [68].

Some years after these results the group of McKellar in Ottawa succeeded in producing molecular beams consisting of small clusters consisting of only a few ^{4}He atoms attached to a chromophore molecule such as OCS [69] and measuring their infra-red absorption spectrum. The many different sized clusters in the beam lead to a multitude of spectral lines which could be unraveled to obtain the spectra for each size selected $OCS(He)_N$ cluster with a specific size of N = 3, 4, 5, ...20 and later with up to about 80 attached ^{4}He atoms [70]. Unambiguous assignments were aided by confirming the infra-red transitions with fine structures detected with a microwave spectrometer. Figure 6 shows the change of the measured rotational energy constant B ($B = \hbar^2/(2\pi I)$), where I is the MOI for end-over-end rotations versus the number of attached He atoms. For $N \leq 8$ B falls off with increasing N as expected for the increased MOI and mass of the rotating cluster. Then for $N \geq 9$, surprisingly, the rotational energy constant does not fall further, but instead increases. This phenomenon was attributed to the onset of a superfluid induced slippage between the rotating cluster with a few attached He atoms and the other He atoms of the cluster. At about N = 22 the rotational energy constant increases almost two-fold from its minimum at N = 9 and with further increasing N it oscillates slowly and at N = 72 is only about 4 % larger than the value measured inside the droplets. Thus these new experiments provide the missing link between the small clusters and the droplets. They also give evidence that, depending on the chromophore, between only 5 and 9 ^{4}He atoms are sufficient for another type of microscopic superfluidity [71].

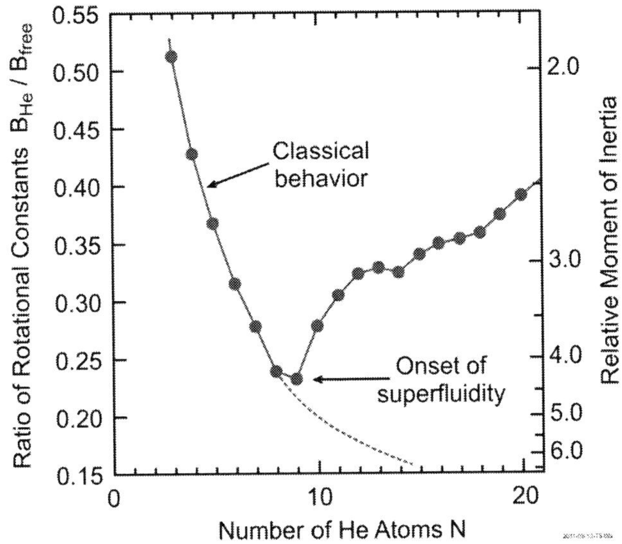

Figure 6. The rotational energy constant obtained from the rotational spectrum of OCS in a beam seeded with ^{4}He is plotted as a function of the number of attached ^{4}He atoms. The moment of inertia relative to the free molecule is plotted on the left ordinate.

The droplet spectroscopies and these new small cluster spectroscopic experiments have attracted innumerable theoretical investigations, which have been summarized in several recent reviews [14, 72, 73]. In order to simulate both the structures and energies as well as the dynamics it has proven important to accurately know the intermolecular potentials between all the particles. As discussed above the He-He potential is now very well known so the recent attention has been directed at applying quantum chemical methods such as advanced density functional techniques to either calculate for the first time or refine earlier calculations of the chromophore-He atom potentials. [14] Accurate results are presently available for the following small molecules: CO, HF, HCN, CO_2, N_2O, OCS, HCCH (acetylene), HCCCN (cyanoacytelene). To simulate the effective MOI and other properties such as line shifts and widths sophisticated dynamical theories, able to handle the ultra-low temperature strong quantum effects in a system with many degrees of freedom, are required. These usually involve stochastic Monte-Carlo methods such as Variational Monte Carlo (VMC), Diffusion Monte Carlo (DMC), Reptation Quantum Monte Carlo (RQMC) and the projection operator Imaginary Time Spectral Evolution (POITSE) method [73]. All these methods are, however, restricted to zero-temperature ($T = 0$). Only the Path Integral Monte Carlo (PIMC) method can cope with finite temperatures. This powerful computational technique is based on one of Feymann's ideas of describing a system of quantum particles in terms of the same number of interacting classical ring polymers each with a large number of elements [74]. In the superfluid state the polymers become interlinked thereby mimicing the exchanges between indistinguishable bosons. The PIMC method is also the only one able to calculate the superfluid fraction within the two fluid model of helium. The superfluid density is defined in terms of the response of the system to an imposed rotation about an axis $\vec{n}$. This leads to the following expression for the superfluid fraction

$$\left.\frac{\rho_{sf}}{\rho_{tot}}\right|_{\vec{n}} = 1 - \frac{I_{qm}(\vec{n})}{I_{cl}(\vec{n})}, \tag{3}$$

where I_{qm} and I_{cl} are the PIMC and classical calculated MOI, respectively [75]. Despite its elegance and remarkable ability to simulate virtually all the properties of superfluid helium [74], it should be realized that the PIMC method is a computational tool in which the real quantum system is projected upon a classical analogy.

The contours in Figure 7(a) illustrate the density distributions calculated with the PIMC method for several small $CO_2(^4He)_N$ clusters at an arbitrarily chosen temperature of 0.15 K [76]. The interaction potential of a single He atom with a linear CO_2 molecule has the deepest well (-45 cm^{-1}) opposite the central C-atom and a smaller minimum (- 24 cm^{-1}) coaxial with the molecular axis at the two O-atom ends. This explains why the total density for N = 5 atoms is located entirely at the center of the molecule where they form a ring about the central waist of the molecule. In the case of the linear OCS molecule this ring has been dubbed a donut ring an expression now generally adopted for other linear molecules. Then with increasing numbers of atoms first the end sites and, when the first shell is filled with 17 atoms, the entire surface of the molecule is coated with He atoms.

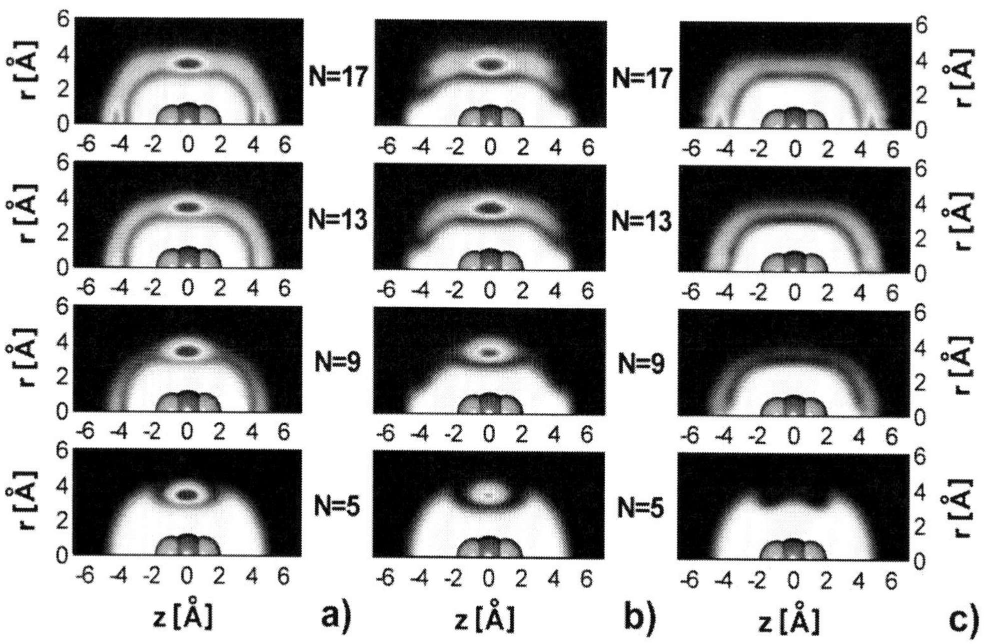

Figure 7. ^4He density distributions for $CO_2(^4He)_N$ clusters for 4 different N (adapted from [73]). The CO_2 molecule is located on the z- axis with the carbon atoms at z = 0.0 Å, and the two oxygen atoms at z = ±1.1615 Å, respectively. Column (a) shows the total density $\rho(z,r)$, column (b) the parallel superfluid density $\rho_{\parallel}^2(z,r)$ and column (c) the perpendicular superfluid density $\rho_{\perp}^2(z,r)$. The regions colored in red indicate a high density and those in dark blue a low density. The repulsive potential excludes helium from the regions shown in white.

In Figure 7 the series of contour maps in (b) and (c) show the superfluid densities obtained from calculations of the axial (parallel) superfluid response and for end-over-end (perpendicular) superfluid response, respectively. The parallel superfluid response is localized exactly in the position of the donut ring and is attributed to cyclic permutation exchanges, which correspond to the K = 0 ground state for axial rotations of the 5 atoms in the ring. The perpendicular superfluid response only sets in after the donut ring is filled. In the contours of Figure 7(c) the superfluid fraction is only apparent for N = 9 but, in fact, starts at a somewhat smaller value of N. Thus the addition of the He atoms to the donut ring explains the decrease in the rotational constant (Figure 6). When the ring is filled then the added atoms occupy end sites, and the end-over-end superfluidity of the molecules is enabled thereby explaining the turnaround in the rotational energy constant.

As noted earlier, the moments of inertia in the large droplets are also significantly larger than for the free molecule. Depending on the molecule the increase can be as much as a factor five for heavy extended molecules and approaches unity for fast rotating light molecules. Initially this effect was attributed to the rigid attachment of some of the atoms of the helium environment to the chromophore so that they rotate with the chromophore [60]. This model, which first was applied to SF_6, when applied to the linear OCS molecule predicted that about 6 He atoms would be attached to form a donut ring around the molecule at its waist where the van der Waals attraction is greatest [77] in good agreement with the observed increase in the moment of inertia. This simple model preceded the detailed confirming calculations presented in Figure 7. A cluster with such a structure is classified as a symmetric top and is expected to exhibit a Q-branch which is related to rotations about the symmetry axis of the top. The lack of a Q-branch in the infra-red spectrum (Figure 5 (b)) was explained by the fact that the six-fold symmetry of the indistinguishable bosonic atoms only permits axial rotations with quantum numbers K = 0, as mentioned in connection with Figure 7 (b), or multiples of 6. The lowest excited state with K = 6 has an energy of 2.1 cm^{-1} (3.2 K) which is too high for excitation at the droplet temperature of 0.37 K. Thus the chromophore with its donut ring behaves as if it is a diatomic molecule and therefore does not exhibit the Q-branch expected for a symmetric top.

The increase in the moment of inertia of chromophores has been investigated with PIMC and related theoretical methods. The PIMC and related calculations, which so far have not been possible for large droplets are limited to sizes of the order of about 40 atoms. These calculations have shown that the effect is related to the superfluid nature of the droplets [72]. As a result of the strong local interaction with the superfluid some of the atoms become non-superfluid enabling them to *adiabatically follow* the rotations of the molecule. Referring to the relatively new results in Figure 7 (b) this description is somewhat misleading. The adiabatic following concept applies to the end-over-end rotations , but the coaxial rotations of the donut can be regarded as very small microscopic superfluids as indicated in Fig 7 (b).

Very recently the broadened lines observed in pure ^{3}He droplets (Fig 5 (c)) have been carefully analyzed [78]. The analysis indicates that in these non-superfluid fermionic droplets a similar effect applies. In this case the increased MOI indicates that about 11 atoms rotate with the OCS molecule. The analysis also reveals that the ^{3}He droplets are much colder with a temperature of 0.07 $\pm$ 0.06 K. Related Kohn-Sham density functional calculations for OCS in small clusters reveal the existence of magic numbers at N= 8, 18 and 40. The difference with respect to the experimental number of 11 suggests that in the large ^{3}He droplets other effects must play a role [79]. The effect of small numbers of ^4He atoms on the properties of ^3He droplets is under investigation [63].

In summary, the PIMC calculations are able to simulate and explain many of the observations made with the small clusters as well as in the large droplets.

V. Helium Droplets: The Ultimate Spectroscopic Matrix

The droplet experiments described above have established helium droplets as the ultimate spectroscopic matrix [80]. They are far colder than can be achieved with ordinary helium cryostats and are more gentle and less perturbative than all other cryogenic liquid and solid matrices Their very low heat of evaporation make helium droplets especially sensitive for spectroscopic photon absorption by the frequently used depletion method. New species with unusual properties can be assembled in their interiors and made available for spectroscopic interrogation [11]. Finally because of their large size and large cross sections for pick-up only moderate vapor pressures of only 10^{-5} mbar are required so that refractory metals, biological molecules and other fragile species can readily be picked up and investigated.

These attractive features have lead to a wide variety of innovative chemical and technologically interesting applications: (1) The low temperatures eliminate the hot bands and the spectroscopy of large molecules is greatly simplified. Recent examples include the spectroscopy of the organic semiconductor PTCDA ($C_{24}H_8O_6$), a flat 2D molecule, which is widely used in light emitting devices and other photonic applications [81], or the spectroscopy of the succinamide thread of the rotaxane molecule [82]. (2) At the low temperatures of the droplets even very weakly bound van der Waals clusters and in some instances clusters with unusual structures have been synthesized. Examples are the creation of non-equilibrium linear chains of with up to 9 cyanoacetylene molecules [83] and the creation of OCS molecules with several rings of para-H_2 molecules with superfluid properties [84]. (3) He droplets have also been used for the study of chemical reactions. Initially it was felt that chemical reactions because of their activation barriers would not occur at the extremely low temperatures of the droplets, but in recent years a number of examples have lead to the realization that there are many chemical reactions which can occur inside He droplets. These include $Ba + N_2O \rightarrow BaO + N_2$, which was the first to be studied [85]. Recently the reaction between Mg and O_2 molecules was reported [86]. Also the reaction occurring within clusters of $(HCl)_m(H_2O)_n$ inside He droplets has received considerable attention [87, 88]. (4) He droplets can serve as cushions for soft landing large metal clusters on solid substrates. This was recently demonstrated with Ag clusters with mean sizes between $N = 300$ and $N = 6 \cdot 10^3$ which were found to undergo little fragmentation upon deposition [89]. In the future the deposition of taylor-made mixed metal or even semiconductor clusters on optimal substrates can be envisaged to serve as catalysts or in plasmonic devices.

The above examples are just a few of the new interesting applications of helium droplets in chemical physics which have been reported in recent years. Several recent reviews describe in more detail these and other applications of He droplets [11, 12, 90].

ACKNOWLEDGMENTS

We thank Oleg Kornilov for valuable comments on an earlier version.

REFERENCES

[1] Lai, T.Y., et al., Benchmark of density functional theory methods on the prediction of bond energies and bond distances of noble-gas containing molecules. *Journal of Chemical Physics,* 2011. *134*(24): p. 244110-110-1-10-8.

[2] Allen, J.F. and A.D. Misener, Flow of Liquid Helium II. *Nature,* 1938. *141*: p. 75-75.

[3] Kapitza, P.L., Viscosity of Liquid Helium below the Lambda-Point. *Nature, 1938. 141*: p. 74-74.

[4] London, F., The Lambda-Phenomenon of Liquid Helium and the Bose-Einstein Degeneracy. *Nature,* 1938. *141*: p. 643-644.

[5] Bose-Einstein Condensation in Atomic Gases. in *Enrico Fermi International School of Physics.* 1999. IOS Press, Amsterdam.

[6] Dalfovo, F., et al., Theory of Bose-Einstein condensation in trapped gases. *Reviews of Modern Physics,* 1999. *71*(3): p. 463-512.

[7] Tilley, D.R. and J. Tilley, eds. Superfluidity and Superconductivity. *Graduate Student Series in Physics,* ed. D.F. Brewer1990, Adam Hilger: Bristol & New York.

[8] Migdal, A.B., Superfluidity and the Moments of Inertia of Nuclei. *Soviet Physics JETP,* 1960. *37*(10): p. 176-185.

[9] Sindzingre, P., M.L. Klein, and D.M. Ceperley, Path-Integral Monte-Carlo Study of Low-Temperature He-4 Clusters. *Physical Review Letters,* 1989. *63*(15): p. 1601-1604.

[10] Northby, J.A., Experimental studies of helium droplets. *Journal of Chemical Physics,* 2001. *115*(22): p. 10065-10077.

[11] Toennies, J.P. and A.F. Vilesov, Superfluid helium droplets: A uniquely cold nanomatrix for molecules and molecular complexes. *Angewandte Chemie-International Edition,* 2004. *43*(20): p. 2622-2648.

[12] Choi, M.Y., et al., Infrared spectroscopy of helium nanodroplets: novel methods for physics and chemistry. I*nternational Reviews in Physical Chemistry,* 2006. *25*(1-2): p. 15-75.

[13] Barranco, M., et al., Helium nanodroplets: An overview. *Journal of Low Temperature Physics,* 2006. *142*(1-2): p. 1-81.

[14] Szalewicz, K., Interplay between theory and experiment in investigations of molecules embedded in superfluid helium nanodroplets. *International Reviews in Physical Chemistry,* 2008. *27*(2): p. 273-316.

[15] Callegari, C. and W.E. Ernst, Helium Droplets as Nanocryostats for Molecular Spectroscopy - from the Vacuum Ultraviolet to the Microwave Regime, in *Handbook of High-resolution Spectroscopy,* M. Quack and F. Merkt, Editors. 2011, John Wiley & Sons Ltd: Chichester, West Sussex, UK. p. 1551-1594.

[16] Luo, F., et al., Influence of Retardation on the Vibrational Wave-Function and Binding-Energy of the Helium Dimer. *Journal of Chemical Physics,* 1993. *98*(12): p. 9687-9690.

[17] Schollkopf, W., L.W. Bruch, and J.P. Toennies, The formation of dimers and trimers in free jet He-4 cryogenic expansions. *Journal of Chemical Physics,* 2002. *117*(4): p. 1544-1566.
[18] Buchenau, H., et al., Mass-Spectra and Time-of-Flight Distributions of Helium Cluster Beams. *Journal of Chemical Physics,* 1990. *92*(11): p. 6875-6889.
[19] Gomez, L.F., et al., Sizes of large He droplets. *Journal of Chemical Physics,* 2011. *135*(15): p. 154201-1 -01-9.
[20] Grisenti, R.E. and J.P. Toennies, Cryogenic microjet source for orthotropic beams of ultralarge superfluid helium droplets. *Physical Review Letters,* 2003. *90*(23): p. 234501-1-01-4.
[21] Meyer, E.S., J.C. Mester, and I.F. Silvera, Comment on "The weakest bond: Experimental observation of helium dimer" [J. Chem. Phys. 98(4), 3564 (1993)]. *Journal of Chemical Physics,* 1994. *100*(5): p. 4021-4022.
[22] Luo, F., et al., The Weakest Bond - Experimental-Observation of Helium Dimer - Response. *Journal of Chemical Physics,* 1994. *100*(5): p. 4023-4024.
[23] Efimov, V., Energy Levels Arising from Resonant 2-Body Forces in a 3-Body System. *Physics Letters B,* 1970. *33*(8): p. 563-564.
[24] Schollkopf, W. and J.P. Toennies, Nondestructive Mass Selection of Small Van-Der-Waals Clusters. *Science,* 1994. *266*(5189): p. 1345-1348.
[25] Grisenti, R.E., et al., Determination of the bond length and binding energy of the helium dimer by diffraction from a transmission grating. *Physical Review Letters,* 2000. *85*(11): p. 2284-2287.
[26] Grisenti, R.E., et al., He-atom diffraction from nanostructure transmission gratings: The role of imperfections. *Physical Review A,* 2000. *61*(3): p. 033608-1-08-15.
[27] Kornilov, O. and J.P. Toennies, Matter-wave diffraction of quantum magical helium clusters. *Europhysics News,* 2007. *38*(1): p. 22-27.
[28] Przybytek, M., et al., Relativistic and Quantum Electrodynamics Effects in the Helium Pair Potential. *Physical Review Letters,* 2010. *104*(18): p. 183003-1-03-4.
[29] Przybytek, M., et al., Onset of Casimir-Polder Retardation in a Long-Range Molecular Quantum State. *Physical Review Letters,* 2012. *108*(18): p. 183201-1-01-4.
[30] Grisenti, R.E., et al., Determination of atom-surface van der Waals potentials from transmission-grating diffraction intensities. *Physical Review Letters,* 1999. *83*(9): p. 1755-1758.
[31] Bruhl, R., et al., The van der Waals potential between metastable atoms and solid surfaces: Novel diffraction experiments vs. theory. *Europhysics Letters,* 2002. *59*(3): p. 357-363.
[32] Lonij, V.P.A., et al., Atom Diffraction Reveals the Impact of Atomic Core Electrons on Atom-Surface Potentials. *Physical Review Letters,* 2010. *105*(23): p. 233202-1 -02-4.
[33] Perreault, J.D. and A.D. Cronin, Observation of atom wave phase shifts induced by van der Waals atom-surface interactions. *Physical Review Letters,* 2005. *95*(13): p. 133201-1 -01-4.
[34] Lepoutre, S., et al., Atom interferometry measurement of the atom-surface van der Waals interaction. *European Physical Journal D,* 2011. *62*(3): p. 309-325.
[35] Lonij, V.P.A., et al., Can Atom-Surface Potential Measurements Test Atomic Structure Models? *Journal of Physical Chemistry A,* 2011. *115*(25): p. 7134-7140.

[36] Kraemer, T., et al., Evidence for Efimov quantum states in an ultracold gas of caesium atoms. *Nature,* 2006. *440*(7082): p. 315-318.
[37] Barletta, P. and A. Kievsky, Variational description of the helium trimer using correlated hyperspherical harmonic basis functions. *Physical Review A,* 2001. *64*(4): p. 042514-1 -14-9.
[38] Kolganova, E.A., A.K. Motovilov, and W. Sandhas, Ultracold collisions in the system of three helium atoms. *Physics of Particles and Nuclei,* 2009. *40*(2): p. 206-235.
[39] Bruhl, R., et al., Matter wave diffraction from an inclined transmission grating: Searching for the elusive He-4 trimer Efimov state. *Physical Review Letters,* 2005. *95*(6): p. 063002-1-02-4.
[40] Knoop, S., et al., Universal three-body parameter in ultracold He-4. *Physical Review* A, 2012. *86*(6): p. 062705-1 -05-4.
[41] Kornilov, O., et al., Observation of mixed fermionic-bosonic helium clusters by transmission grating diffraction. *Physical Review Letters,* 2005. *95*(11): p. 113402-1-02-4.
[42] Friedrich, B., M. Gupta, and D. Herschbach, Probing weakly-bound species with nonresonant light: Dissociation of He-2 induced by rotational hybridization. *Collection of Czechoslovak Chemical Communications,* 1998. *63*(8): p. 1089-1093.
[43] Lemeshko, M. and B. Friedrich, Probing Weakly Bound Molecules with Nonresonant Light. *Physical Review Letters,* 2009. *103*(5): p. 053003-1 -03-4.
[44] Balanarayan, P. and N. Moiseyev, Strong chemical bond of stable He-2 in strong linearly polarized laser fields. *Physical Review A,* 2012. *85*(3): p. 032516-1 -16-6.
[45] Cederbaum, L.S., J. Zobeley, and F. Tarantelli, Giant intermolecular decay and fragmentation of clusters. *Physical Review Letters,* 1997. *79*(24): p. 4778-4781.
[46] Sisourat, N., et al., Ultralong-range energy transfer by interatomic Coulombic decay in an extreme quantum system. *Nature Physics,* 2010. *6*(7): p. 508-511.
[47] Havermeier, T., et al., Single Photon Double Ionization of the Helium Dimer. *Physical Review Letters,* 2010. *104*(15): p. 153401-1 -01-4.
[48] Havermeier, T., et al., Interatomic Coulombic Decay following Photoionization of the Helium Dimer: Observation of Vibrational Structure. *Physical Review Letters,* 2010. *104*(13): p. 133401-1 -01-4.
[49] Bruhl, R., et al., Diffraction of neutral helium clusters: Evidence for "magic numbers". *Physical Review Letters,* 2004. *92*(18): p. 185301-1-01-4.
[50] Guardiola, R., et al., Magic numbers, excitation levels, and other properties of small neutral He-4 clusters (N <= 50). *Journal of Chemical Physics,* 2006. *124*(8): p. 084307-1-07-16.
[51] Melzer, R. and J.G. Zabolitzky, No Magic Numbers in Neutral He-4 Clusters. *Journal of Physics A-Mathematical and General,* 1984. *17*(11): p. L565-L568.
[52] Sola, E., J. Casulleras, and J. Boronat, Theoretical search for shell effects in He-4 clusters. *Journal of Low Temperature Physics,* 2004. *134*(1-2): p. 787-792.
[53] Moroshkin, P., A. Hofer, and A. Weis, Atomic and molecular defects in solid (4)He. *Physics Reports-Review Section of Physics Letters,* 2008. *469*(1): p. 1-57.
[54] Lebedev, V., et al., Spectroscopy of the copper dimer in normal fluid, superfluid, and solid He-4. *Journal of Chemical Physics,* 2010. *133*(15): p. 154508-1 -08-8.
[55] Becker, E.W., R. Klingelhöfer, and P. Lohse, *Strahlen aus kondensiertem Helium im Hochvakuum.* Z. Naturforschg. A, 1961. *16*(11): p. 1259-1259.

[56] Becker, E.W., On the History of Cluster Beams. *Zeitschrift Fur Physik D-Atoms Molecules and Clusters,* 1986. *3*(2-3): p. 101-107.
[57] Toennies, J.P. Helium Clusters. in *Enrico Fermi International School of Physics.* 1988. Varenna: Elsevier Science Publishers B. V. North Holland.
[58] Scheidemann, A., J.P. Toennies, and J.A. Northby, Capture of Neon Atoms by He-4 Clusters. *Physical Review Letters,* 1990. *64*(16): p. 1899-1902.
[59] Goyal, S., D.L. Schutt, and G. Scoles, Vibrational Spectroscopy of Sulfur-Hexafluoride Attached to Helium Clusters. *Physical Review Letters,* 1992. *69*(6): p. 933-936.
[60] Hartmann, M., et al., Rotationally Resolved Spectroscopy of SF_6 in Liquid-Helium Clusters - a Molecular Probe of Cluster Temperature. *Physical Review Letters,* 1995. *75*(8): p. 1566-1569.
[61] Dalfovo, F., Atomic and Molecular Impurities in He-4 Clusters. *Zeitschrift Fur Physik D-Atoms Molecules and Clusters,* 1994. *29*(1): p. 61-66.
[62] Toennies, J.P., S. Grebenev, and A.F. Vilesov, Superfluidity within a small helium-4 cluster: The microscopic Andronikashvili experiment. *Science,* 1998. *279*(5359): p. 2083-2086.
[63] Sartakov, B., J.P. Toennies, and A. Vilesov, *Infrared spectroscopy inside pure ^{3}He and mixed ^{4}He/^{3}He droplets.* submitted, 2013.
[64] Guirao, A., M. Pi, and M. Barranco, Finite Size Effects in the Evaporation Rate of He-3 Clusters. *Zeitschrift Fur Physik D-Atoms Molecules and Clusters,* 1991. *21*(2): p. 185-188.
[65] Hartmann, M., et al., Direct spectroscopic observation of elementary excitations in superfluid He droplets. *Physical Review Letters,* 1996. *76*(24): p. 4560-4563.
[66] Vilesov, A.F., et al., Spectroscopy of molecules in helium droplets. *Physica* B, 2000. *280*(1-4): p. 65-72.
[67] Poertner, N., et al., Anomalously sharp phonon excitations in He-3 droplets. *Europhysics Letters,* 2009. *88*(2): p. 26007.
[68] Grebenev, S., J.P. Toennies, and A.F. Vilesov, Superfluidity within a samll helium-4 cluster:The microscopic Andronokashvili experiment. *Science,* 1998. *279*(10): p. 2083-2086.
[69] McKellar, A.R.W., et al., Quantum solvation of carbonyl sulfide with helium atoms. *Science,* 2002. *297*(5589): p. 2030-2033.
[70] McKellar, A.R.W., Y.J. Xu, and W. Jager, Spectroscopic studies of OCS-doped He-4 clusters with 9-72 helium atoms: Observation of broad oscillations in the rotational moment of inertia. *Journal of Physical Chemistry A,* 2007. *111*(31): p. 7329-7337.
[71] McKellar, A.R.W., Tunable diode laser spectroscopy of helium clusters. *Applied Physics B-Lasers and Optics,* 2008. *90*: p. 213-225.
[72] Whaley, K.B., et al., Quantum solvation and molecular rotations in superfluid helium clusters. *Journal of Chemical Physics,* 2000. *113*(16): p. 6469-6501.
[73] Paesani, F., Superfluidity of Clusters, in *Handbook of Nanophysics,* K.D. Sattler, Editor 2010, Taylor & Francis. p. 12-1-12-16.
[74] Ceperley, D.M., Path-Integrals in the Theory of Condensed Helium. *Reviews of Modern Physics,* 1995. *67*(2): p. 279-355.
[75] Draeger, E.W. and D.M. Ceperley, Superfluidity in a doped helium droplet. *Physical Review Letters,* 2003. *90*(6): p. 065301-1-01-4.

[76] Paesani, F., Y. Kwon, and K.B. Whaley, Onset of superfluidity in small CO(2)((4)He)(N) clusters. *Physical Review Letters,* 2005. *94*(15): p. 153401-1-01-4.
[77] Grebenev, S., et al., The rotational spectrum of single OCS molecules in liquid He-4 droplets. *Journal of Chemical Physics,* 2000. *112*(10): p. 4485-4495.
[78] Sartakov, B.G., J.P. Toennies, and A.F. Vilesov, Infrared spectroscopy of carbonyl sulfide inside a pure He-3 droplet. *Journal of Chemical Physics,* 2012. *136*(13): p. 134316-1 -16-8.
[79] Mateo, D., et al., A density functional study of the structure of small OCS@He-3(N) clusters. *Journal of Chemical Physics,* 2013. *138*(4): p. 044321-1 -21-8.
[80] Lehmann, K.K. and G. Scoles, Superfluid Helium - The Ultimate Spectroscopic Matrix? *Science,* 1998. *279*(5359): p. 2065-2066.
[81] Wewer, M. and F. Stienkemeier, Molecular versus excitonic transitions in PTCDA dimers and oligomers studied by helium nanodroplet isolation spectroscopy. *Physical Review B,* 2003. *67*(12): p. 125201-1-01-5.
[82] Smolarek, S., et al., Conformational flexibility of a rotaxane thread probed by electronic spectroscopy in helium nanodroplets. *J. Am. Chem. Soc.,* 2009. *131*(36): p. 12902-12903.
[83] Nauta, K. and R.E. Miller, Nonequilibrium self-assembly of long chains of polar molecules in superfluid helium. *Science,* 1999. *283*(5409): p. 1895-1897.
[84] Grebenev, S., et al., Spectroscopic investigation of OCS (p-H(2))(n) (n=1-16) complexes inside helium droplets: Evidence for superfluid behavior. *Journal of Chemical Physics,* 2010. *132*(6): p. 064501-1-01-19.
[85] Lugovoi, E., J.P. Toennies, and A. Vilesov, Manipulating and enhancing chemical reactions in helium droplets. *Journal of Chemical Physics,* 2000. *112*(19): p. 8217-8220.
[86] Krasnokutski, S.A. and F. Huisken, Ultra-low-temperature reactions of Mg atoms with O_2 moelcules in helium droplets. *J. Phys. Chem.,* 2010. *114*: p. 7292-7300.
[87] Gutberlet, A., et al., Aggregation-Induced Dissociation of $HCl(H_2O)_4$ Below 1 K: The Smallest Droplet of Acid. *Science,* 2009. *324*(5934): p. 1545-1548.
[88] Flynn, S.D., et al., Infrared Spectra of HCl-H_2O Clusters in Helium Nanodroplets. *Journal of Physical Chemistry Letters,* 2010. *1*(15): p. 2233-2238.
[89] Loginov, E., L.F. Gomez, and A.F. Vilesov, Surface Deposition and Imaging of Large Ag Clusters Formed in He Droplets. *Journal of Physical Chemistry A,* 2011. *115*(25): p. 7199-7204.
[90] Slenczka, A. and J.P. Toennies, Chemical Dynamics Inside Superfluid Helium Nanodroplets at 0.37 K, in *Low Temperatures and Cold Molecules,* I.W.M. Smith, Editor 2008, Imperial College Press: Birmingham and Cambridge. p. 345-392.

In: Helium: Occurrence, Applications and Biological Effects ISBN: 978-1-62808-612-6
Editor: Simon B. Davis

Chapter 4

INTERACTION OF HELIUM ATOMS AND IONS WITH MATTER

Ya. A. Teplova**, N. V. Novikov and Yu. A. Belkova
Scobeltsyn Institute of Nuclear Physics,
Lomonosov Moscow State University, Moscow, Russia

ABSTRACT

The electron loss and electron capture cross sections, equilibrium target thickness, energy loss and ranges are studied experimentally for helium atoms and helium ions with velocities $(2.6 - 20)\times10^8$ cm/s in various targets. The energy, angular and charge distributions of helium ions and atoms reflected from a metal surface at grazing incidence are measured at the energy 250 and 300 keV. On the bases of experimental data some effects of ion interaction with matter are discussed theoretically (nonmonotonic dependence of range, energy loss, and charge-changing cross sections on nuclear charges of the colliding particles, density effects in gaseous and solid targets etc.). The experimental values are given by figures and by tables in appendix

PACS number: 34.70+e, 82.30Fa

Keywords: Electron capture cross section, electron loss cross sections, energy loss and ranges for atoms and ions, density effects

1. INTRODUCTION

Helium ions and helium atoms are the traditional objects of investigation in atomic and nuclear physics. Helium is convenient for experimental studies, because natural radioactive sources of helium nucleus (α-particles) can be used. Helium ions can be efficiently

* Corresponding author. e-mail: teplova@anna19.sinp.msu.ru.

accelerated because of the small mass and small amount of charge fractions. At the same conditions the greater velocity can be reached in acceleration of He than for heavier ions.

Besides this, helium is the simple monoatomic gas with filled electron shells in the ground state and the large binding energy. It does not form chemical bonds with atoms of other elements. He has relatively small amount of electrons and only few charge states, which can be rather successfully described theoretically. These features make helium atoms and helium ions the unique objects for investigations.

Our group studied different physical aspects of helium interaction with gaseous and solid matter since 1957 year. The electron loss and electron capture cross sections, charge distributions, equilibrium target thickness, energy loss and ranges are studied experimentally for helium atoms and helium ions, passing with velocities $(2.6 - 20)\times10^8$ cm/s through various targets. The energy, angular and charge distributions of helium ions and atoms reflected from a metal surface at grazing incidence are measured at the energy 250 and 300 keV. On the bases of experimental data some effects of ion interaction with matter are discussed theoretically (nonmonotonic dependence of range, energy loss, and charge-changing cross sections on nuclear charges of the colliding particles, density effects in gaseous and solid targets, the metastable $(1s2s)^{1,3}S$ helium atomic states, etc.).

The investigations of various processes with He ions, including interaction with matter, were performed by some groups, for example [1-4].

2. Experimental Procedure

The technique used in measuring ranges *R*, energy losses *(-dE/dx)*, the range straggling *S* and charge fractions has been described in [5, 6]. A 72-cm cyclotron with a system of magnetic focusing and separating of an energy-homogeneous ion beam was the source of fast particles. The velocity of particles extracted from cyclotron was determined from the field of the focusing magnet, which had been calibrated to within ~ 2%, over the entire interval of magnetic field variation, from the known ranges of protons and α particles with eight different energies. The energy spread of the ion beam did not exceed ±1%. The velocities of swift helium particles can be varied from $V \sim 2.6\times10^8$ cm/s ($E \sim 0.035$ MeV/amu) to $V \sim 20.0\times10^8$ cm/s ($E \sim 2.1$ MeV/amu). The ion charge *i* changed from $i = 0$ to $i = 2$.

The ranges *R* and energy losses *(-dE/dx)* were measured in gases: H_2, He, air, Ar and in solids: Be, C, Al, Ni, Ag, Au, organic films (celluloid). The measurements of *R* and *dE/dx* were not affected by an error associated with scattering of the particles. The mean *R* was determined within 2-3%. Energy losses in solids were calculated from data on the air equivalents of foils for different velocities and from the dependence of *R* on *V* in air. In order to measure partial ionization, a focused ion beam from a 72-cm cyclotron was directed into a deceleration chamber. By moving the detector along the particle run, it was possible to determine the pulse amplitude magnitudes by the beam path, and thus to establish the dependence of partial ionization on the ion run.

The charge fractions of ions were measured in gases: He, N_2, Ne, Ar, Kr, in celluloid and carbon films. Cross sections for loss and capture of electrons σ_{ik}, where *i* and *k* are the initial and final charges of the particles, were determined with an experimental apparatus which differed from the previously described one mainly by using the additional magnetic separator for making measurements with neutral particles. We extracted from the 72-cm cyclotron He^+ ions with $V = 8\times10^8$ cm/s and He^{2+} ions with $V = 1.2\times10^9$ and 2.0×10^9 cm/s. Atoms and ions with other charges at the same velocities V were obtained by passing the ion beam through a celluloid film of thickness 2-3 μg/cm^2. The errors in the resulting cross sections σ_{ik}, which were determined mainly by the errors in determination of the thickness of the gas target and by the statistical spread of the results of individual measurements, amounted as a rule to 10-15% for $\sigma_{ii\pm1}$ and 20-25% for $\sigma_{ii\pm2}$.

The energy and charge distributions of helium ions and atoms with energies E_0 = 250–300 keV reflected from the metal surface for grazing incidence angles α = 1°–3° have been measured. Helium ions were accelerated by an EG_500 cascade generator to energies between 250–300 keV. The helium atom beam formed as a result of electron capture by He^+ ions in a gas with subsequent deflection of the charged component by magnetic field. A Cu polycrystalline target and a semiconductor detector (SD) were fixed onto a multiaxis goniometer making it possible to set the angle α of beam incidence on the target and the scattering angle θ to an accuracy of ±0.2°. The incident beam was monitored using particle backscattering by a rotating plate covered by a thin Au layer. The scattered particle energy E was measured by the SD in the plane of the incident beam and the normal to the target surface with a resolution of approximately 10 keV in the angular ranges α = 1°–5° and θ = 1.5°–10°. The energy spectra of all scattered particles were measured using the SD and an amplitude analyzer. A voltage exceeding 6 kV (sufficient for deflecting the charged component of the scattered beam from the detector input window) was applied to an electrostatic deflector (attached with SD) with an input window ~0.3 mm in diameter. The scattered atom energy spectra were registered under such conditions. The ratio of the distribution of scattered atoms to that of all scattered particles for fixed α, θ, and a fixed number of incident particles makes it possible to obtain the dependence of the relative number of atoms in the scattered particle beam (the neutral fraction) F_0 on E.

3. Ranges and Energy Losses

The mean distance R through which an ion of initial energy E moves through a matter before it is stopped may be found by the relation

$$R(E) = \frac{1}{N}\int_0^E dE' \frac{1}{(-dE'/dx)}, \qquad (1)$$

where N is the number of atoms or molecules per cm^2. In an analogous way we may introduce the range straggling.

$$S = 2\sqrt{\frac{2}{\pi}}\frac{R_{ext} - R}{R}, \qquad (2)$$

where R_{ext} is the extrapolated range. It is necessary to note, that the difference between R and R_{ext} is small.

In order to determine the relative stopping powers of different gases we measured the ranges of He ions in various gases and gaseous mixtures. The ratio of each of these ranges to the corresponding range in air was only slightly dependent on velocity. The ranges *of* He ions in hydrogen are ~ 3.8 times greater, and in helium ~ 5.3 times greater, than in air, while in benzene vapor at 10^5 Pa they are smaller by a factor ~ 3.5 than in air. For He the ranges in methane are - 15% greater than in air. The dependence of R on Z_t was found to be similar to that for protons. The summary error including the errors in measurements of the pressure and foil equivalent fluctuates between 1 and 5 % for range and between 7 and 15% for energy loss.

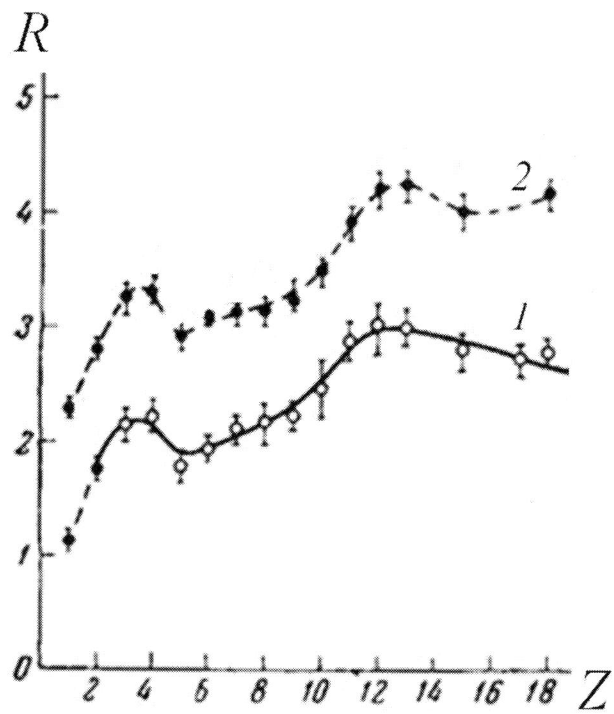

Figure 1. Dependence of ion ranges on Z in air at 10^5 Pa and 19°C. The curves are labeled with the respective ion velocities: (1) $V = 2.6\times 10^8$ cm/s; (2) $V = 4.1\times 10^8$ cm/s . The points are connected to guide the eye.

The experimental measurements of ranges for light ions with various charges Z allowed to obtain the dependence of R on Z. This dependence is nonmonotonic (Figure 1), and is characterized by periodic increases and decreases of the order of 30%. The amplitude of the fluctuations diminishes with increasing velocity. The fluctuations of R indicate on significant role of the electronic structure of the ions. The first maximum corresponds to the ions where L shell begins to be filled. With increasing V the fluctuations diminish and the details of the electronic structure become less important.

A charged particle is slowed down in matter as a result of three basic processes. The energy loss may be written as:

$$(-dE/dx) \approx (-dE/dx)_{el} + (-dE/dx)_n + (-dE/dx)_{ex}, \quad (3)$$

where $(-dE/dx)_{el}$ is the energy loss in inelastic collisions between ions and atomic electrons of medium, induced by ionization and excitation of atoms, $(-dE/dx)_n$ is the energy loss in elastic collisions with atoms, and $(-dE/dx)_{ex}$ is the energy loss in charge exchange process. The relative importance of these processes depends on V, Z, and Z_t.

The energy loss in inelastic collisions $(-dE/dx)_{el}$ comprises 75-90% of the total energy loss at $V>V_0$ ($V_0 = 2.19\times10^8$ cm/s). The calculations of $(-dE/dx)_{el}$ at $V \leq V_0$ can be carried out according to Firsov formula [7]

$$(-dE/dx)_{el} = 2.34\,(Z+Z_t)\,V\times10^{-15}\ \text{eV cm}^2/\text{atom}, \quad (4)$$

where V is in units of 10^8 cm/s, $Z_t \approx (1/4 \div 4)Z$. The analogous dependence is represented by Lindhard's formula [8].

The calculations of $(-dE/dx)_{el}$ at $V>>U$, where U is the mean orbital velocity of the electron being removed, can be carried out according to Bloch's formula [9] with using the effective ion charge taking into account the various charge states of ions, passing through the matter.

The value of $(-dE/dx)_{ex}$ can be evaluated on the basis of the electron capture and loss cross sections, extracted from the experimental data for ions, passing through matter [10, 11]:

$$(-dE/dx)_{ex} = N_t\,\sigma_{i,i\pm1}\,(I+E_k), \quad (5)$$

where N_t is the number of atoms per cm^2 of the stopping material, E_k is the kinetic energy of the electron, that was captured and subsequently lost by an ion, I is the binding energy of electrons in target atom. The presented results coincide with experimental data and qualitatively agree with results of other authors, obtained for other targets and higher projectile energies.

Most of the presented results have to do with the ion velocities in the range $V = (0.5 - 4.0)\,V_0$ ($V_0 = 2.19\times10^8$cm/s) [10]; this corresponds to the energy range $E \leq 0.4$ MeV/amu. This energy range is important in considering practical applications (e.g., ion implantation, surface modification, and medicine).

Figure 2 shows the measuring energy loss dependence of helium ions. Ion energy losses are observed mainly in the low energy range, where $(-dE/dx)$ is proportional to the ion velocity, and partially in the range of maximum energy loss. The position of the energy loss maximum depends on Z_t.

We should note that the traditional Firsov model, in which ion energy losses appear due to electrons transfering the pulse through a plane separating the incident ion and the target atom, it is assumed that energy losses increase monotonously with rising charges of the incident ion and target atom. A modified Firsov theory, however, in which electron flow through the plane is calculated for individual electron shells and then summarized according to all of the shells whose contributions were substantial, allows us to qualitatively describe nonmonotonous dependence $(-dE/dx)$.

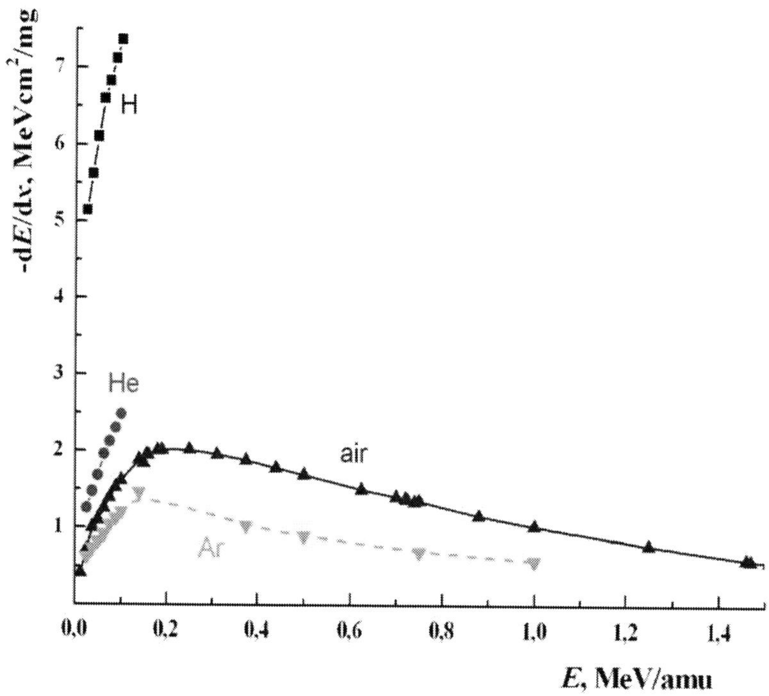

Figure 2. The energy loss dependence of helium ions in hydrogen, helium, air and argon.

4. Charge Exchange Cross Sections and Equilibrium Charge in Gases

A mass-spectrometric method for determining charge fractions Φ_i and effective cross sections of electron loss and capture by ions was used in the experiment and was described in [12, 13]. The cross sections σ_{ik} of the loss and capture of one or several electrons were calculated using the experimental values of Φ_i on the basis of the solution of the system of linear differential equations for the charge–changing process

$$\frac{d\Phi_i}{dt} = \sum_k \Phi_k \sigma_{ki} - \Phi_i \sum_k \sigma_{ik}, \qquad \sum_i \Phi_i = 1, \tag{6}$$

where t is thickness of target. If t is more than an equilibrium target thickness t_{eq}, we obtained

$$\frac{d\Phi_i}{dt} = 0 \text{; at } t \geq t_{eq}, \tag{7}$$

where F_i are equilibrium charge fractions coupled by the relations

$$\sum_k F_k \sigma_{ki} - F_i \sum_k \sigma_{ik} = 0, \qquad \sum_i F_i = 1. \tag{8}$$

If the processes with simultaneous capture or loss of more than one electron can be neglected, the ratio of equilibrium charge fractions are presented as

$$\frac{F_{i+1}}{F_i} = \frac{\sigma_{i,i+1}}{\sigma_{i+1,i}}. \tag{9}$$

The mean charge of ions passing through matter can be written as

$$\bar{i} = \sum_i i\, F_i . \tag{10}$$

Formula (9) taking into account particular corrections for the described experiment was given in [13]. The possibility that the same number of electrons can be lost as a result of several subsequent collisions with the loss of only one electron in each of them was taken into account in determining the cross sections of the simultaneous loss of several electrons. The relations for the calculation of the cross sections σ_{ik} using the experimental values of Φ_i (6) are valid only in the case of a sufficiently low gas pressure where the single-collision conditions are satisfied. An important element of the mass-spectroscopic method is the required control of the absence of errors in the values of cross sections that can be produced by scattering and by the presence of a residual gas in the system and by probable nonlinearity in the dependence of Φ_i on the gas pressure. Thus, the error in the obtained values of cross sections was mainly determined by the errors in the determination of the pressure in the collision chamber and additional volumes located along the ion beam path as well as by the statistical spread of results of individual measurements. It was about 20%.

The equilibrium charge distributions in beams of He ions that have passed through He, N_2, Ne, Ar, Kr and a celluloid foil have been measured for ions with $V \approx (2.6 \div 20.0)\times 10^8$ cm/s. The experimental data showed that the distribution of charge in all media is close to Gaussian for intensive charge group

$$F_i \approx \frac{1}{\sqrt{2\pi d^2}} \exp\left[-\frac{(i-\bar{i})^2}{2d^2} \right] \tag{11}$$

and can be characterized essentially by two parameters: the mean charge (10) and half-width of charge distribution

$$d = \left[\sum_i F_i (i-\bar{i})^2 \right]^{1/2} \tag{12}$$

Exceptions are the cases when the distribution is highly asymmetrical relative to the mean charge, owing to the fact that is close either to zero or to Z. In this case d no longer determines the width of the Gaussian distribution, but yields, as always, the difference between the mean-squared charge which is used in the calculation of energy losses. At a given ion velocity, the value of d in different gases differ as a rule by not more than ten percent. The value of d in celluloid at $V \leq 5 \times 10^8$ cm/s is 10 — 20% greater than in gases. When V is increased, d first increases but starts decreasing at $V > 7 \times 10^8$ cm/s. We find that the maximum value of d is 0.55 for helium ions.

The dependence of on V and Z can be described using the empirical relation [14]

$$\bar{i}/Z = f(V/Z^{\beta}), \tag{13}$$

where $1/3 \leq \beta \leq 2/3$. The function f monotonically increases with V/Z^{β}.

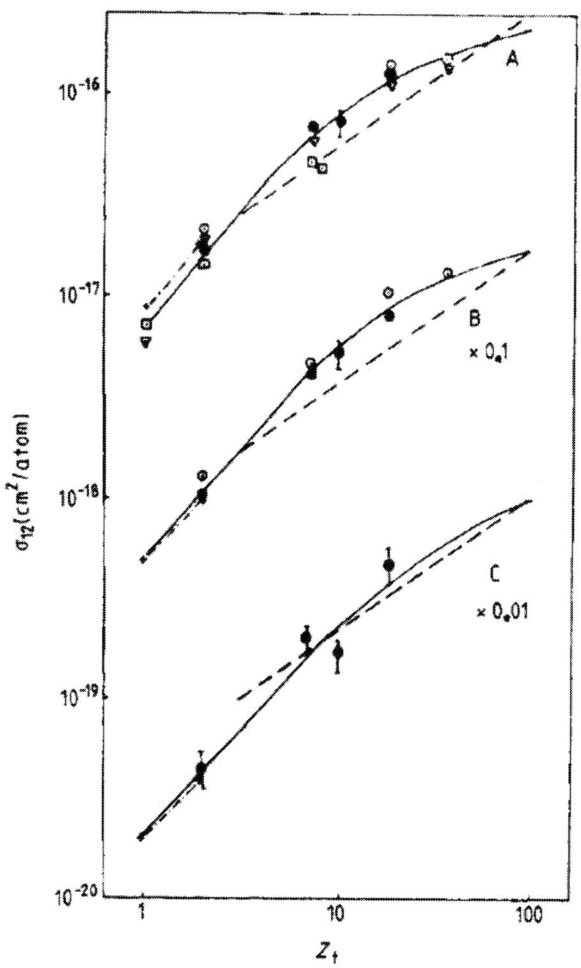

Figure 3. The electron loss cross sections σ_{12} for He^+ ions as a function of Z_t for V=8x10^8, 12x10^8 and 20x10^8 cm/s are shown by A, B and C respectively. Experimental data: ●, [17]; ∇, [18]; O, [19] and □, [20]. --. --, Born approximation according to [21, 22]; - - -, Bohr's formula; ___ , monotonic dependence of σ_{12} on Z_t.

To determine the variation of with the medium, we measured the equilibrium distributions of helium ions in helium, nitrogen, argon, krypton, and celluloid film [15]. The results of the measurements of the equilibrium distribution of the charges of He ions are listed in Tables B.1. In addition the equilibrium distributions of the charges were obtained for approximately the same velocities in a celluloid film, and yielded for $i > 0$ values of Φ_{i+1}/Φ_i which agreed with the results [16].

We observed considerable differences in the values of the mean ion charge in different gases. Although the ratio of the values of in different media does not remain constant with velocity, the following regularities, which are common to all ions, can be noted. In the case of ions passing through a gas, the maximum of is observed in nitrogen or argon, and the values of in these gases do not differ by more than 5%. The mean charge in helium is always smaller than in nitrogen, and the mean charge in krypton is smaller than in argon. In the velocity range (3.5—10) x 10^8 cm/s the minimum of occurs in helium, and at $V \sim 3 \times 10^8$ cm/s and $V \sim$ (11 to 12) x 10^8 cm/s it occurs in krypton. In most cases the difference between the maximum and minimum values of in gases amounts to 20%.

The experimental values of the electron loss cross sections σ_{12} for He^+ ions refer to a target atom and the experimental and theoretical (OBK) values of the electron capture cross section σ_{21} for He^{2+} ions are presented in figures 3 and 4, respectively. The figures also contain the experimental values obtained for σ_{12} and σ_{21} for helium ions in various gases. Also included are the values of σ_{12} for $Z_t = 1$ and 2 obtained in the Born approximation and those for $Z_t \geq 3$ calculated from Bohr's formula which was obtained in the free-collision approximation in a classical treatment of electron scattering in the strongly screened field of the target atom

$$\sigma_{12} = \pi\, a_0^2\, Z_t^{2/3}\left(\frac{V_0}{V}\right)\left(\frac{I_0}{I}\right)^{1/2}, \tag{14}$$

where I is the binding energy of an ejected electron, a_0=5.29 x 10^{-9} cm, V_0= 2.19 x 10^8 cm/s and I_0 = 13.6 eV.

The experimental results have shown that the 20-40% decrease of the electron loss cross section, $\sigma_{i,i+1}$ in neon relative to the corresponding cross sections in nitrogen and argon, which was earlier revealed for nitrogen and neon ions with charges q= 3-6 [23, 24] is practically unobserved for helium atoms and ions (figure 3). From the available experimental data on the electron loss cross section for helium ions it follows that at velocities V= 8 - 20 x 10^8 cm/s these cross sections increase practically monotonically with increasing Z_t. The experimental values of σ_{12} in hydrogen and helium coincide to within 20% with the values calculated in the Born approximation [21] for $Z_t = 1$ and 2. The experimental values of σ_{12} for Z_t = 7-36 are the same as, or larger by no more than 1.5 times than, the values calculated by Bohr's formula (figure 3).

The experimental values of the electron capture cross sections σ_{10}, σ_{21} and σ_{20} for helium ions in neon at V = 8 x 10^8 cm/s are larger by a factor of two to three than in nitrogen and argon (figure 4). Thus, the increase of the electron capture cross sections in neon revealed earlier [24] for nitrogen and neon ions with charges i= 3-6, is completely confirmed for all helium ions. Therefore, this increase can be assumed to be general for all ions of light elements.

The calculated values of the cross sections multiplied by the same coefficient k = 0.18 for all media, and for velocities V = 8 x 10^8, 1.2 x 10^9 and 2 x 10^9 cm/s are practically coincident with the experimental cross sections σ_{21} (figure 4).

At $V = 8 \times 10^8$ cm/s the values of $0.18\,\sigma_{10}^{OBK}$ in nitrogen and neon are close to the experimental values and those in hydrogen and helium are smaller by a factor of two to three, and in argon and krypton they are larger by a factor of three to five than the experimental values. Hence, with an increasing Z_t the experimental values of σ_{10} increase more strongly than the values of σ_{10}^{OBK} .

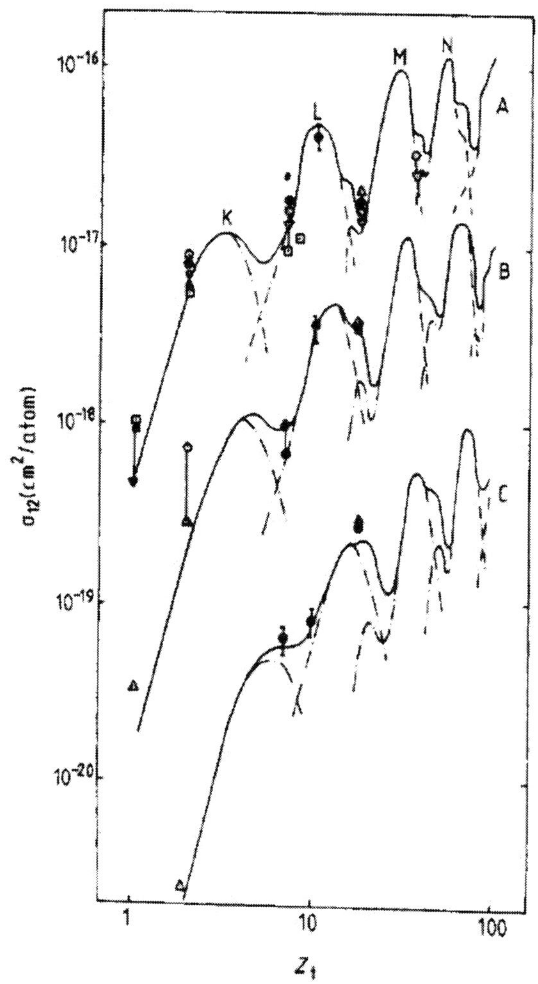

Figure 4. The electron capture cross sections σ_{21} for He^+ ions as a function of Z_t for $V = 8 \times 10^8$, 1.2×10^9 and 2×10^9 cm/s are shown by A, B and C respectively. Experimental values: ●, [17]; O, [25]; ∇, [18]; Δ, [26] and □, [27]. The lines are OBK calculations.

At all the velocities examined the dependence of the calculated cross sections and on Z_t is of an oscillating character. An analogous oscillation in the dependence of the cross sections on the nuclear charge Z_t of the target atoms was previously obtained [33] when calculating the cross section for electron capture to the K shell of Cl ions with an energy of 120 MeV using the Brinkman-Kramers formula. These calculations have confirmed the experimental non-monotonic dependence of the x-ray yield on Z_t for Cl ions. Maxima in the cross sections and are observed for the target atoms with electron binding energy I_s close to $I_V/3$ where s is K, L, M or N shell and $I_V = mV^2/2$ is the transfer energy of the electron. As the ion velocity increases, the maxima shift to larger Z_t for $V = 8 \times 10^8$ cm/s, when $I_V = 180$ eV, they should be observed at $Z_t = 3$, 10, 30 and 55 and for $V = 2 \times 10^9$ cm/s, when $I_V = 1.1$ keV at $Z_t = 6$, 16, 36 and 70. It will be noted that the relations between the calculated values of σ_{21}^{OBK} for $Z_t = 7$, 10 and 18, which are qualitatively different at $V = 8 \times 10^8$, 12×10^8 and 2×10^9 cm/s, are well substantiated by experiment.

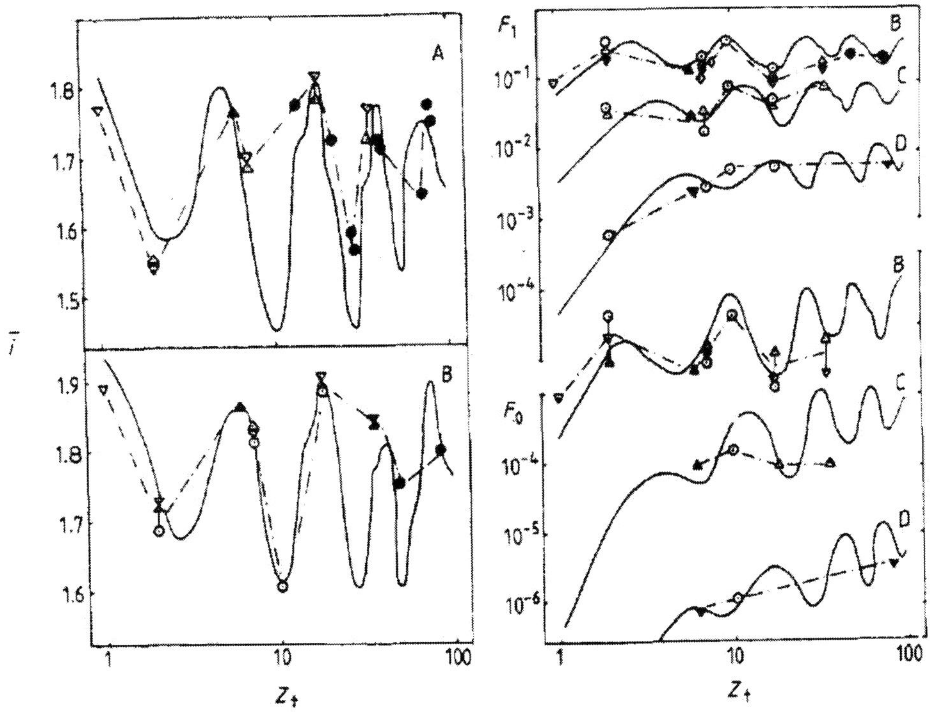

Figure 5. The equilibrium charge fractions F_0 and F_1 and mean equilibrium charge for helium ions as a function of Z_t, at $V = 6.9 \times 10^8$, 8×10^8, 1.2×10^9 and 2×10^9 cm/s are shown by A, B, C and D respectively. O, from experimental cross sections [17], ___ from values of $\sigma_{i,i-1}^{OBK}$ and the experimental monotonic dependence of $\sigma_{i,i+1}$ on Z_t, (full curve in figure 3). The experimental results for gaseous targets: Δ, [13]; ∇, [28]; □, [29]; and for solid targets: ▲, [30]; ●, [31] and θ, [32]. The experimental values of F_q and are connected by a chain-like line to guide the eye.

Oscillations of the electron capture cross sections with Z_t must give rise to oscillations of the equilibrium charge fractions F_i and in the mean equilibrium charge $=\sum i\, F_i$ of helium ions. Really, the experimental values of F_i and [15, 28-32] show the oscillating Z_t dependences which are qualitatively the same as those calculated from the experimental monotonic dependences of the electron loss cross sections σ_{12} on Z_t, and also from the nonmonotonic dependences of the theoretical electron capture cross sections σ_{21}^{OBK} on Z_t which are normalised to the experimental values of σ_{21} at $Z_t = 7$. The experimental values of F_i and obtained for helium ions passing through the gaseous media H_2, He, N_2, O_2, Ne, Ar, Kr [15, 28, 29] and through solid films [30, 32] and the values of F_i and obtained for helium ions scattered in the opposite direction by various solid targets heated up to 800°C [31] lie on the same Z_t dependence curves and supplement each other (figure 5). These facts have finally convinced us about the oscillating character of the dependence of the electron capture cross section on Z_t for helium ions at ion velocities $V = 8$ - 20×10^8 cm/s. According to the OBK calculation these oscillations in the capture cross sections and σ_{10}^{OBK} must also be observed at smaller velocities down to $V \sim V_0$.

5. Metastable States

When fast helium atoms pass through matter, many of them are in excited states and a part of them, η, is in metastable states. The experiments described involved the excitement of fast ions when they passed through thin celluloid films and recording these ions on the basis of increased probability of electron loss in a mass spectrometer installation intended for measuring the cross-sections of loss and capture of electrons by ions. From the results of measurements [34] we have determined the cross sections for loss and capture of one and two electrons by unexcited helium-like ions and the cross sections for loss of one electron by metastable particles. We have estimated the relative number of metastable ions in beams of helium-like particles for the two indicated methods of formation. A comparison has been made of the cross sections for unexcited and metastable helium-like particles with the corresponding cross sections for other very simple ionic systems.

Since in beams of fast helium ions a certain fraction of them η is in metastable states, the values σ_{01} and σ_{02} obtained from experiments should differ somewhat from the cross sections and for unexcited atoms: $\sigma_{0k} / \sigma_{0k}^0 = 1 + \eta\left[\sigma_{0k}^m / \sigma_{0k}^0 - 1\right]$, where is the cross section for metastable particles. The values of σ_{01} for $Z_t = 2$, 7, and 10 should exceed the values of σ_{01}^0 by 10-15% at $V = 8\times10^8$ cm/s and by 6-8% at $V = 2\times10^9$ cm/s. The excess of the values of σ_{01} in argon is respectively 12-18% and 7-10%. The relative error in the cross section σ_{02} turns out to be the same if in accordance with the estimates obtained in [34] we assume that $\left|\sigma_{02}^m / \sigma_{02}^0 - 1\right| \leq 0.8$. Thus, the deviation of the values found for σ_{01} and σ_{02}, from and σ_{02}^0, in most cases does not exceed the error given above for the cross sections.

It was found that lifetime of $(1s2s)^1S$ and of $(1s2s)^3S$ state is about 1.4×10^8 and 3.8×10^{12} ns, respectively.

6. Equilibrium Charge Fractions in Dense Media

The nonmonotonic character of the variations in F_i and with increasing Z_t, was found experimentally [35] to also occur in the case of He ion beams after passage through solid targets. It is obvious that it is not possible to study the Z_t-dependence of the charge composition of an ion beam traversing through solid media without allowing for a high density of target atoms. The interatomic distances in some solid targets (for example, Be, B, C, Al, etc.) do not exceed 2.7 Å thereby suggesting that all the excited states of fast atoms and He ions whose characteristics linear dimensions are no less then 2.7 Å are unlikely to exist in such targets. In this case the Z_t -variations of the charge distribution of the He ion beam scattered by solid targets found in [35] are actually due to the regularities governing the Z_t-dependence of partial, rather than total, cross sections of single electrons capture to the He^0 and He^+ ground states and must differ from the variations of the charge distribution in rarefied media.

The equations [17] are used to calculate the equilibrium charge composition F_i and of He ion beams after the passage through dense media with $Z_t \leq 24$ at collision velocities of V= 1.0, 1.83 and 3.16 a.u. on the assumption that only the ground state incident particles can exist inside a target [36]. The cross sections for the electrons capture to the ground state by He^{2+} and He^+ ions were calculated as

$$\sigma_{i,i-1} = K\sigma_{i,i-1}^{OBK} , \tag{15}$$

where $\sigma_{i,i-1}$ is the total cross section for the electron capture by an ion of charge i; the superscripts EXP and OBK correspond to the experimental cross sections and to those calculated in the Oppenheimer-Brinkman-Kramers approximation. The values are not known for some of the media examined. The $\sigma_{i,i-1}$ in (15) were calculated using the K-values obtained by linear interpolation of the $K(Z_t)$ values at the points where the $\sigma_{i,i-1}^{EXP}$ values are known. Simultaneously with the F_i and values for dense media we have calculated the equilibrium charge distribution of He ion beams after passage through gas targets with $Z_t \leq 24$. The F_i and values for gaseous media were obtained by solving the same set of equations as in the case of dense targets.

The results of calculations [36] of the equilibrium charge fractions and the mean charge of He ions emerging from dense and gas targets are shown in figure 6 together with the available experimental data. It can be seen that the charge fractions and the mean charges of the equilibrium ion beams that have passed through dense targets ($F_i^{sol}, \bar{i}^{sol}$) and gases ($F_i^{gas}, \bar{i}^{gas}$), are characterized by an oscillating dependence on nuclear charge Z_t. The charge fractions, as compared with are more sensitive to variations in V. At V= 3.16 a.u. the difference between F_i^{sol} and in most of the targets does not exceed 20%, and the character of the variations in $F_i^{sol}(Z_t)$ and is the same; namely, the extrema of the two functions occur approximately the same Z_t values and, besides, the oscillation range is about the same. At V= 1.83 a.u., not only the differences between the and values increase, but also the positions of the extrema of the two functions shift in such a way that the positions of the $F_i^{sol}(Z_t)$ minima coincide with the positions of the local maxima of $F_i^{gas}(Z_t)$. At V= 1 a.u. the differences between the dependences $F_i^{sol}(Z_t)$ and are even more pronounced, with the $F_i^{sol}(Z_t)$ and $F_i^{gas}(Z_t)$ values being different by an order of magnitude in some media. Besides, it is clear that the variations of the F_2^{sol} and F_2^{gas} values as functions of Z_t are practically opposite in phase.

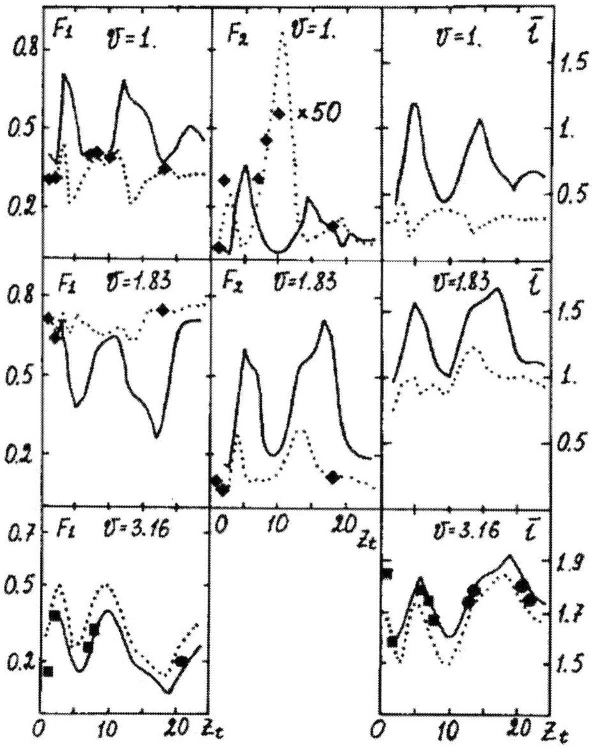

Figure 6. The equilibrium charge fractions F_i (F_1 and F_2) and the mean charge of fast He ion beams in dense targets (the solid line) and gases (the dotted line) as functions of nuclear charge Z_t of the target atoms at collision velocities V= 1.0, 1.83, and 3.16 a.u. The F_2-values at V = 3.16 are $F_2 = 1 - F_1$. The experimental data are from [35] (the full circles), from [29] (the full squares), and from [16] (the full diamonds).

Thus, the assumption that the fast excited particles, whose characteristic dimension exceeds the mean interatomic distance in a medium, are absent in such media has led us to conclude that the Z_t -dependences of the equilibrium beam charge and of the equilibrium charge fractions must be very different in rarefied gaseous and dense targets. This circumstance may be used to carry out direct experimental verification of the very hypothesis that fast excited particles, whose characteristic dimensions exceed the mean interatomic distance in a given target, cannot occur in solid media.

7. Energy and Angular Distributions of Ions Scattered by Surface

On the base of measured energetic and angular distributions of particles the scattering processes for H^+ and He^+ ions at incidence on a metallic surface under small angles were compared. The experimental measurements [37] were performed for particle energies of 300 and 500 keV in incidence angle interval of $\alpha = 2°\text{-}6°$ and scattering angle interval of θ from $\theta = \alpha$ to $\theta = 30°$.

The obtained energetic distributions of He^+ ions are shown on figure 7. The form of distribution curves is close to dome shaped. It was determined that half-width of energetic distribution of ΔE for He^+ ions is greater than for H^+ ions for 300 keV energy. If the incidence angle of α increases, the maximum of angle distribution for He is displaced to greater scattering angles stronger than for H. The most symmetric distribution is for $\theta \approx 2\alpha$. For α=const the maximum of the energetic distribution E_{max} *moves to the smaller energy with increasing of* θ, and the effective half-width of the energetic distribution of scattered particles ΔE decreases.

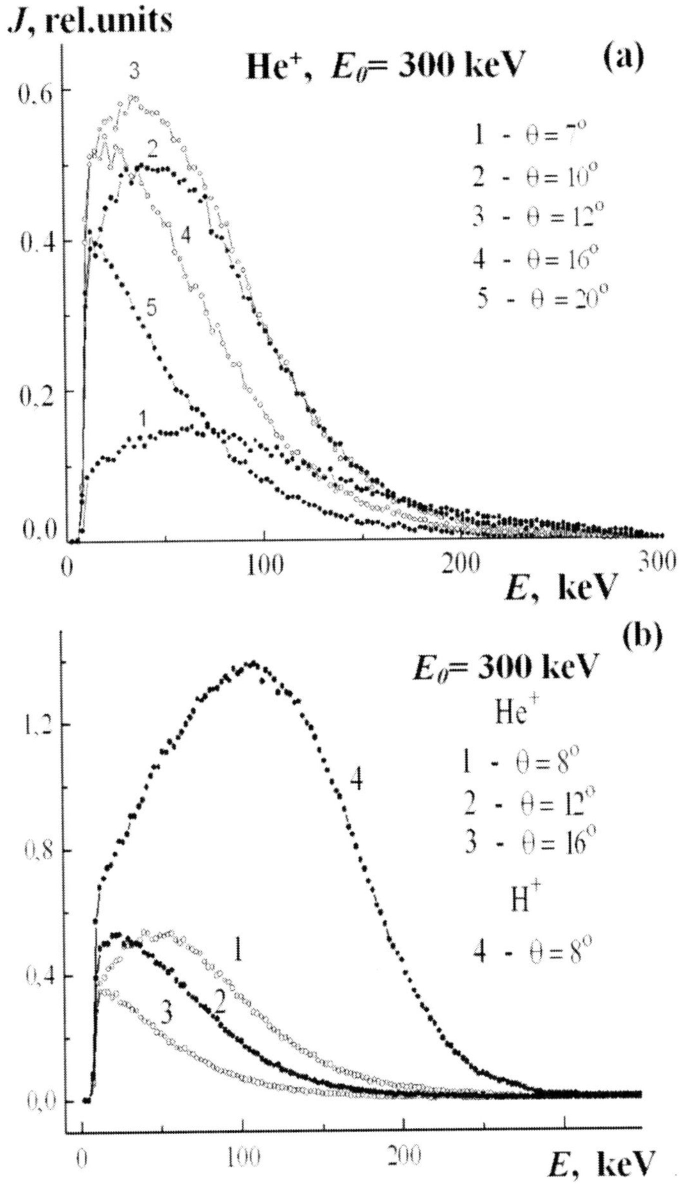

Figure 7. (Continued)

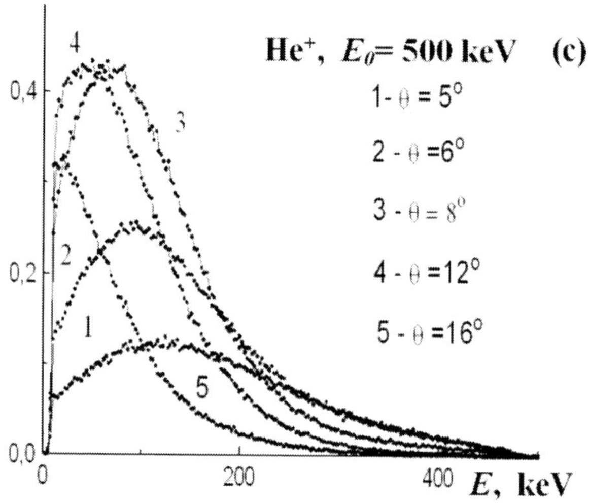

Figure 7. Relative number of particles scattered by a steel surface: (a) at E_0 = 300 keV , α=6° ; (b) at E_0 = 300 keV , α=4° (c) E_0 = 500 keV, α=4°.

With the increasing of the energy of He^+ ions the half-width ΔE increases, and its value depends on the relations of angles α and θ. For particle energy E=500 keV and $\theta < 2\alpha$ the maximum value $E_{max} < \Delta E$. For E=300 keV the relation between $\boldsymbol{E_{max}}$ and ΔE can be different: for $\theta > 2\alpha$ and $E_{max} > \Delta E$, but for $\theta < 2\alpha$ and $E_{max} < \Delta E$.

The angular distribution of helium ions reflected from surface is asymmetrical (Figure 8). Experimental results demonstrate the maximum in angular distribution when the reflecting angle (θ-α) is equal of the grazing angle α or $\theta \approx 2\alpha$. The number of reflected particles decreases with increasing of scattering angle θ and the dependence on α becomes weak at large θ. If the incidence angle increases the maximum of angle distribution for He is displaced to greater scattering angles stronger than for H (Figure 8).

8. Energy and Charge Distributions of Ions Scattered by Surface

The relative number of helium atoms in the scattered beam upon He ion and atom incidence on the Cu surface is studied [38]. Theoretical estimation of the charged state of an ion reflected from a surface is complicated because there is hardly any information concerning cross sections for ion atom interaction in solids; in this case, estimation of the charge exchange cross section for solids must differ from that for gases. In the case of gases, the charge exchange cross section $\sigma_{i,j}$ characterizes the interaction of the projectile in the ground state with one isolated atom of the medium, while, in the case of solids, the projectile can interact with several of the medium's atoms at the same time, because interatomic distances are small. In addition, because the time interval between collisions is small, there is no time for the excited scattered particle to make the transition to the ground state, and, being in the excited state, it participates in a subsequent collision with another atom of the medium.

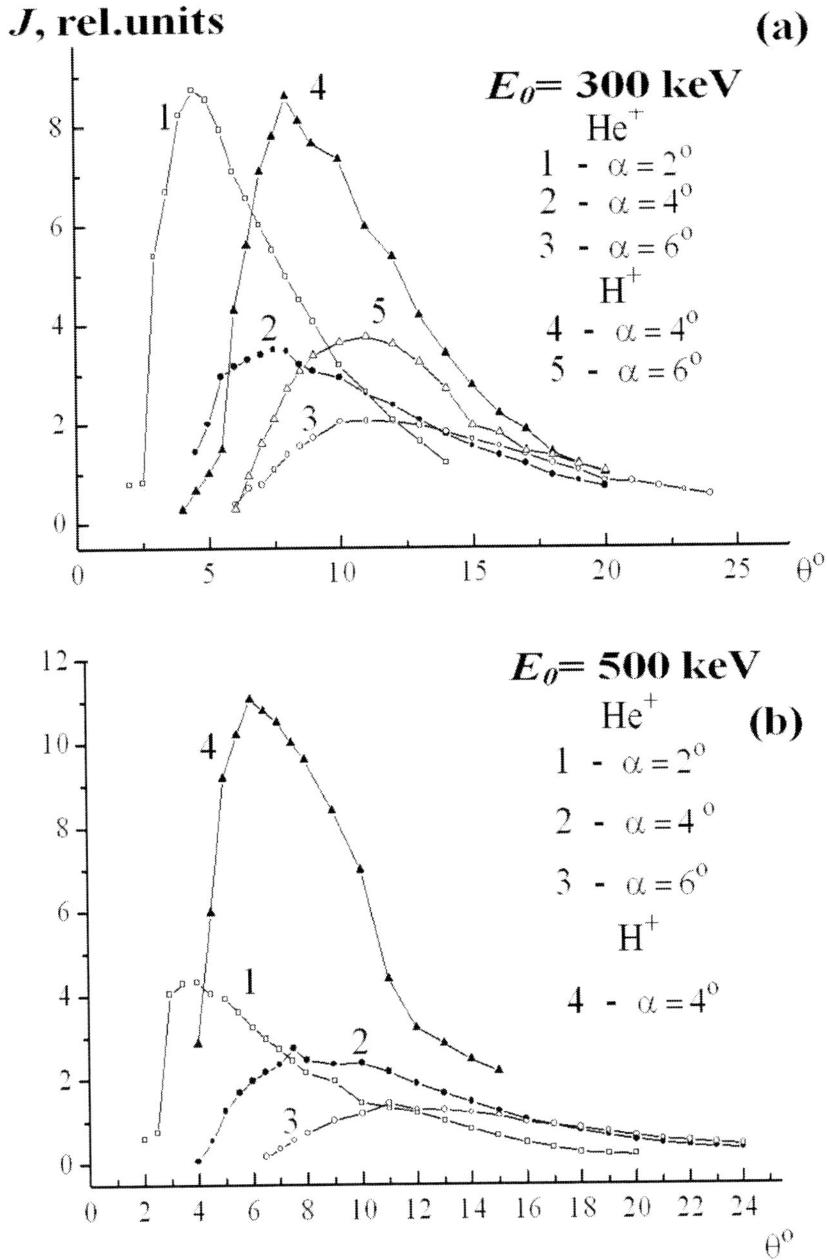

Figure 8. Relative number of particles scattered by a steel surface: (a) at E_0 = 300 keV; (b) at E_0 = 500 keV.

Theoretical estimation of the charge fractions F_j in the He ion–Cu atom collisions is based on solving a set of charge exchange equations [16] with the interpolation of experimental electron loss and capture cross sections $\sigma_{i,j}(Z_t, E)$ in the range Z_t = 18–36:

$$\sigma_{i,j}(Z_t,E) = \sigma_{i,j}(Z_t=18,E) + A_{i,j}(Z_t)\,[\sigma_{i,j}(Z_t=36,E) - \sigma_{i,j}(Z_t=18,E)], \tag{16}$$

where $A_{i,j}(Z_t)$ are the interpolation coefficients. It should be mentioned that the dependence of the charge exchange cross sections on Z_t can be not only nonlinear, but also nonmonotonous due to the target atom's shell structure [17]. In connection with this, to estimate $A_{i,j}(Z_t)$, theoretical calculations are used:

$$A_{i,j}(Z_t) = [\sigma^c_{i,j}(Z_t) - \sigma^c_{i,j}(Z_t=18)] / [\sigma^c_{i,j}(Z_t=36) - \sigma^c_{i,j}(Z_t=18)] , \qquad (17)$$

where are calculated charge exchange cross sections. The coefficients $A_{i,i+1}(Z_t)$ were calculated in the first Born approximation [39], and $A_{i,i-1}(Z_t)$, in the Oppenheimer–Brinkman–Kramers approximation [13]. Both approximations take the shell structures of the projectile and target atom into account. It was assumed for multiple processes that

$$A_{i,i+k}(Z_t) = A_{i,i+1}(Z_t) \; ; \; A_{i,i-k}(Z_t) = A_{i,i-1}(Z_t) \text{ for } k \geq 2. \qquad (18)$$

The difference between the theoretical and experimental values which can be significant for the single electron capture process, does not affect the quantity $A_{i,j}(Z_t)$, because expression (17) contains the ratio of cross sections in the same approximation. The empirical coefficient $C_{g-s}(Z,Z_t) > 1$ was used to estimate the cross sections in solids; all electron loss cross sections for gases were multiplied by it and those of electron capture for gases were divided by it:

$$\sigma^s_{i,j}(E) = \sigma_{i,j}(E) * C_{g-s}(Z,Z_t) \text{ for } i<j; \; \sigma^s_{i,j}(E) = \sigma_{i,j}(E) / C_{g-s}(Z,Z_t) \text{ for } i>j. \qquad (19)$$

It was assumed that $C_{g-s}(Z,Z_t)$ weakly depends on E and i.

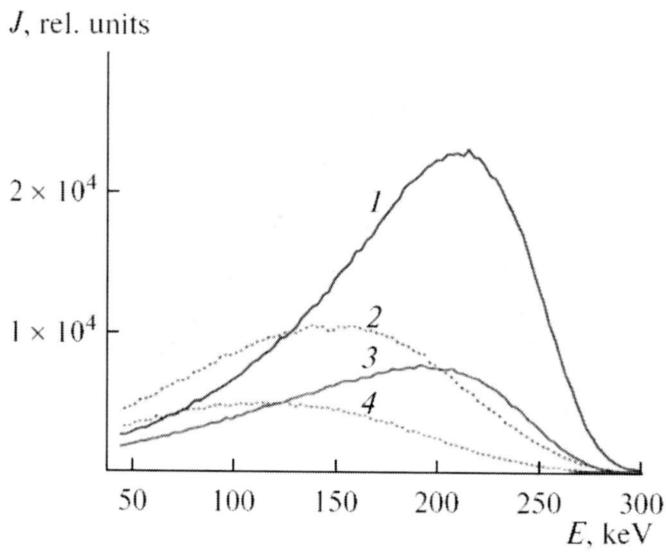

Figure 9. Relative number J of particles scattered by the Cu surface at $E_0 = 300$ keV and $\alpha = 1°$. The following particles were registered: (*1*) He^+ and He^0 for $\theta = 2°$, (*2*) He^+ and He^0 for $\theta = 6°$, (*3*) He^0 for $\theta = 2°$, and (*4*) He^0 for $\theta = 6°$.

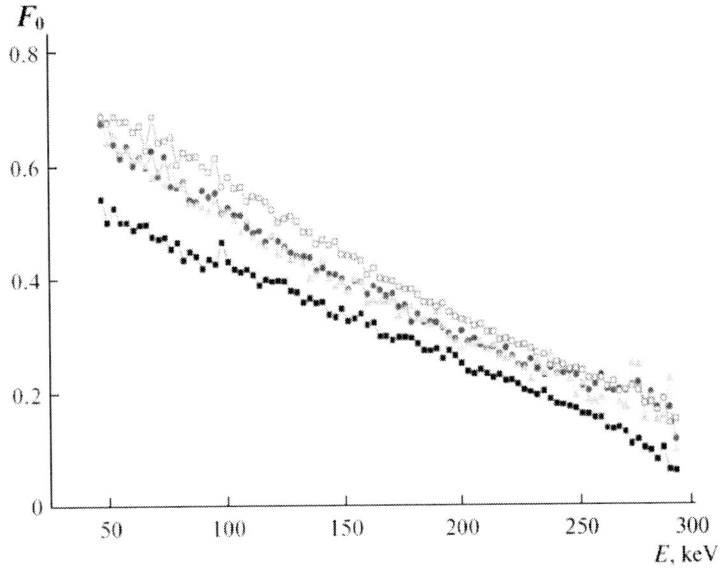

Figure 10. Relative number of helium atoms reflected from the Cu surface in the case of He^0 collisions with E_0 = 300 keV for α = 1° and different scattering angles θ: 2° (■), 3° (●), 5° (▲), and 6° (□).

The energy distribution with a broad maximum (Figure 9) is typical of the number J of particles reflected from amorphous and polycrystalline surfaces for the grazing incidence angle α. The projectile interacting with the target atoms experiences a rather large number of collisions and loses a large portion of its energy. The measured results (Figure 10) show that, for $\theta \geq 3°$, the quantity F_0 varies with θ within the limits of experimental error in the entire range E. In the range $E < 150$ keV, the number of atoms in the scattered particle beam for θ = 2° is noticeably smaller than that for $\theta \geq 3°$, and F_0 is almost independent of θ for $E \geq 150$ keV. This feature is related to the interaction between scattered ions and target electrons.

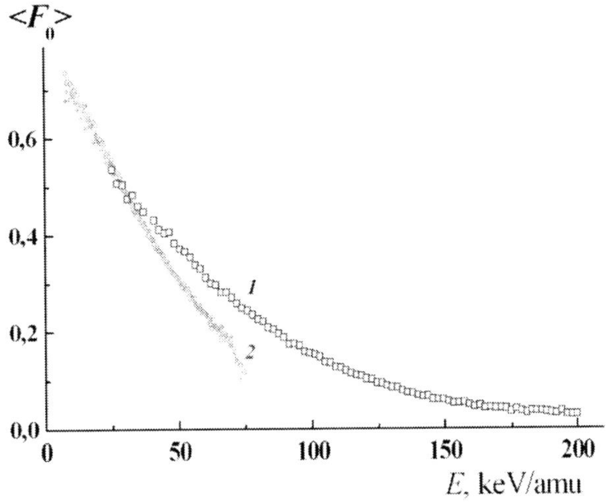

Figure 11. Dependence of the neutral fraction $\langle\Phi_0\rangle$ on the energy averaged over θ in the H^0, He^0, and He^+ beams scattered by the Cu surface for α = 1°. Projectiles: He^0 for E_0 = 300 keV (o), He^+ at E_0 = 300 keV (+), and H^0 for E_0 = 200 keV (□) [40].

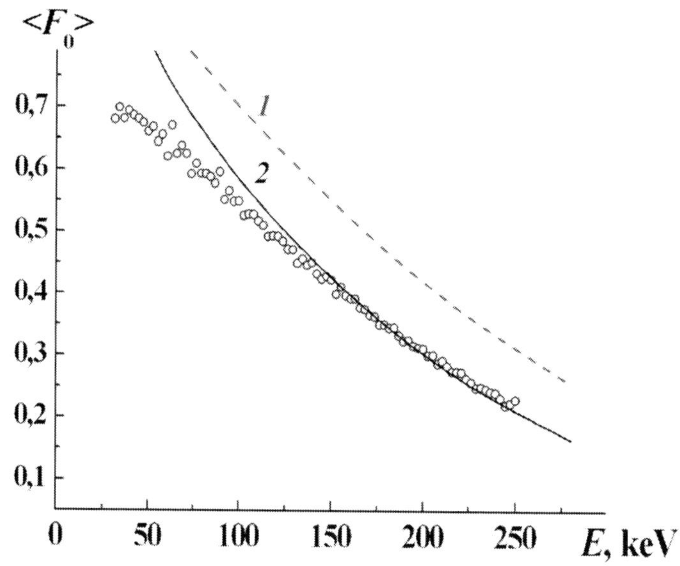

Figure 12. Relative number of atoms reflected from the Cu surface for $E_0 = 300$ keV and $\alpha = 1°$. The experimental data for the He^0 beams incident on the surface are averaged over θ. Theoretical estimates: (*1*) interpolation of charge exchange cross sections in gases, $C_{g\text{-}s} = 1$; and (*2*) data with the gas–solid correction taken into account for $C_{g\text{-}s} = 1.3$.

If the perpendicular velocity component for scattered particles is small, then interaction with the target leads to particle redistribution in the scattered beam; as a result of this, the number of charged particles increases for small scattering angles and the relative number of reflected atoms decreases. Analysis of the results for different α showed that the θ dependence of F_0 vanishes as the incidence angle increases to $\alpha \geq 2°$. This effect in the charge distributions of scattered helium ions is manifested in the same α range and scattered particle velocities as in experiments with protons and hydrogen atoms [40], but, for helium atoms, the decrease in F_0 for $\theta = 2°$ is considerably larger than that for $\theta \geq 3°$.

Figure 11 shows the neutral fraction $< F_0>$ averaged over the scattering angle upon proton and helium atom reflection from the Cu surface. In the range $E = 30$–40 keV/amu, the neutral fraction value obtained in experiments with helium atoms is close to that obtained in experiments with hydrogen atoms. As E increases, the fraction of helium atoms in the scattered beam turns out to be considerably smaller than the fraction of hydrogen atoms for the same scattered particle velocity. This is related to the presence of He^{2+} ions in the scattered beam, which decreases the fraction of He^+ ions and the probability of He atom formation. We also note that the quantity $< F_0>$ for incident He^+ and He^0 coincides within the limits of experimental error in the entire E range (Figure 11). The lack of a dependence of the number of helium atoms on the charge projectile state can be explained by the fact that a scattered particle experiences a number of collisions which is sufficient for establishing equilibrium charge distribution. Theoretical estimates in the gas approximation ($C_{g\text{-}s} = 1$) give a neutral fraction value that is higher than the experimental value in the entire E range (Figure 12).

The increase in $C_{g\text{-}s}$ decreases the fraction of atoms in the particle beam scattered by the surface. The best agreement between theoretical and experimental data regarding He atom reflection from the Cu surface is achieved for $C_{g\text{-}s}(Z = 2, Z_t = 29) = 1.3$ (Figure 12). The obtained coefficient $C_{g\text{-}s}$ can be used to theoretically estimate He^+ and He^{2+} ion charge fractions in the E range, where satisfactory agreement between the experimental and theoretical distributions $<F_0>$ is achieved. The quantitative difference between the theoretical estimate and the experimental data for $E < 100$ keV is probably caused by the different dependences of $\sigma_{i,i-1}$ on E for gases and solids at low velocities, which is not taken into account in (19).

CONCLUSION

The experimental measurements of ranges for helium ions confirmed the nonmonotonic dependence of R on Z, obtained earlier, which is characterized by periodic increases and decreases of the order of 30%. The amplitude of the fluctuations diminishes with increasing velocity. The fluctuations of R indicate on significant role of the electronic structure of the ions. The dependence of $(-dE/dx)$ on V has the maximum, and its position depends strongly on Z and to a less extent on Z_t. The dependence can be represented qualitatively by $dE/dx \sim V^m$, where $m \approx 1$ at $V < 6\times10^8$cm/s, m diminishes to zero at higher velocities, and is negative approaching to $m \approx -2$ for $V \geq 12\times10^8$cm/s.

The electron capture and loss cross sections for helium ions at velocities $V = (8\text{-}20) \times 10^8$ cm/s in helium, nitrogen, neon and argon have been measured. The performed theoretical calculations of the electron capture cross sections for He^+ and He^{2+} ions in various media are in qualitative agreement with experimental results. At all the velocities examined the dependence of the electron capture cross sections on Z_t has an oscillating character. Oscillations of the electron capture cross sections with Z_t give rise to oscillations of the equilibrium charge fractions F_i and in the mean equilibrium charge of helium ions.

The presence of metastable He ions in ionic beam can significantly influence on the measured charge exchange cross sections. For example, the electron loss cross section can increase in several times.

The nonmonotonic character of the variations in F_i and with increasing Z_t, was also found experimentally in the case of He ion beams passing through solid targets. The assumption that the fast excited particles, whose characteristic dimension exceeds the mean interatomic distance in dense medium, are absent in ionic beam, has led us to conclude that the Z_t - dependences of the equilibrium charge and charge fractions are different in gaseous and dense targets.

The energy and charge distributions of helium ions and atoms reflected from a metallic surface at grazing incidence have been measured at the energies 250 and 300 keV. It was shown that for $E \leq 250$ keV neutral fraction in the reflected He beam exceeds 25% and must be taken into account in further studies. For increasing energy the neutral fraction decreases to 10%.

REFERENCES

[1] H.B. Gilbody, K.F. Dunn, R. Browning, and C.J. Latimer, *J. Phys.* B V.4 p.800 (1971).
[2] P. Hvelplund and A. Andersen, *Phys. Scripta,* 26, 370 (1982).
[3] M.E. Rudd, T.V. Goffe, A. Itoh, and R.D. DuBois, *Phys. Rev.* A 32 829 (1985).
[4] P.Sigmund, *Particle penetration and radiation effects general aspects and stopping of swift point charges*" Springer, 2008, 437p.
[5] I. S. Dmitriev, *Ya. A. Teplova*, Yu.A. Belkova, N. V. Novikov, and Yu. A. Fainberg, *Atomic Data Nucl. Data Tables.* 96, 85, (2010).
[6] *Ya. A. Teplova*, I. S. Dmitriev, Yu.A. Belkova, and N. V. Novikov, *Boron beam interaction with matter. Boron: Compounds, Production and Application.* Editor. G.L. Perkins. Chapter 12. P.1-36, Nova Science Publishers, Inc. (2011) .
[7] O.B. Firsov, *Zh. Eksp Teor. Fiz.* 36, 1517 (1959); Sov.Phys.JETP 9, 1076 (1959).
[8] J. Lindhard and M. Scharff, *Phys. Rev.* 124, 128 (1961). J. Lindhard, M. Scharff, and H.E.Schiott, *Mat. Fys. Medd. Dan. Vid. Selsk.* 33, 14 (1963).
[9] F. Bloch, *Ann. Phys. (Leipzig)* 16, 285 (1933).
[10] Ya.A.Teplova, V.S.Nikolaev, I.S.Dmitriev, and L.N.Fateeva, Zh. Eksp Teor. Fiz. 42, 44 (1962) [Sov. Phys. JETP 15, 31 (1962)].
[11] Ya.A. Teplova, I.S. Dmitriev, and Yu.A. Belkova, *Nucl. Instr. and Meth.* B 164-165, 291 (2000).
[12] *V.S.Nikolaev*, I. S. Dmitriev, Ya.A. Teplova, and **L.N.Fateyeva**. *Accelerators*, Moscow, Atomizdat, (1960) p.90-96.
[13] *V.S.Nikolaev*, *Usp. Fiz. Nauk* 85, 679 (1965) [Sov. Phys. Usp. 8, 269 (1965)].
[14] V.S. Nikolaev and I.S. Dmitriev, *Phys. Let.* A, 28, 122 (1968).
[15] V.S. Nikolaev, I.S. Dmitriev, L.H. Fateeva, Ya.A. Teplova, *Zh. Eksp Teor. Fiz.* 39, 905 (1960) [*Sov. Phys. JETP* 12, 627 (1960)].
[16] ***S***. K. Allison, Rev. Mod. Phys. 30, 1137 (1958).
[17] I.S. Dmitriev, N.F. Vorob'ev, Zh.M. Konovalova, et al., *Zh. Eksp. Teor. Fiz.* 84, 1987 (1983) [Sov. Phys. JETP 57, 1157 (1983)].
[18] L.I. Pivovar, V.M. Tubaev, and M.I. Novikov, *Zh. Eksp. Teor. Fiz.* 41, 26 (1961).
[19] I.S. Dmitriev, V. S. Nikolaev, L.N. Fateeva, and Ya. A. Teplova *Zh. Eksp. Teor. Fiz.* 42, 16 (1962) [Sov. Phys. JETP 15, 11 (1962)].
[20] A. Itoh, K. Ohnishi, and F. Fukusawa F *J. Phys. Soc. Japan* 49, 1513 (1980).
[21] I.S. Dmitriev, Ya. M. Zhileikin, and V.S. Nikolaev, *Zh. Eksp. Teor. Fiz.* 49, 500 (1965) [Sov. Phys. JETP 20, 352 (1966)].
[22] V.S. Nikolaev, V.S. Senashenko, V.A. Sidorovich, and V. Yu. Shafer, *Zh. Tech. Fiz.* 48, 1399 (1978).
[23] Yu. S. Volodyagin, I.S. Dmitriev, V.S. Nikolaev, Yu.A. Tashaev, and Ya. A. Teplova, *J. Phys.* B, 6, L171 (1973).
[24] I.S. Dmitriev, Yu.A. Tashaev, V.S. Nikolaev, Ya.A. Teplova, and B.M. Popov, *Zh. Eksp. Teor. Fiz.* 73, 1684 (1977) [Sov. Phys. JETP 46, 884 (1977)].
[25] V.S. Nikolaev, I.S. Dmitriev, L.N. Fateeva, and Ya.A. Teplova, *Zh. Eksp. Teor. Fiz.* 40, 989 (1961) [Sov. Phys. JETP 13, 695 (1961)].
[26] P. Hvelplund, J. Heinemeier, E.H. Pedersen, and F.R. Simpson *J. Phys. B.* 9, 491 (1976).

[27] A. Itoh, M. Asari, and F. Fukusawa, *J. Phys. Soc. Japan* 48, 943 (1980).
[28] L.I. Pivovar, M.I. Novikov, and V.M. Tubaev, *Zh. Eksp. Teor. Fiz.* 42, 1490 (1962).
[29] A. Itoh and F. Fukusawa *J. Phys. Soc. Japan* 50, 632 (1981).
[30] J.C. Armstrong, J.V. Mullendore, W.R. Harris, and J.B. Marion *Proc. Phys. Soc.* 86, 1283 (1965).
[31] Y. Kido, Y. Kanamori, and F. Fukusawa *Nucl. Instrum. Methods* 164, 565 (1979).
[32] A. Gladieux, A. Chateau-Therry, and B. Delaunay *J. Phys. E* 12, 3591 (1979).
[33] F. Bell and H.D. Betz *J. Phys.* B 10, 483 (1977).
[34] I.S. Dmitriev, V.S. Nikolaev,Yu.A. Tashaev, Ya.A. Teplova, and B.M. Popov, *Zh. Eksp. Teor. Fiz.* 73 2047 (1974) [Sov. Phys. JETP 40, 1017 (1975)].
[35] Y. Haruyama, Y. Kanamory, T. Kido, A. Itoh, and F.Fukuzawa, *J. Phys.* B 16, 1225 (1983).
[36] N.F. Vorobjev, V. P. Zaikov, E. A. Kralkina, and E.I. Sirotinin. // *Nucl. Instrum. Methods* B, 33, 207 (1988).
[37] Yu. A. Fainberg, Ya. A. Teplova, and V. S. Kulikauskas, *Poverxnost'*, N4, 43, 5 (2002).
[38] Yu. A. Fainberg, Ya. A. Teplova, and N. V. Novikov, *J. Surf. Invest.* 3, 5 (2012).
[39] N. V. Novikov and Ya. A. Teplova, *Phys. Lett.* A 373, 550 (2009).
[40] Yu. A. Fainberg, Ya. A. Teplova, and N. V. Novikov, *J. Surf. Invest.* 4, 494 (2010).

APPENDIX A: Ranges of Helium Ions in Various Gases

Table A.1.

target	*V*, 10^8 cm/s	*E*,MeV/amu	*R*, mm	*S/R*, %
air	5.3	0.702	3.4 ± 0.1	3.5
	16.9	7.14	45.1 ± 0.2	2.0
	19.4	9.41	70.4 ± 1.2	1.5

APPENDIX B: Energy Losses of Helium Ions

Table B.1. Helium ions in Hydrogen

E ,	V ,	-dE/dx	
MeV/amu	10^8 cm/s	keV cm^2/μg	10^{-15} eV cm^2/atom
0.025	2.19	5.14	8.6
0.038	2.69	5.62	9.4
0.050	3.11	6.10	10.2
0.063	3.47	6.59	11.0
0.075	3.80	6.825	11.4
0.088	4.12	7.12	11.9
0.100	4.39	7.36	12.3

Table B.2. Helium ions in HELIUM

E,	V,	-dE/dx	
MeV/amu	10^8 cm/s	keV cm^2/μg	10^{-15} eV cm^2/atom
0.025	2.19	1.25	8.33
0.038	2.69	1.47	9.8
0.050	3.11	1.68	11.2
0.063	3.47	1.96	13.07
0.075	3.80	2.12	14.13
0.088	4.12	2.30	15.3
0.100	4.39	2.48	16.53

Table B.3. Helium ions in AIR

E,	V,	-dE/dx	
MeV/amu	10^8 cm/s	keV cm^2/μg	10^{-15} eV cm^2/atom
0.0125	1.55	0.41	9.9
0.025	2.19	0.72	17.3
0.038	2.69	1.00	24.0
0.050	3.11	1.10	26.4
0.063	3.47	1.25	30.0
0.075	3.80	1.39	33.4
0.088	4.12	1.525	36.6
0.10	4.39	1.615	38.8
0.14	5.22	1.89	45.4
0.15	5.38	1.844	44.3
0.157	5.49	1.955	47.0
0.16	5.57	1.95	46.9
0.18	5.94	2.01	48.3
E,	V,	-dE/dx	
MeV/amu	10^8 cm/s	keV cm^2/μg	10^{-15} eV cm^2/atom
0.19	6.02	2.015	48.4
0.25	6.95	2.02	48.55
0.31	7.77	1.958	47.1
0.375	8.5	1.885	45.3
0.44	9.2	1.786	43.0
0.50	9.82	1.696	40.8
0.625	10.97	1.50	36.05
0.70	11.58	1.412	33.9
0.72	11.75	1.386	33.3
0.724	11.80	1.372	33.0
0.74	11.91	1.347	32.4
0.75	12.04	1.358	32.6
0.88	13.0	1.159	27.9
1.00	13.9	1.028	24.7
1.25	15.54	0.783	18.8
1.46	16.75	0.6025	14.5
1.47	16.83	0.588	14.1
1.51	17.03	0.571	13.7

Table B.4. Helium ions in Argon

E ,	V ,	-dE/dx	
MeV/amu	10^8 cm/s	keV cm²/μg	10^{-15} eV cm²/atom
0.025	2.19	0.626	41.5
0.038	2.69	0.735	48.7
0.050	3.11	0.825	54.6
0.063	3.47	0.918	60.8
0.075	3.80	1.043	69.1
0.088	4.12	1.127	74.6
0.100	4.39	1.205	79.8
0.14	5.22	1.458	96.6
0.375	8.5	1.03	68.2
0.50	9.82	0.90	59.6
0.75	12.04	0.70	46.4
1.00	13.9	0.58	38.4

APPENDIX C:
Equilibrium Charge Distributions of Helium Ions

Table C.1.

target	*V*, 10^8 cm/s	F_0, %	F_1, %	F_2, %	$\bar{i}$
He	5.67	6.4	61.4	32.2	1.26
	7.92	0.3	27.7	72.0	1.72
	11.67	___	4.0	96.0	1.96
	16.86	___	0.6	99.4	1.99
N_2	5.67	2.6	51.2	46.2	1.44
	7.92	0.05	17.3	82.7	1.83
	11.67	___	3.8	96.2	1.96
	16.86	___	0.75	99.25	1.99
Ar	5.67	1.3	42.9	55.8	1.54
	7.92	0.04	10.1	89.9	1.90
	11.67	0.01	4.4	95.6	1.96
	16.86	___	1.15	98.85	1.99
Kr	3.89	14.0	78.0	8.0	0.93
	5.67	1.4	49.1	49.5	1.48
	7.92	0.06	16.2	83.7	1.84
	11.67	0.01	7.6	92.4	1.92
	16.86	___	1.6	98.4	1.98

APPENDIX D:
Electron Capture and Loss Cross Section

Table D.1. $He^{i+} + He \rightarrow He^{(i-1)+}$

E, MeV/amu	V, 10^8 cm/s	$\sigma_{i,i-1}$, 10^{-16} cm^2	
		i=1	i=2
0.091	4.1	0.37	2.2
0.164	5.6	0.089	0.49
0.334	8.0	0.018	0.095
0.752	12.0	—	0.0083 ± 0.0040

Table D.2. $He^{i+} + He \rightarrow He^{(i-1)+}$

E, MeV/amu	V, 10^8 cm/s	$\sigma_{i,i+1}$, 10^{-16} cm^2	
		i=0	i=1
0.334	8.0	0.57	0.18 ± 0.03
0.752	12.0	0.35	0.11
2.1	20.0	0.14	0.044 ± 0.007

Table D.3. Hei+ + He → He(i-2)+

E, MeV/amu	V, 10^8 cm/s	$\sigma_{i,i-2}$, 10^{-16} cm^2
		i=2
0.091	4.1	0.17
0.164	5.6	0.012
0.334	8.0	0.00027 ± 0.00025

Table D.4 $He^{i+} + He \rightarrow He^{(i+2)+}$

E, MeV/amu	V, 10^8 cm/s	$\sigma_{i,i+2}$, 10^{-16} cm^2
		i=0
0.334	8.0	0.14 ± 0.04
0.752	12.0	0.0105 ± 0.0020
2.1	20.0	0.0034 ± 0.0012

Table D.5. $He^{i+} + N_2 \rightarrow He^{(i-1)+}$

E, MeV/amu	V, 10^8 cm/s	$\sigma_{i,i-1}$, 10^{-16} cm^2/atom	
		i=1	i=2
0.091	4.1	0.60	2.9
0.164	5.6	0.15	0.96
0.334	8.0	0.017 ± 0.003	0.15
0.752	12.0	—	0.0071
2.1	20.0	—	0.00064 ± 0.00014

Table D.6. $He^{i+} + N_2 \rightarrow He^{(i+1)+}$

E, MeV/amu	V, 10^8 cm/s	$\sigma_{i,i+1}$, 10^{-16} cm^2/atom	
		i=0	i=1
0.334	8.0	1.8	0.74
0.752	12.0	1.3	0.44
2.1	20.0	0.60 ± 0.12	0.21

Table D.7. $He^{i+} + N_2 \rightarrow He^{(i-2)+}$

E, MeV/amu	V, 10^8 cm/s	$\sigma_{i,i-2}$, 10^{-16} cm^2/atom
		i=2
0.091	4.1	0.22
0.334	8.0	0.0015 ± 0.0007

Table D.8. $He^{i+} + N_2 \rightarrow He^{(i+2)+}$

E, MeV/amu	V, 10^8 cm/s	$\sigma_{i,i+2}$, 10^{-16} cm^2/atom
		i=0
0.334	8.0	0.20
0.752	12.0	0.095
2.1	20.0	0.035 ± 0.008

Table D.9. $He^{i+} + Ne \rightarrow He^{(i-1)+}$

E, MeV/amu	V, 10^8 cm/s	$\sigma_{i,i-1}$, 10^{-16} cm^2	
		i=1	i=2
0.334	8.0	0.076 ± 0.014	0.42 ± 0.07
0.752	12.0	0.0040 ± 0.0006	0.038 ± 0.010
2.1	20.0	0.00025 ± 0.00008	0.00080 ± 0.00020

Table D.10. $He^{i+} + Ne \rightarrow He^{(i+1)+}$

E, MeV/amu	V, 10^8 cm/s	$\sigma_{i,i+1}$, 10^{-16} cm^2	
		i=0	i=1
0.334	8.0	2.1	0.77 ± 0.12
0.752	12.0	1.4	0.57 ± 0.11
2.1	20.0	0.88	0.18 ± 0.04

Table D.11. $He^{i+} + Ne \rightarrow He^{(i-2)+}$

E, MeV/amu	V, 10^8 cm/s	$\sigma_{i,i-2}$, 10^{-16} cm^2
		i=2
0.334	8.0	0.010 ± 0.002

Table D.12. $He^{i+} + Ne \rightarrow He^{(i+2)+}$

E, MeV/amu	V, 10^8 cm/s	$\sigma_{i,i+2}$, 10^{-16} cm^2
		i=0
0.334	8.0	0.19 ± 0.05
0.752	12.0	0.19 ± 0.04
2.1	20.0	0.071 ± 0.020

Table D.13. $He^{i+} + Ar \rightarrow He^{(i-1)+}$

E, MeV/amu	V, 10^8 cm/s	$\sigma_{i,i-1}$, 10^{-16} cm^2	
		i=1	i=2
0.334	8.0	0.018 ± 0.005	0.18
0.752	12.0	—	0.036
2.1	20.0	—	0.0027

Table D.14. $He^{i+} + Ar \rightarrow He^{(i+1)+}$

E, MeV/amu	V, 10^8 cm/s	$\sigma_{i,i+1}$, 10^{-16} cm^2	
		i=0	i=1
0.334	8.0	3.4	1.2
0.752	12.0	2.5	0.84
2.1	20.0	1.5	0.48 ± 0.11

Table D.15. $He^{i+} + Ar \rightarrow He^{(i-2)+}$

E, MeV/amu	V, 10^8 cm/s	$\sigma_{i,i-2}$, 10^{-16} cm^2
		i=2
0.091	4.1	0.32
0.164	5.6	0.012 ± 0.004

Table D.16. $He^{i+} + Ar \rightarrow He^{(i+2)+}$

E, MeV/amu	V, 10 8 cm/s	σ i,i+2 , 10 -16 cm2
		i=0
0.334	8.0	0.35
0.752	12.0	0.39
2.1	20.0	0.12 ± 0.03

Table D.17. $He^{i+} + Kr \rightarrow He^{(i-2)+}$

E, MeV/amu	V, 10^8 cm/s	$\sigma_{i,i-2}$, 10^{-16} cm^2
		i=2
0.091	4.1	0.34
0.164	5.6	0.020 ± 0.008
0.334	8.0	0.0030 ± 0.0015

In: Helium: Occurrence, Applications and Biological Effects ISBN: 978-1-62808-612-6
Editor: Simon B. Davis

Chapter 5

POSTSEISMIC LEAKAGE OF MANTLE AND CRUSTAL HELIUM FROM SEISMICALLY ACTIVE REGIONS

Koji Umeda**, Koichi Asamori, Ryo Komatsu, Chifumi Kakuta, Sunao Kanazawa, Atusi Ninomiya, Tomohiro Kusano and Kazuo Kobori
Tono Geoscientific Research Unit,
Geological Isolation Research and Development Directorate,
Japan Atomic Energy Agency, Jorinji, Izumi, Toki, Japan

ABSTRACT

It is well known that mantle degassing does not occur homogeneously over the Earth's surface. The elevated $^3He/^4He$ ratios found in volcanic regions and tectonically active areas are higher than the atmospheric values. The anomalous helium isotopic ratios have been interpreted to transfer mantle volatiles into the crust by processes or mechanisms such as magmatic intrusion, continental underplating and lithospheric rifting. This study was undertaken to elucidate the geographic distribution of $^3He/^4He$ ratios around seismically active regions in Japan, using helium isotope data obtained from gas samples. Several case studies suggest that there seem to be a significant trend of high ^3He emanations around the source regions following inland earthquakes, resulting in leakage of mantle volatiles through crustal pathways (faults) due to more frequent development of higher permeability pathways and/or upwelling of mantle fluids through the ductile lower crust. The other case is significant decrease of $^3He/^4He$ ratios found in groundwaters before and after a large inland earthquake. Episodic faulting could release accumulated crustal (radiogenic) helium from host rocks, or enhance the transfer of mantle volatiles through permeable fault zones, such that subsequent fluid flow proximal to the source region could then explain the spatio-temporal variations in $^3He/^4He$ ratios.

* Corresponding author's e-mail : umeda.koji@jaea.go.jp.

1. INTRODUCTION

The Earth has been continually losing helium through the degassing of rocks from its interior over the past of 4.5 billion years. Any helium left on the Earth at all is largely due to its replenishment in the interior through radioactive decay. The decay of uranium and thorium produces ^{4}He, whereas ^{3}He in the Earth's mantle is not produced by radioactive decay and was only incorporated during accretion, that is, it is primordial (Ozima and Podosek, 2002). Since the discovery of mantle helium in oceanic waters over the East Pacific Rise in 1968 (Clarke et al., 1969), considerable efforts have been made to identify, quantify and elucidate the variation of helium isotope ratios ($^3He/^4He_{air}$ = 1.4×10^{-6}; $^3He/^4He_{mantle}$ = $1\text{-}1.4\times10^{-5}$; $^3He/^4He_{radiogenic}$ = $1\text{-}5\times10^{-8}$: Lupton, 1983). In the most widely accepted mantle helium evolution model, the ^{3}He/^{4}He ratios observed in the mid-ocean ridge basalts (MORBs) (Craig et al., 1975) and ocean island basalts (OIBs) (Kurz et al., 1983) are interpreted in terms of mixing two reservoirs (e.g., Porcelli and Ballentine, 2002). OIBs are thought to sample undegassed, high ^{3}He/^{4}He mantle that has retained most of its ^{3}He since the origin of the Earth and are often associated with the lower mantle. In contrast, MORBs are thought to sample degassed, low ^{3}He/^{4}He mantle that has lost essentially all of its ^{3}He through melting (Kurz et al., 1982) and is associated with the upper mantle.

It is common knowledge that mantle degassing does not occur homogeneously over the Earth's surface. In continental settings, the flux of radiogenic ^{4}He from the crust has been investigated and the distribution of mantle helium and its manifestation in tectonically active areas has been mapped (e.g., Marty et al., 1989). The continental crust is a low permeability barrier to mantle degassing, and radiogenic helium production within the crust dilutes any trans-crustal mantle helium. This is reflected in the low ^{3}He/^{4}He ratios observed in crustal fluids from stable platforms (e.g., Torgersen and Clarke, 1987). In contrast, ^{3}He/^{4}He ratios higher than the typical crustal values found in volcanic regions and tectonically active regions have been believed to transfer mantle volatiles into the crust by processes or mechanisms such as magmatic intrusion, continental underplating and lithospheric rifting (O'Nions and Oxburgh, 1988; Torgersen, 1993). Mantle melting is considered to be the most likely mechanism responsible for the transfer of mantle helium from the subcrustal lithosphere where the volatiles would be released directly from a magma body into the crustal fluids (e.g. Torgersen, 1993). Significantly higher ^{3}He/^{4}He ratios have been observed in seismically active zones compared to non-active zones (e.g., Güleç et al., 2002). Consequently, the presence or the absence of mantle helium in crustal fluids could be efficiently used to trace the inter-relationship of tectonics, magmatism and fluids circulating in active fault zones (e.g, Kennedy and van Soest, 2007). In addition, temporal changes in the helium isotope ratios have been examined as a predictive indicator in studies monitoring earthquakes and/or volcanic eruptions (e.g., Sano et al., 1998; Shimoike and Notsu, 2000).

For the last decade, many researchers have made efforts to acquire helium isotope data from hot springs and drinking water wells throughout the Japanese Islands, which are above active subduction zones in this island are system (e.g., Umeda et al., 2012). These activities led to the elucidation of the geographical distribution of ^{3}He/^{4}He ratios, and revealed a relatively high contribution of mantle helium in tectonically active regions. In this contribution, we provide an overview of the geographic distribution of the ^{3}He/^{4}He ratios throughout the Japanese Islands, and indicate the spatio-temporal variations in ^{3}He/^{4}He ratios around the source regions of recent large inland earthquakes, based on the helium isotope data set from previously published reports.

2. Helium Isotope Data in Japan

Helium isotope data used in this study were compiled from 108 reports that were previously published. The majority of samples used in the studies were collected from wells and bubbling natural springs, the remainder are from hot spring waters. The synthesized data were compiled from a total of 1728 samples (Table 1), which are available in the auxiliary material (http://jolissrch-inter.tokai-sc.jaea.go.jp/pdfdata/JAEA-Data-Code-2012-017-appendix (CD-ROM).zip).

Table 1. Isotopic compositions of gas samples compiled from previous reports

No.	Latitude	Longitude	Feature	$^3He/^4He(x10^{-6})$	±	1σ	$^4He/^{20}Ne$	Ref.
1	45.42	141.67	Gas	0.35	±	0.05	7.1	Wakita et al. (1990)
2	45.12	142.36	Gas	0.93	±	0.04	3760	Ueda et al. (2010)
3	45.10	141.78	Gas	0.53	±	0.01	723	Ueda et al. (2010)
4	45.10	141.78	Gas	0.56	±	0.01	519	Ueda et al. (2010)
5	45.07	141.83	Gas	0.40	±	0.01	>1200	Wakita et al. (1990)
6	45.04	141.77	Gas	0.11	±	0.01	178	Ueda et al. (2010)
7	45.03	141.88	Water	1.61			6.52	Nagao(2004)
8	45.03	141.88	Water	0.59			4.25	Nagao(2004)
9	45.03	141.88	Water	0.98			1.23	Nagao(2004)
10	45.03	141.88	Water	0.41			25.7	Nagao(2004)
11	45.03	141.88	Water	0.90			5.66	Nagao(2004)
12	45.03	141.88	Water	0.59			2.90	Nagao(2004)
13	45.03	141.88	Water	3.10			1.77	Nagao(2004)
14	45.03	141.88	Water	1.19			0.46	Nagao(2004)
15	45.03	141.88	Water	1.05			0.53	Nagao(2004)
16	44.89	141.75	Gas	0.40	±	0.02	31.2	Ueda et al. (2010)
17	44.66	141.87	Gas	1.98	±	0.06	3090	Ueda et al. (2010)
18	44.66	141.87	Gas	1.96	±	0.04	274	Ueda et al. (2010)
19	44.65	141.86	Gas	1.70	±	0.06	>770	Wakita et al. (1990)
20	44.36	141.70	Gas	1.43	±	0.03	684	Ueda et al. (2010)
21	44.36	141.70	Gas	1.44	±	0.02	246	Ueda et al. (2010)
22	44.31	141.65	Gas	2.35	±	0.08	1830	Ueda et al. (2010)
23	44.31	141.65	Gas	2.41	±	0.03	6.09	Ueda et al. (2010)
24	44.20	145.33	Gas	9.34	±	0.24	71	Sano and Wakita (1988)
25	44.18	142.40	Gas	1.38	±	0.04	8.77	Ueda et al. (2010)
26	44.18	142.65	Gas	2.07	±	0.05	29.9	Ueda et al. (2010)
27	44.04	145.16	Gas	2.77	±	0.08	0.53	Sano and Wakita (1988)
28	43.90	145.01	Gas	9.57	±	0.17	13	NEDO (2001a)

Table 1. (Continued)

No.	Latitude	Longitude	Feature	$^{3}He/^{4}He(x10^{-6})$	±	1σ	$^{4}He/^{20}Ne$	Ref.
29	43.89	144.91	Gas	10.09	±	0.16	18	NEDO (2001a)
30	43.80	144.95	Gas	9.08	±	0.19	51	NEDO (2001a)
31	43.74	144.83	Gas	2.60			1	NEDO (2001b)
32	43.72	142.96	Gas	7.45	±	0.34	2.2	Sano and Wakita (1988)
33	43.72	142.91	Gas	4.43			6.5	NEDO (1990a)
34	43.72	142.91	Gas	4.17			2.3	NEDO (1990a)
35	43.70	144.92	Gas	6.59	±	0.10	1.51	NEDO (1997)
36	43.70	144.91	Gas	7.80			36	NEDO (2001b)
37	43.68	144.94	Gas	7.70			10	NEDO (2001b)
38	43.67	144.85	Gas	3.73			0.39	NEDO (2001b)
39	43.67	142.85	Gas	6.48			6.0	NEDO (1990a)
40	43.67	142.85	Gas	10.14			57	NEDO (1990a)
41	43.66	145.03	Gas	3.00			18	NEDO (2001b)
42	43.65	142.80	Gas	9.20	±	0.08	21	Sano and Wakita (1988)
43	43.62	144.45	Gas	9.58	±	0.48	63	Sano and Wakita (1988)
44	43.62	142.93	Gas	7.76			4.7	NEDO (1990a)
45	43.61	144.44	Gas	3.14	±	0.12	0.465	Marty et al. (1989)
46	43.59	144.71	Gas	7.73	±	0.09	59	Sano and Wakita (1988)
47	43.59	144.31	Gas	7.62	±	0.12	3.0	Sano and Wakita (1988)
48	43.59	144.31	Gas	3.13	±	0.12	0.53	Sano and Wakita (1988)
49	43.58	141.40	Water	5.30	±	0.08	55.9	Ueda et al. (2010)
50	43.56	141.91	Water	3.92	±	0.09	52.5	Ueda et al. (2010)
51	43.55	142.24	Gas	2.46	±	0.08	2470	Ueda et al. (2010)
52	43.55	142.24	Water	2.57	±	0.05	19.1	Ueda et al. (2010)
53	43.53	142.90	Gas	5.98	±	0.12		NEDO (1988)
54	43.53	142.90	Gas	6.54	±	0.13		NEDO (1988)
55	43.53	142.90	Gas	6.26	±	0.08		NEDO (1988)
56	43.52	142.19	Gas	1.03	±	0.02	0.529	Ueda et al. (2010)
57	43.50	142.90	Gas	7.09	±	0.16		NEDO (1988)
58	43.50	142.90	Gas	6.75	±	0.13		NEDO (1988)
59	43.50	142.90	Gas	6.64	±	0.08		NEDO (1988)
60	43.50	142.00	Water	1.52	±	0.07	2.79	Ueda et al. (2010)
61	43.49	142.96	Gas	2.75	±	0.07		NEDO (1988)
62	43.49	142.96	Gas	1.67	±	0.05		NEDO (1988)
63	43.47	142.65	Gas	10.24	±	0.14		NEDO (1988)
64	43.47	142.65	Gas	9.78	±	0.11		NEDO (1988)
65	43.47	142.65	Gas	9.15	±	0.06		NEDO (1988)
66	43.47	142.65	Gas	10.41	±	0.28		NEDO (1988)
67	43.47	143.11	Gas	4.92	±	0.31	2.9	Sano and Wakita (1988)
68	43.46	142.88	Gas	2.63	±	0.11		NEDO (1988)

No.	Latitude	Longitude	Feature	$^{3}He/^{4}He(x10^{-6})$	±	1σ	$^{4}He/^{20}Ne$	Ref.
69	43.46	142.88	Gas	4.99	±	0.10		NEDO (1988)
70	43.42	141.97	Gas	0.15	±	0.01	4.83	Ueda et al. (2010)
71	43.42	142.69	Gas	9.19	±	0.55	35	Nagao et al. (1981)
72	43.42	144.15	Gas	8.00	±	0.17	5.6	Sano and Wakita (1988)
73	43.42	142.68	Gas	8.61	±	0.11	3.2	NEDO (1990a)
74	43.42	142.68	Gas	9.83	±	0.26	55	NEDO (1990a)
75	43.41	142.66	Gas	9.22	±	0.15		NEDO (1988)
76	43.41	142.66	Gas	9.32	±	0.15		NEDO (1988)
77	43.41	142.66	Gas	8.50	±	0.18		NEDO (1988)
78	43.41	142.66	Gas	9.67	±	0.20		NEDO (1988)
79	43.41	142.66	Gas	7.89	±	0.13	58.824	Marty et al. (1989)
80	43.40	142.66	Gas	9.89	±	0.14	11	NEDO (1990a)
81	43.40	142.66	Gas	9.23	±	0.11	18	NEDO (1990a)
82	43.40	142.66	Gas	8.81	±	0.21	13	NEDO (1990a)
83	43.39	142.77	Gas	3.24	±	0.07		NEDO (1988)
84	43.39	144.01	Gas	8.72	±	0.20	22.222	Marty et al. (1989)
85	43.37	143.18	Gas	2.81	±	0.02	9.6	Sano and Wakita (1988)
86	43.29	141.60	Gas	1.19	±	0.02	26	Wakita et al. (1990)
87	43.28	141.95	Water	0.27	±	0.02	1.98	Ueda et al. (2010)
88	43.19	144.50	Gas	0.24	±	0.01	7.7	Sano and Wakita (1988)
89	43.05	142.40	Gas	9.80	±	0.17		NEDO (1988)
90	43.05	142.40	Gas	10.30	±	0.15		NEDO (1988)
91	42.91	141.72	Gas	1.62	±	0.02	21	Wakita et al. (1990)
92	42.90	144.32	Water	0.02				Mahara and Nakata (2003)
93	42.90	144.32	Water	0.01				Mahara and Nakata (2003)
94	42.90	144.32	Water	0.01				Mahara and Nakata (2003)
95	42.90	144.32	Water	0.28				Mahara and Nakata (2003)
96	42.90	144.32	Water	0.02				Mahara and Nakata (2003)
97	42.90	144.32	Water	0.26				Mahara and Nakata (2003)
98	42.90	144.32	Water	0.02				Mahara and Nakata (2003)
99	42.90	144.32	Water	0.01				Mahara and Nakata (2003)
100	42.90	144.32	Water	0.02				Mahara and Nakata (2003)
101	42.90	140.41	Gas	2.66	±	0.06	1.1	Sano and Wakita (1985)
102	42.87	140.63	Gas	7.64	±	0.19	20	Sano and Wakita (1985)
103	42.77	141.28	Water	4.36	±	0.18	1.4	Sano and Wakita (1985)
104	42.62	141.03	Gas	2.95	±	0.04	2.5	Sano and Wakita (1985)
105	42.57	140.23	Gas	1.20	±	0.04	0.47	Sano and Wakita (1985)

Table 1. (Continued)

No.	Latitude	Longitude	Feature	$^3He/^4He(x10^{-6})$	±	1σ	$^4He/^{20}Ne$	Ref.
106	42.54	140.84	Gas	7.91	±	0.05	55	Scott et al. (1994)
107	42.54	140.84	Gas	6.24	±	0.06	35	Scott et al. (1994)
108	42.54	140.83	Gas	7.43	±	0.06	66.667	Marty et al. (1989)
109	42.54	140.83	Gas	7.10	±	0.20	26.316	Marty et al. (1989)
110	42.54	140.83	Gas	6.89	±	0.07		Marty et al. (1989)
111	42.54	140.86	Gas	7.40	±	0.30	120	Nagao et al. (1980b)
112	42.54	140.86	Gas	6.70	±	0.40	2.1	Nagao et al. (1980b)
113	42.54	140.86	Gas	7.00	±	0.20	10	Nagao et al. (1980b)
114	42.54	140.86	Gas	7.00	±	0.50	75	Nagao et al. (1980b)
115	42.54	140.86	Gas	7.00	±	0.50	86	Nagao et al. (1980b)
116	42.54	140.86	Gas	3.90	±	0.20	2.0	Nagao et al. (1980b)
117	42.54	140.86	Gas	3.10	±	0.30	0.61	Nagao et al. (1980b)
118	42.54	140.83	Gas	9.60	±	0.40	650	Nagao et al. (1980b)
119	42.54	140.83	Gas	6.90	±	0.29	9.4	Sano and Wakita (1988)
120	42.54	140.83	Gas	6.80	±	0.25	12	Sano and Wakita (1988)
121	42.54	140.83	Gas	6.98	±	0.16	23	Sano and Wakita (1988)
122	42.54	140.83	Gas	6.93	±	0.16	1.5	Sano and Wakita (1988)
123	42.54	140.83	Gas	6.84	±	0.29	20	Sano and Wakita (1988)
124	42.54	140.83	Gas	3.62	±	0.06	0.63	Sano and Wakita (1988)
125	42.50	141.11	Gas	5.20	±	0.03	1.8	Sano and Wakita (1985)
126	42.50	140.36	Gas	1.88	±	0.08	9.1	Sano and Wakita (1985)
127	42.49	141.15	Gas	5.30	±	0.18	11	Sano and Wakita (1985)
128	42.45	141.21	Water	1.72	±	0.05	12	Sano and Wakita (1985)
129	42.19	140.13	Gas	9.50	±	0.10	12	NEDO (1999c)
130	42.17	140.00	Gas	7.90	±	0.10	57	NEDO (1999c)
131	42.15	140.05	Gas	4.10	±	0.30	1.6	NEDO (1999c)
132	42.15	140.04	Gas	6.80	±	0.20	2000	NEDO (1999c)
133	42.14	140.45	Gas	9.95	±	0.25		Nagao et al. (1979)
134	42.14	140.21	Gas	5.10	±	0.17	4.3	Sano and Wakita (1985)
135	42.14	140.21	Gas	6.05	±	0.11	75	Sano and Wakita (1988)
136	42.13	140.45	Gas	3.00	±	0.15		Nagao et al. (1979)
137	42.13	140.45	Gas	2.16	±	0.25		Nagao et al. (1979)
138	42.13	140.45	Gas	4.55	±	0.27		Nagao et al. (1979)
139	42.13	140.45	Gas	4.38	±	0.29		Nagao et al. (1979)
140	42.13	140.45	Gas	1.60	±	0.07		Nagao et al. (1979)
141	42.12	140.43	Gas	2.67	±	0.05	0.38	Sano and Wakita (1985)
142	42.02	140.81	Gas	3.75	±	0.04	16	Sano and Wakita (1985)
143	42.02	140.81	Gas	4.03	±	0.03	12	Sano and Wakita (1988)

No.	Latitude	Longitude	Feature	$^3He/^4He(x10^{-6})$	±	1σ	$^4He/^{20}Ne$	Ref.
144	41.95	140.83	Gas	4.40	±	0.10	10	Sano and Wakita (1985)
145	41.80	141.17	Gas	4.55	±	0.02	152	Scott et al. (1994)
146	41.77	140.78	Gas	7.28	±	0.03	59	Sano and Wakita (1985)
147	41.77	140.78	Gas	7.97	±	0.10	96	Sano and Wakita (1988)
148	40.97	141.37	Water	1.58	±	0.01	0.217	Mahara and Igarashi (2003)
149	40.97	141.37	Water	2.12	±	0.02	0.336	Mahara and Igarashi (2003)
150	40.97	141.37	Water	2.14	±	0.02	0.243	Mahara and Igarashi (2003)
151	40.97	141.37	Water	1.54	±	0.01	0.246	Mahara and Igarashi (2003)
152	40.97	141.37	Water	1.37	±	0.01	0.229	Mahara and Igarashi (2003)
153	40.97	141.37	Water	1.28	±	0.01	0.222	Mahara and Igarashi (2003)
154	40.97	141.37	Water	1.60	±	0.01	0.265	Mahara and Igarashi (2003)
155	40.97	141.37	Water	1.70	±	0.01	0.242	Mahara and Igarashi (2003)
156	40.97	141.37	Water	1.59	±	0.01	0.256	Mahara and Igarashi (2003)
157	40.97	141.37	Water	1.35	±	0.01	0.248	Mahara and Igarashi (2003)
158	40.97	141.37	Water	2.56	±	0.02	0.268	Mahara and Igarashi (2003)
159	40.97	141.37	Water	2.60	±	0.02	0.238	Mahara and Igarashi (2003)
160	40.97	141.37	Water	1.66	±	0.01	0.236	Mahara and Igarashi (2003)
161	40.97	141.37	Water	2.14	±	0.02	0.239	Mahara and Igarashi (2003)
162	40.97	141.37	Water	1.34	±	0.01	0.267	Mahara and Igarashi (2003)
163	40.97	141.37	Water	1.41	±	0.01	0.279	Mahara and Igarashi (2003)
164	40.97	141.37	Water	1.94	±	0.02	0.244	Mahara and Igarashi (2003)
165	40.97	141.37	Water	2.17	±	0.02	0.236	Mahara and Igarashi (2003)
166	40.97	141.37	Water	4.47	±	0.03	0.246	Mahara and Igarashi (2003)
167	40.97	141.37	Water	2.51	±	0.02	0.211	Mahara and Igarashi (2003)
168	40.97	141.37	Water	3.82	±	0.03	0.238	Mahara and Igarashi (2003)
169	40.97	141.37	Water	1.46	±	0.01	0.247	Mahara and Igarashi (2003)
170	40.97	141.37	Water	1.48	±	0.01	0.195	Mahara and Igarashi (2003)
171	40.97	141.37	Water	1.22	±	0.01	0.327	Mahara and Igarashi (2003)
172	40.97	141.37	Water	0.67	±	0.01	1.087	Mahara and Igarashi (2003)
173	40.97	141.37	Water	0.69	±	0.01	1.515	Mahara and Igarashi (2003)
174	40.97	141.37	Water	1.08	±	0.01	0.382	Mahara and Igarashi (2003)

Table 1. (Continued)

No.	Latitude	Longitude	Feature	$^3He/^4He(x10^{-6})$	±	1σ	$^4He/^{20}Ne$	Ref.
175	40.97	141.37	Water	1.03	±	0.01	0.493	Mahara and Igarashi (2003)
176	40.97	141.37	Water	0.55	±	0.00	1.818	Mahara and Igarashi (2003)
177	40.97	141.37	Water	1.42	±	0.01	0.288	Mahara and Igarashi (2003)
178	40.97	141.37	Water	1.40	±	0.01	0.248	Mahara and Igarashi (2003)
179	40.97	141.37	Water	1.37	±	0.01	0.217	Mahara and Igarashi (2003)
180	40.97	141.37	Water	1.41	±	0.01	0.243	Mahara and Igarashi (2003)
181	40.97	141.37	Water	1.43	±	0.01	0.269	Mahara and Igarashi (2003)
182	40.97	141.37	Water	1.10	±	0.01	2.725	Mahara and Igarashi (2003)
183	40.97	141.37	Water	0.64	±	0.01	16.667	Mahara and Igarashi (2003)
184	40.97	141.37	Water	1.61	±	0.01	0.245	Mahara and Igarashi (2003)
185	40.97	141.37	Water	1.43	±	0.01	0.285	Mahara and Igarashi (2003)
186	40.97	141.37	Water	1.57	±	0.01	0.307	Mahara and Igarashi (2003)
187	40.97	141.37	Water	1.72	±	0.01	0.210	Mahara and Igarashi (2003)
188	40.97	141.37	Water	1.33	±	0.01	0.331	Mahara and Igarashi (2003)
189	40.97	141.37	Water	0.58	±	0.00	16.667	Mahara and Igarashi (2003)
190	40.97	141.37	Water	1.63	±	0.01	0.196	Mahara and Igarashi (2003)
191	40.97	141.37	Water	1.43	±	0.01	0.221	Mahara and Igarashi (2003)
192	40.97	141.37	Water	1.44	±	0.01	0.238	Mahara and Igarashi (2003)
193	40.97	141.37	Water	1.45	±	0.01	0.238	Mahara and Igarashi (2003)
194	40.97	141.37	Water	1.37	±	0.01	2.381	Mahara and Igarashi (2003)
195	40.97	141.37	Water	2.01	±	0.02	0.239	Mahara and Igarashi (2003)
196	40.97	141.37	Water	1.58	±	0.01	0.287	Mahara and Igarashi (2003)
197	40.97	141.37	Water	0.80	±	0.01	2.083	Mahara and Igarashi (2003)
198	40.97	141.37	Water	1.37	±	0.01	0.247	Mahara and Igarashi (2003)
199	40.97	141.37	Water	1.42	±	0.01	0.217	Mahara and Igarashi (2003)
200	40.97	141.37	Water	1.35	±	0.01	0.334	Mahara and Igarashi (2003)
201	40.97	141.37	Water	1.14	±	0.01	0.510	Mahara and Igarashi (2003)
202	40.97	141.37	Water	0.78	±	0.01	4.000	Mahara and Igarashi (2003)
203	40.97	141.37	Water	1.34	±	0.01	0.272	Mahara and Igarashi (2003)

No.	Latitude	Longitude	Feature	$^3He/^4He(x10^{-6})$	±	1σ	$^4He/^{20}Ne$	Ref.
204	40.97	141.37	Water	1.16	±	0.01	0.613	Mahara and Igarashi (2003)
205	40.97	141.37	Water	1.62	±	0.01	7.692	Mahara and Igarashi (2003)
206	40.97	141.37	Water	0.75	±	0.01	333.333	Mahara and Igarashi (2003)
207	40.64	140.85	Gas	1.87	±	0.04	0.30	Sano and Wakita (1985)
208	40.60	139.86	Gas	8.84	±	0.45	91	Sano and Wakita (1985)
209	40.57	140.03	Gas	5.59	±	0.24	2.8	Sano and Wakita (1985)
210	40.05	139.93	Gas	4.12	±	0.03		Wakita and Sano (1983)
211	40.02	139.90	Gas	4.76	±	0.08	160	Wakita et al. (1990)
212	40.00	139.89	Gas	6.60			240	Sakata et al.(1986)
213	40.00	139.93	Gas	6.24	±	0.07		Wakita and Sano (1983)
214	40.00	139.93	Gas	5.96	±	0.12		Wakita and Sano (1983)
215	39.99	140.88	Gas	5.70			8	NEDO (1999b)
216	39.99	140.80	Gas	15.40	±	0.40	2.3	Nagao et al. (1981)
217	39.96	139.93	Gas	7.20			21	Sakata et al.(1986)
218	39.95	139.92	Gas	6.66	±	0.20	71	Wakita et al. (1990)
219	39.94	139.98	Gas	6.45	±	0.24	380	Wakita et al. (1990)
220	39.87	140.92	Gas	4.30	±	0.14	10	NEDO (1991a)
221	39.85	140.08	Gas	4.73	±	0.08	11	Wakita et al. (1990)
222	39.85	140.97	Gas	7.15	±	0.15	56	NEDO (1991a)
223	39.84	140.94	Gas	2.83	±	0.08	0.58	NEDO (1991a)
224	39.84	140.87	Gas	1.33	±	0.04	4.4	NEDO (1991a)
225	39.84	140.89	Gas	3.64	±	0.10	1.3	NEDO (1991a)
226	39.83	140.86	Gas	6.86	±	0.14	3.3	NEDO (1991a)
227	39.83	140.13	Gas	4.77	±	0.10	5.9	Wakita et al. (1990)
228	39.83	140.86	Gas	2.04	±	0.09	2.5	NEDO (1991a)
229	39.83	140.86	Gas	5.04	±	0.14	2.9	NEDO (1991a)
230	39.83	140.87	Gas	0.38	±	0.02	52	NEDO (1991a)
231	39.82	140.87	Gas	5.71	±	0.15	1.8	NEDO (1991a)
232	39.82	140.89	Gas	0.15	±	0.00	58	NEDO (1991a)
233	39.80	140.81	Gas	9.28	±	0.15	26	NEDO (1991a)
234	39.80	140.81	Gas	8.91	±	0.08	39	NEDO (1991a)
235	39.80	140.80	Gas	9.18	±	0.25	55	NEDO (1991a)
236	39.80	140.81	Gas	9.51	±	0.14	78	NEDO (1991a)
237	39.75	140.10	Gas	1.68				Sakata (1991)
238	39.74	140.08	Gas	1.18	±	0.04		Wakita and Sano (1983)
239	39.74	140.08	Gas	0.71	±	0.04	330	Wakita et al. (1990)
240	39.69	140.10	Gas	1.42	±	0.08		Wakita and Sano (1983)
241	39.69	140.10	Gas	0.85	±	0.06	450	Wakita et al. (1990)
242	39.60	139.97	Gas	1.20	±	0.10	3	Kita and Nagao (1996)
243	39.60	139.97	Gas	1.80	±	0.10	101	Kita and Nagao (1996)

Table 1. (Continued)

No.	Latitude	Longitude	Feature	$^3He/^4He(x10^{-6})$	±	1σ	$^4He/^{20}Ne$	Ref.
244	39.54	140.83	Water	1.43	±	0.04	0.637	Horiguchi and Matsuda (2008)
245	39.54	140.83	Water	0.96	±	0.04		Horiguchi and Matsuda (2008)
246	39.53	141.94	Water	1.41	±	0.03	0.29	Horiguchi et al. (2010)
247	39.53	141.95	Water	1.32	±	0.04	0.2	Horiguchi et al. (2010)
248	39.50	141.97	Water	1.44	±	0.03	0.25	Horiguchi et al. (2010)
249	39.39	140.77	Water	0.99	±	0.03	6.9	Horiguchi et al. (2010)
250	39.38	141.22	Water	0.39	±	0.01	1.4	Horiguchi et al. (2010)
251	39.34	141.88	Water	1.26	±	0.01	0.29	Horiguchi et al. (2010)
252	39.33	139.88	Gas	4.70	±	0.10	18	Kita and Nagao (1996)
253	39.33	141.56	Water	1.54	±	0.01	0.39	Horiguchi et al. (2010)
254	39.31	140.93	Water	1.20	±	0.01	22	Horiguchi et al. (2010)
255	39.31	140.93	Water	1.33	±	0.06	70.4	Horiguchi and Matsuda (2008)
256	39.31	140.93	Water	1.19	±	0.03		Horiguchi and Matsuda (2008)
257	39.27	140.72	Water	1.16	±	0.03	1.09	Horiguchi and Matsuda (2008)
258	39.27	140.72	Water	1.20	±	0.07		Horiguchi and Matsuda (2008)
259	39.27	140.07	Gas	7.57	±	0.23	625	Wakita et al. (1990)
260	39.27	140.07	Gas	2.66				Sakata (1991)
261	39.27	140.07	Gas	<4.1			153	Sakata et al.(1986)
262	39.27	140.07	Gas	9.30			72	Sakata et al.(1986)
263	39.27	140.72	Water	1.29	±	0.03	0.83	Horiguchi et al. (2010)
264	39.25	140.08	Gas	7.25	±	0.13		Wakita and Sano (1983)
265	39.25	140.08	Gas	7.13	±	0.08		Wakita and Sano (1983)
266	39.25	140.08	Gas	7.35	±	0.20	500	Wakita et al. (1990)
267	39.24	140.08	Gas	2.45	±	0.05		Wakita and Sano (1983)
268	39.24	140.08	Gas	2.54	±	0.05		Wakita and Sano (1983)
269	39.24	140.08	Gas	2.39	±	0.04		Wakita and Sano (1983)
270	39.24	140.08	Gas	2.69	±	0.06	100	Wakita et al. (1990)
271	39.24	140.08	Gas	3.74	±	0.20	140	Wakita et al. (1990)
272	39.24	139.91	Gas	2.42	±	0.03		Wakita and Sano (1983)
273	39.24	139.91	Gas	2.42	±	0.03	0.36	Wakita et al. (1990)
274	39.17	139.93	Gas	9.37	±	0.34	1700	Wakita et al. (1990)
275	39.17	139.93	Gas	9.13	±	0.15	1700	Wakita et al. (1990)
276	39.17	141.72	Water	0.07	±	0.01	20.2	Horiguchi et al. (2010)

No.	Latitude	Longitude	Feature	$^{3}He/^{4}He(x10^{-6})$	±	1σ	$^{4}He/^{20}Ne$	Ref.
277	39.17	140.50	Water	2.73	±	0.03	14.7	Horiguchi et al. (2010)
278	39.17	140.50	Water	1.76	±	0.02	16.5	Horiguchi and Matsuda (2008)
279	39.17	140.50	Water	2.72	±	0.04		Horiguchi and Matsuda (2008)
280	39.15	141.54	Water	1.40	±	0.03	0.34	Horiguchi et al. (2010)
281	39.12	140.65	Water	0.38	±	0.02	2.98	Horiguchi and Matsuda (2008)
282	39.12	140.65	Water	1.43	±	0.04		Horiguchi and Matsuda (2008)
283	39.12	140.65	Water	0.40	±	0.02	2.72	Horiguchi and Matsuda (2008)
284	39.12	140.65	Water	0.14	±	0.04		Horiguchi and Matsuda (2008)
285	39.12	140.65	Water	0.48	±	0.01	1.2	Horiguchi et al. (2010)
286	39.11	141.72	Water	1.47	±	0.03	0.31	Horiguchi et al. (2010)
287	39.08	140.99	Water	0.42	±	0.01	26.4	Horiguchi et al. (2010)
288	39.07	139.88	Gas	6.87	±	0.17		Wakita and Sano (1983)
289	39.07	139.90	Gas	8.98	±	0.15	1700	Wakita et al. (1990)
290	39.07	139.88	Gas	6.87	±	0.17	1.923	Sano et al. (1985)
291	39.00	140.62	Water	1.48	±	0.03	0.34	Horiguchi et al. (2010)
292	38.99	140.62	Gas	6.40				NEDO (1998)
293	38.99	140.60	Gas	9.40				NEDO (1998)
294	38.99	140.91	Water	0.45	±	0.01	3.81	Horiguchi et al. (2010)
295	38.99	140.60	Gas	9.50			180	Kita et al. (1992)
296	38.99	140.57	Gas	0.30				NEDO (1998)
297	38.99	140.57	Gas	0.37			420	NEDO (1998)
298	38.99	140.57	Gas	2.90			51	NEDO (1998)
299	38.99	140.57	Gas	3.80			51	NEDO (1998)
300	38.99	140.57	Gas	4.40			46	NEDO (1998)
301	38.99	140.57	Gas	4.40			46	NEDO (1998)
302	38.99	140.57	Gas	5.50			43	NEDO (1998)
303	38.99	140.57	Gas	5.70			36	NEDO (1998)
304	38.98	140.58	Gas	8.20				NEDO (1998)
305	38.98	140.58	Gas	8.30				NEDO (1998)
306	38.98	140.77	Gas	1.41	±	0.03	0.38	Horiguchi et al. (2010)
307	38.98	140.93	Water	0.51	±	0.03	4.26	Horiguchi and Matsuda (2008)
308	38.98	140.93	Water	0.36	±	0.03		Horiguchi and Matsuda (2008)
309	38.97	140.56	Gas	5.52			37	NEDO (2001c)
310	38.97	140.56	Gas	0.68			450	NEDO (2001c)
311	38.96	140.56	Gas	6.80				NEDO (1998)
312	38.96	140.57	Gas	7.80			4.3	NEDO (2001c)
313	38.96	140.57	Gas	8.47			25	NEDO (2001c)
314	38.96	140.55	Gas	0.55			540	NEDO (2001c)
315	38.95	140.63	Gas	7.70				NEDO (1998)

Table 1. (Continued)

No.	Latitude	Longitude	Feature	$^3He/^4He(x10^{-6})$	±	1σ	$^4He/^{20}Ne$	Ref.
317	38.92	141.32	Water	1.39	±	0.03	0.29	Horiguchi et al. (2010)
318	38.92	141.51	Water	1.16	±	0.01	0.27	Horiguchi et al. (2010)
319	38.90	141.58	Water	0.13			44.5	Horiguchi et al. (2010)
320	38.85	139.93	Gas	7.77	±	0.15		Wakita and Sano (1983)
321	38.85	139.93	Gas	7.14	±	0.22	120	Wakita et al. (1990)
322	38.85	139.90	Gas	6.39	±	0.13		Wakita and Sano (1983)
323	38.85	139.90	Gas	7.53	±	0.13	290	Wakita et al. (1990)
324	38.85	139.90	Gas	7.53	±	0.32	710	Wakita et al. (1990)
325	38.85	139.90	Gas	7.90	±	0.10	1100	Wakita et al. (1990)
326	38.83	141.08	Water	0.18			6.9	Horiguchi et al. (2010)
327	38.83	141.08	Water	0.13	±	0.01	21.1	Horiguchi et al. (2010)
328	38.81	140.67	Gas	3.62	±	0.10	1.3	Asamori et al. (2010)
329	38.81	140.67	Gas	2.07	±	0.03	0.84	Horiguchi et al. (2010)
330	38.81	140.67	Gas	3.25	±	0.09	0.872	Horiguchi and Matsuda (2008)
331	38.81	140.67	Gas	3.81	±	0.02		Horiguchi and Matsuda (2008)
332	38.81	140.67	Gas	2.78	±	0.06	0.663	Horiguchi and Matsuda (2008)
333	38.81	140.67	Gas	3.81	±	0.02		Horiguchi and Matsuda (2008)
334	38.80	140.67	Gas	3.09	±	0.04	1.07	Horiguchi et al. (2010)
335	38.75	140.73	Gas	5.19	±	0.08	3.29	Horiguchi et al. (2010)
336	38.74	140.72	Gas	1.36	±	0.01	0.29	Horiguchi et al. (2010)
337	38.74	140.71	Gas	5.79	±	0.15	12.821	Sano et al. (1985)
338	38.72	140.68	Gas	4.86	±	0.17	6.3	Asamori et al. (2010)
339	38.71	140.56	Gas	4.62	±	0.56	1.6	Takaoka and Imada (1989)
340	38.70	141.29	Water	0.21	±	0.01	4.6	Horiguchi et al. (2010)
341	38.66	141.45	Water	0.10	±	0.01	273	Horiguchi et al. (2010)
342	38.64	139.69	Water	1.22	±	0.04	4.5	Umeda et al. (2008a)
343	38.61	139.61	Water	3.23	±	0.04	140	Umeda et al. (2008a)
344	38.60	140.16	Gas	7.93	±	0.20	3.846	Sano et al. (1985)
345	38.57	140.53	Gas	6.44	±	0.42	44	Takaoka and Imada (1989)
346	38.57	140.53	Gas	6.30	±	0.56	27	Takaoka and Imada (1989)
347	38.55	140.78	Water	1.71	±	0.03	12.7	Horiguchi et al. (2010)
348	38.54	141.15	Water	0.78	±	0.03	2.6	Asamori et al. (2010)
349	38.49	139.52	Water	4.35	±	0.10	300	Umeda et al. (2008a)

No.	Latitude	Longitude	Feature	$^3He/^4He(x10^{-6})$	±	1σ	$^4He/^{20}Ne$	Ref.
350	38.48	140.35	Gas	6.86	±	0.98	79	Takaoka and Imada (1989)
351	38.48	141.05	Water	0.13	±	0.01	26.8	Horiguchi et al. (2010)
352	38.46	140.40	Gas	6.83	±	0.18	33	Takaoka and Imada (1989)
353	38.45	141.44	Water	0.03			35.6	Horiguchi et al. (2010)
354	38.41	140.77	Water	2.23	±	0.03	0.34	Horiguchi et al. (2010)
355	38.37	141.16	Water	0.25			126	Horiguchi et al. (2010)
356	38.37	140.28	Gas	7.26	±	0.04	20.000	Sano et al. (1982)
357	38.33	140.61	Water	1.41	±	0.01	0.34	Horiguchi et al. (2010)
358	38.30	139.97	Gas	8.07	±	0.09	48.0	Asamori et al. (2011)
359	38.30	140.78	Gas	3.55	±	0.04	17.857	Sano et al. (1985)
360	38.30	140.11	Water	1.46	±	0.03	0.32	Asamori et al. (2011)
361	38.29	140.74	Water	2.91	±	0.03	137	Horiguchi et al. (2010)
362	38.28	139.52	Water	2.99	±	0.05	62	Umeda et al. (2008a)
363	38.28	140.28	Water	7.74	±	0.11	16	Asamori et al. (2011)
364	38.26	140.32	Gas	7.66	±	0.12	62.0	Asamori et al. (2011)
365	38.25	140.25	Water	8.70	±	0.45	98.1	Takaoka and Mizutani (1987)
366	38.25	140.88	Gas	0.46	±	0.01	591	Asamori et al. (2010)
367	38.25	140.88	Water	0.41	±	0.01	410	Asamori et al. (2010)
368	38.22	140.06	Gas	1.61	±	0.05	1.6	Asamori et al. (2011)
369	38.22	140.74	Water	1.71	±	0.03	1.47	Horiguchi et al. (2010)
370	38.21	139.44	Water	1.45	±	0.04	1.3	Umeda et al. (2008a)
371	38.15	140.28	Gas	7.88	±	0.09	39.0	Asamori et al. (2011)
372	38.14	140.46	Gas	11.20	±	0.20		Takaoka (1985)
373	38.14	140.46	Gas	10.30	±	0.30		Takaoka (1985)
374	38.14	140.46	Gas	11.00	±	0.20		Takaoka (1985)
375	38.14	140.46	Gas	10.60	±	0.30		Takaoka (1985)
376	38.14	140.46	Gas	11.20	±	0.40		Takaoka (1985)
377	38.14	140.46	Gas	10.40	±	0.60		Takaoka (1985)
378	38.14	140.46	Gas	10.70	±	0.40		Takaoka (1985)
379	38.14	140.46	Gas	10.50	±	0.70		Takaoka (1985)
380	38.14	140.46	Gas	11.00	±	1.10		Takaoka (1985)
381	38.14	140.46	Gas	10.20	±	0.70		Takaoka (1985)
382	38.14	140.46	Gas	10.90	±	0.30		Takaoka (1985)
383	38.14	140.46	Gas	10.40	±	0.70		Takaoka (1985)
384	38.14	140.46	Gas	10.30	±	0.80		Takaoka (1985)
385	38.14	140.46	Gas	10.60	±	0.40		Takaoka (1985)
386	38.14	140.46	Gas	11.00	±	0.60		Takaoka (1985)
387	38.14	140.46	Gas	10.50	±	0.50		Takaoka (1985)
388	38.14	140.46	Gas	11.20	±	1.00		Takaoka (1985)
389	38.14	140.46	Gas	11.80	±	0.50		Takaoka (1985)
390	38.14	140.46	Gas	11.10	±	0.90		Takaoka (1985)
391	38.14	140.46	Gas	10.20	±	0.60		Takaoka (1985)

Table 1. (Continued)

No.	Latitude	Longitude	Feature	$^3He/^4He(x10^{-6})$	±	1σ	$^4He/^{20}Ne$	Ref.
392	38.14	140.46	Gas	9.38	±	0.86		Takaoka (1985)
393	38.14	140.46	Gas	9.99	±	0.72		Takaoka (1985)
394	38.14	140.46	Gas	9.78	±	0.89		Takaoka (1985)
395	38.14	140.46	Gas	9.70	±	0.84		Takaoka (1985)
396	38.14	140.46	Gas	10.50	±	0.70		Takaoka (1985)
397	38.14	140.46	Gas	9.55	±	0.51		Takaoka (1985)
398	38.14	140.46	Gas	10.30	±	0.70		Takaoka (1985)
399	38.14	140.46	Gas	9.62	±	0.53		Takaoka (1985)
400	38.14	140.46	Gas	11.20	±	0.30		Takaoka (1985)
401	38.14	140.46	Gas	9.70	±	1.00		Takaoka (1985)
402	38.14	140.46	Gas	11.00	±	1.00		Takaoka (1985)
403	38.14	140.46	Gas	10.30	±	0.60		Takaoka (1985)
404	38.12	140.54	Water	8.25	±	0.10	7.97	Horiguchi et al. (2010)
405	38.10	139.11	Gas	0.14	±	0.01	110	Wakita et al. (1990)
406	38.10	139.11	Gas	0.76	±	0.18	100	Wakita et al. (1990)
407	38.10	139.11	Gas	1.62	±	0.08	260	Wakita et al. (1990)
408	38.10	139.11	Gas	0.93	±	0.06	48	Wakita et al. (1990)
409	38.10	139.11	Gas	1.90	±	0.02		Wakita and Sano (1983)
410	38.10	139.11	Gas	0.75	±	0.04		Wakita and Sano (1983)
411	38.10	139.11	Gas	0.34	±	0.02		Wakita and Sano (1983)
412	38.10	139.58	Gas	3.42	±	0.13	111	Umeda et al. (2007b)
413	38.09	139.37	Gas	4.20			12	Sakata et al. (1996)
414	38.09	139.57	Gas	3.28	±	0.10	65	Umeda et al. (2007b)
415	38.08	139.58	Water	4.03	±	0.06	260	Umeda et al. (2008a)
416	38.07	139.60	Water	6.04	±	0.07	100	Umeda et al. (2008a)
417	38.05	139.76	Gas	3.76	±	0.08	210	Umeda et al. (2007b)
418	38.04	140.92	Water	0.49	±	0.01	15.9	Horiguchi et al. (2010)
419	38.02	139.50	Gas	0.31	±	0.01	48	Umeda et al. (2007b)
420	38.02	139.31	Gas	<1.4			54	Sakata et al.(1986)
421	38.02	139.31	Gas	6.70			201	Sakata et al.(1986)
422	38.02	140.58	Gas	5.00	±	0.08	24.390	Sano et al. (1985)
423	37.98	140.56	Water	2.98	±	0.06	7.74	Horiguchi et al. (2010)
424	37.97	139.17	Gas	2.24				Sakata (1991)
425	37.97	139.17	Gas	4.48				Sakata (1991)
426	37.97	139.17	Gas	<1.4				Sakata et al. (1989)
427	37.97	139.17	Gas	1.30				Sakata et al. (1989)
428	37.97	139.17	Gas	<1.5				Sakata et al. (1989)
429	37.97	139.17	Gas	<1.1				Sakata et al. (1989)
430	37.97	139.17	Gas	2.20				Sakata et al. (1989)
431	37.97	139.17	Gas	4.40				Sakata et al. (1989)
432	37.97	139.17	Gas	5.90			63	Sakata et al.(1986)
No.	Latitude	Longitude	Feature	$^3He/^4He(x10^{-6})$	±	1σ	$^4He/^{20}Ne$	Ref.
433	37.97	139.17	Gas	3.00			4.3	Sakata et al. (1996)

No.	Latitude	Longitude	Feature	$^3He/^4He(x10^{-6})$	±	1σ	$^4He/^{20}Ne$	Ref.
434	37.97	139.17	Gas	4.20			6.6	Sakata et al. (1996)
435	37.95	139.36	Gas	7.39	±	0.23	592	Umeda et al. (2007b)
436	37.95	139.70	Gas	9.01	±	0.09	169.492	Sano et al. (1985)
437	37.95	139.10	Gas	3.22	±	0.09	91	Wakita et al. (1990)
438	37.95	139.10	Gas	3.23	±	0.03	77	Wakita et al. (1990)
439	37.95	139.10	Gas	3.51	±	0.06	110	Wakita et al. (1990)
440	37.95	139.10	Gas	3.85	±	0.09	22	Wakita et al. (1990)
441	37.94	139.12	Gas	4.30			96	Sakata et al.(1986)
442	37.93	139.11	Gas	3.22	±	0.09		Wakita and Sano (1983)
443	37.93	139.11	Gas	3.23	±	0.03		Wakita and Sano (1983)
444	37.93	139.42	Gas	6.72	±	0.10	66	Umeda et al. (2007b)
445	37.92	139.66	Gas	11.12	±	0.11	53	Umeda et al. (2007b)
446	37.88	139.31	Gas	6.12	±	0.11	330	Umeda et al. (2007b)
447	37.86	138.99	Water	1.82	±	0.04	0.54	Umeda et al. (2008b)
448	37.84	139.16	Gas	3.76	±	0.08		Wakita and Sano (1983)
449	37.84	139.16	Gas	3.08	±	0.02		Wakita and Sano (1983)
450	37.79	139.78	Gas	8.05	±	0.13	2.1	Umeda et al. (2007b)
451	37.74	138.84	Gas	10.08	±	0.13	4200	Umeda et al. (2008b)
452	37.74	139.39	Water	5.84	±	0.12	34	Umeda et al. (2008b)
453	37.73	139.89	Gas	9.95	±	0.11	270	Umeda et al. (2008a)
454	37.73	140.25	Gas	10.27	±	0.15		NEDO (1991b)
455	37.73	140.25	Gas	12.07	±	0.45		NEDO (1991b)
456	37.73	140.25	Gas	11.76	±	0.22		NEDO (1991b)
457	37.73	140.25	Gas	11.03	±	0.13		NEDO (1991b)
458	37.73	140.25	Gas	12.43	±	0.15		NEDO (1991b)
459	37.73	140.25	Gas	12.09	±	0.30		NEDO (1991b)
460	37.73	140.25	Gas	11.42	±	0.12		NEDO (1991b)
461	37.73	140.25	Gas	12.14	±	0.30		NEDO (1991b)
462	37.72	140.25	Gas	11.28	±	0.13		NEDO (1991b)
463	37.71	139.47	Gas	9.13	±	0.29	194	Umeda et al. (2007b)
464	37.70	140.98	Gas	0.48	±	0.02	344.828	Sano et al. (1985)
465	37.70	140.98	Gas	0.47	±	0.01	114.943	Sano et al. (1985)
466	37.70	140.98	Gas	0.47	±	0.02	555.556	Sano et al. (1985)
467	37.70	140.98	Gas	0.47	±	0.02	500.000	Sano et al. (1985)
468	37.70	140.98	Gas	0.50	±	0.02	1000.000	Sano et al. (1985)
469	37.70	140.98	Gas	0.46	±	0.02	434.783	Sano et al. (1985)
470	37.70	140.98	Gas	0.48	±	0.02	555.556	Sano et al. (1985)
471	37.67	140.25	Gas	1.35	±	0.08		NEDO (1991b)
472	37.62	140.06	Gas	7.41	±	0.14	4.444	Marty et al. (1989)
473	37.61	140.06	Gas	4.42	±	0.21	0.800	Marty et al. (1989)
474	37.54	138.68	Gas	9.10			38	Sakata et al.(1986)
475	37.53	138.93	Gas	10.50				Sakata (1991)
476	37.53	138.93	Gas	10.90			810	Sakata et al.(1986)

Table 1. (Continued)

No.	Latitude	Longitude	Feature	$^3He/^4He(x10^{-6})$	±	1σ	$^4He/^{20}Ne$	Ref.
477	37.52	138.85	Gas	8.61	±	0.11		Wakita and Sano (1983)
478	37.52	138.85	Gas	8.67	±	0.28	>2500	Wakita et al. (1990)
479	37.52	138.85	Gas	8.16	±	0.19	1700	Wakita et al. (1990)
480	37.51	137.34	Water	4.13	±	0.06	1.9	Umeda et al. (2009)
481	37.50	138.80	Gas	0.28	±	0.01	468	Kita and Nagao (1996)
482	37.50	138.80	Gas	0.25	±	0.02	22	Kita and Nagao (1996)
483	37.50	138.80	Gas	7.00	±	0.10	79	Kita and Nagao (1996)
484	37.50	138.80	Gas	7.10	±	0.10	76	Kita and Nagao (1996)
485	37.46	138.76	Gas	7.10			85	Sakata et al.(1986)
486	37.46	138.90	Gas	8.35	±	0.12	52	Umeda et al. (2008b)
487	37.45	138.90	Gas	9.16	±	0.14	17	Umeda et al. (2008b)
488	37.44	138.71	Gas	2.41	±	0.04	1.1	Umeda et al. (2008b)
489	37.42	138.80	Gas	9.38				Sakata (1991)
490	37.42	138.80	Gas	9.50			1140	Sakata et al.(1986)
491	37.42	139.54	Water	8.75	±	0.06	4.39	Horiguchi et al. (2010)
492	37.42	138.86	Gas	9.40	±	0.14	22	Umeda et al. (2008b)
493	37.41	138.65	Gas	9.20				Sakata et al. (1989)
494	37.40	141.84	Gas	0.23	±	0.01	3650	Kita and Nagao (1996)
495	37.40	141.84	Gas	0.22	±	0.01	410	Kita and Nagao (1996)
496	37.40	141.84	Gas	0.25	±	0.01	430	Kita and Nagao (1996)
497	37.40	141.84	Gas	0.25	±	0.01	580	Kita and Nagao (1996)
498	37.40	136.94	Gas	1.38	±	0.04	0.87	Umeda et al. (2008a)
499	37.40	136.94	Gas	1.38	±	0.04	0.87	Umeda et al. (2009)
500	37.39	138.78	Gas	8.65	±	0.08		Wakita and Sano (1983)
501	37.39	138.78	Gas	7.20	±	0.23	2000	Wakita et al. (1990)
502	37.39	138.78	Gas	8.57	±	0.26	2500	Wakita et al. (1990)
503	37.39	138.84	Gas	9.18	±	0.11	11	Umeda et al. (2008b)
504	37.38	138.56	Gas	1.41	±	0.03	25	Umeda et al. (2008b)
505	37.37	138.77	Gas	8.45	±	0.28	2500	Wakita et al. (1990)
506	37.37	138.56	Gas	7.69	±	0.08		Wakita and Sano (1983)
507	37.37	138.56	Gas	6.98	±	0.10		Wakita and Sano (1983)
508	37.37	138.56	Gas	7.33	±	0.09		Wakita and Sano (1983)
509	37.37	138.56	Gas	6.90	±	0.10		Wakita and Sano (1983)
510	37.37	138.78	Gas	5.59	±	0.08		Wakita and Sano (1983)
511	37.37	138.78	Gas	8.54	±	0.39	670	Wakita et al. (1990)
512	37.37	138.61	Gas	6.93	±	0.14	830	Wakita et al. (1990)
513	37.37	138.76	Gas	9.38				Sakata (1991)

No.	Latitude	Longitude	Feature	$^3He/^4He(x10^{-6})$	±	1σ	$^4He/^{20}Ne$	Ref.
514	37.37	138.90	Gas	5.71	±	0.07	11	Umeda et al. (2008a)
515	37.37	138.90	Gas	5.71	±	0.07	11	Umeda et al. (2008b)
516	37.37	137.09	Gas	0.04	±	0.00	13	Umeda et al. (2009)
517	37.35	138.62	Gas	9.00			1960	Sakata et al.(1986)
518	37.35	138.75	Gas	5.41	±	0.06	370	Wakita et al. (1990)
519	37.33	138.75	Gas	6.90	±	0.13	830	Wakita et al. (1990)
520	37.31	137.20	Gas	1.73	±	0.02	720	Umeda et al. (2008a)
521	37.31	137.20	Gas	1.73	±	0.02	720	Umeda et al. (2009)
522	37.30	138.82	Gas	0.48	±	0.01	6	Umeda et al. (2008b)
523	37.29	139.00	Gas	2.90	±	0.06	0.73	Umeda et al. (2008b)
524	37.28	138.44	Gas	1.02	±	0.02	0.67	Umeda et al. (2008b)
525	37.27	136.73	Gas	3.25	±	0.04	320	Umeda et al. (2009)
526	37.26	138.87	Gas	5.19	±	0.08	2.2	Umeda et al. (2008b)
527	37.25	140.98	Gas	0.36	±	0.02	27.027	Sano et al. (1985)
528	37.25	140.98	Gas	0.33	±	0.01	22.727	Sano et al. (1985)
529	37.25	140.98	Gas	0.38	±	0.01	33.333	Sano et al. (1985)
530	37.25	140.98	Gas	0.35	±	0.01	27.027	Sano et al. (1985)
531	37.25	140.90	Gas	0.38	±	0.01	33.333	Sano et al. (1982)
532	37.23	138.77	Water	1.41	±	0.04	0.35	Umeda et al. (2008b)
533	37.21	138.63	Gas	0.20	±	0.01	13	Umeda et al. (2008b)
534	37.20	138.33	Gas	3.31	±	0.10	480	Wakita et al. (1990)
535	37.20	138.33	Gas	3.10	±	0.06		Wakita and Sano (1983)
536	37.20	136.90	Gas	1.63	±	0.03	0.38	Umeda et al. (2009)
537	37.19	139.08	Water	1.76	±	0.03	0.59	Umeda et al. (2008b)
538	37.19	138.27	Gas	4.20				Sakata (1991)
539	37.16	138.74	Gas	1.34	±	0.02	0.48	Umeda et al. (2008b)
540	37.16	138.92	Gas	3.29	±	0.05	80	Umeda et al. (2008a)
541	37.16	138.92	Gas	3.29	±	0.05	80	Umeda et al. (2008b)
542	37.14	136.88	Gas	1.66	±	0.03	8.0	Umeda et al. (2009)
543	37.13	139.98	Gas	5.07	±	0.11	26	NEDO (1990b)
544	37.13	139.96	Gas	5.60	±	0.17	11	NEDO (1990b)
545	37.13	138.90	Gas	1.58	±	0.03	0.34	Umeda et al. (2008b)
546	37.13	139.96	Gas	10.58	±	0.24	26	NEDO (1990b)
547	37.13	139.95	Gas	6.66	±	0.29	97	Nagao et al. (1981)
548	37.12	139.96	Gas	6.59	±	0.12	34	NEDO (1990b)
549	37.11	140.00	Gas	4.01	±	0.17	5.5	NEDO (1990b)
550	37.09	136.92	Gas	0.96	±	0.02	5.4	Umeda et al. (2008a)
551	37.09	136.92	Gas	0.96	±	0.02	5.4	Umeda et al. (2009)
552	37.08	137.04	Gas	0.47	±	0.02	16000	Umeda et al. (2008a)
553	37.08	137.04	Gas	0.47	±	0.02	1600	Umeda et al. (2009)
554	37.07	138.78	Gas	3.78	±	0.07	1.2	Umeda et al. (2008b)
555	37.06	138.60	Gas	1.75	±	0.03	0.37	Umeda et al. (2008a)
556	37.06	138.60	Gas	1.75	±	0.03	0.37	Umeda et al. (2008b)
557	37.06	138.88	Gas	6.95	±	0.09	34	Umeda et al. (2008b)
558	37.06	138.93	Gas	2.62	±	0.04	2.1	Umeda et al. (2008a)
559	37.06	138.93	Gas	2.62	±	0.04	2.1	Umeda et al. (2008b)

Table 1. (Continued)

No.	Latitude	Longitude	Feature	$^{3}He/^{4}He(x10^{-6})$	±	1σ	$^{4}He/^{20}Ne$	Ref.
560	37.04	138.33	Gas	1.99	±	0.07	2000	Wakita et al. (1990)
561	37.04	138.33	Gas	2.28	±	0.07	330	Wakita et al. (1990)
562	37.04	138.33	Gas	2.09	±	0.08		Wakita and Sano (1983)
563	36.98	138.80	Gas	1.51	±	0.04	0.32	Umeda et al. (2008b)
564	36.96	139.78	Gas	8.36	±	0.31	120	NEDO (1990b)
565	36.90	140.76	Gas	0.16	±	0.01	1000	Sano and Wakita (1985)
566	36.89	136.77	Gas	0.83	±	0.02	300	Umeda et al. (2008a)
567	36.89	136.77	Gas	0.83	±	0.02	300	Umeda et al. (2009)
568	36.88	137.87	Gas	1.21	±	0.13	0.26	Kusakabe et al. (2003)
569	36.87	136.98	Gas	0.42	±	0.02	640	Umeda et al. (2008a)
570	36.87	136.98	Gas	0.42	±	0.02	640	Umeda et al. (2009)
571	36.87	137.88	Gas	8.27	±	0.11	4.7	NEDO (1999a)
572	36.86	136.82	Gas	0.75	±	0.01	20	Umeda et al. (2009)
573	36.84	136.87	Water	0.51	±	0.01	14	Umeda et al. (2009)
574	36.82	137.93	Water	1.43	±	0.19	1.12	Kusakabe et al. (2003)
575	36.82	137.58	Gas	5.21	±	0.07	1.1	Umeda et al. (2008a)
576	36.81	137.04	Water	0.20			200	Umeda et al. (2009)
577	36.80	139.41	Gas	4.09	±	0.11	7.1	Sano and Wakita (1985)
578	36.80	137.87	Water	5.63	±	0.14	2.91	Kusakabe et al. (2003)
579	36.80	137.83	Gas	9.73	±	0.14	2.9	NEDO (1999a)
580	36.80	137.83	Gas	10.61	±	0.15	36	NEDO (1999a)
581	36.80	137.80	Gas	8.35	±	0.22	31.7	Kusakabe et al. (2003)
582	36.80	137.80	Gas	9.24	±	0.16	3.6	NEDO (1999a)
583	36.80	137.80	Gas	9.72	±	0.15	14	NEDO (1999a)
584	36.75	137.77	Water	9.57	±	0.25	3.27	Kusakabe et al. (2003)
585	36.73	137.82	Water	2.11	±	0.10	2.41	Kusakabe et al. (2003)
586	36.73	137.00	Water	4.77	±	0.14	0.23	Takaoka and Mizutani (1987)
587	36.73	137.00	Water	2.18	±	0.10	0.22	Takaoka and Mizutani (1987)
588	36.72	137.00	Water	1.80	±	0.13	0.21	Takaoka and Mizutani (1987)
589	36.70	136.97	Water	1.47	±	0.11	0.22	Takaoka and Mizutani (1987)
590	36.70	136.98	Water	1.41	±	0.11	0.22	Takaoka and Mizutani (1987)
591	36.68	136.86	Water	1.21	±	0.12	0.22	Takaoka and Mizutani (1987)
592	36.65	138.65	Gas	5.04			1.53	Sano et al. (1994)
593	36.65	138.63	Gas	5.17				Sano and Marty (1995)
594	36.65	138.63	Gas	5.32			0.97	Sano et al. (1994)
595	36.65	138.58	Gas	9.09			2.03	Sano et al. (1994)
596	36.65	138.58	Gas	9.14			2.89	Sano et al. (1994)

No.	Latitude	Longitude	Feature	$^3He/^4He(x10^{-6})$	±	1σ	$^4He/^{20}Ne$	Ref.
597	36.65	138.64	Gas	5.30	±	0.09	1.650	Marty et al. (1989)
598	36.65	138.54	Gas	9.58	±	0.12	500.000	Marty et al. (1989)
599	36.65	138.54	Gas	9.06	±	0.20	500.000	Marty et al. (1989)
600	36.65	138.54	Gas	10.09			2.54	Sano et al. (1994)
601	36.65	138.65	Gas	5.16	±	0.03	2.778	Marty et al. (1989)
602	36.64	138.54	Gas	10.61			12.0	Sano et al. (1994)
603	36.64	138.51	Gas	2.11			0.35	Sano et al. (1994)
604	36.63	138.51	Gas	8.06	±	0.25	30.864	Marty et al. (1989)
605	36.63	138.52	Gas	9.39			1.89	Sano et al. (1994)
606	36.63	138.51	Gas	8.75			1.27	Sano et al. (1994)
607	36.63	136.85	Water	1.33	±	0.10	0.23	Takaoka and Mizutani (1987)
608	36.63	138.51	Gas	5.74			0.62	Sano et al. (1994)
609	36.63	138.56	Gas	9.34	±	0.12	20.408	Marty et al. (1989)
610	36.63	138.56	Gas	9.15	±	0.40	250.000	Marty et al. (1989)
611	36.63	138.56	Gas	9.55	±	0.26	250.000	Marty et al. (1989)
612	36.63	138.56	Gas	8.12	±	0.14	58.824	Marty et al. (1989)
613	36.63	138.56	Gas	10.57				Sano and Marty (1995)
614	36.63	138.56	Gas	10.23			11.8	Sano et al. (1994)
615	36.62	138.60	Gas	7.74			1.27	Sano et al. (1994)
616	36.62	138.60	Gas	7.01	±	0.21	1.333	Marty et al. (1989)
617	36.62	138.60	Gas	7.22	±	0.17	1.227	Marty et al. (1989)
618	36.62	138.60	Gas	6.80	±	0.13	2.315	Marty et al. (1989)
619	36.62	138.60	Gas	7.17				Sano and Marty (1995)
620	36.62	138.60	Gas	7.71			1.28	Sano et al. (1994)
621	36.62	138.60	Gas	7.03			1.02	Sano et al. (1994)
622	36.61	138.64	Gas	4.47	±	0.10	23.810	Marty et al. (1989)
623	36.59	137.61	Gas	10.90	±	0.80	22	Nagao et al. (1981)
624	36.57	138.21	Gas	9.12	±	0.58		Nagao et al. (1980a)
625	36.57	138.21	Gas	9.10	±	0.69		Nagao et al. (1980a)
626	36.57	138.20	Gas	6.41	±	0.10	29	Sano and Wakita (1985)
627	36.53	137.82	Gas	0.70	±	0.07	0.55	Kusakabe et al. (2003)
628	36.52	137.75	Water	3.52	±	0.09	3.69	Kusakabe et al. (2003)
629	36.52	137.72	Water	1.26	±	0.01	0.41	Kusakabe et al. (2003)
630	36.49	136.91	Water	6.05	±	0.59	3.89	Ohwada et al. (2007)
631	36.49	137.91	Water	4.18	±	0.07	1.06	Umeda et al. (2011b)
632	36.49	138.38	Gas	4.62	±	0.09	5.3	Sano and Wakita (1985)
633	36.48	136.76	Water	7.17	±	0.64	93.9	Ohwada et al. (2007)
634	36.48	136.76	Water	7.13	±	0.20	19.8	Ohwada et al. (2007)
635	36.45	137.98	Water	1.26	±	0.02	0.67	Umeda et al. (2011b)
636	36.44	138.07	Water	2.41	±	0.04	0.52	Umeda et al. (2011b)
637	36.44	138.41	Gas	4.36	±	0.07	1.887	Sano et al. (1982)
638	36.44	136.55	Water	5.33	±	0.45	2.18	Ohwada et al. (2007)
639	36.44	136.55	Water	5.53	±	0.14	19.0	Ohwada et al. (2007)

Table 1. (Continued)

No.	Latitude	Longitude	Feature	$^3He/^4He(x10^{-6})$	±	1σ	$^4He/^{20}Ne$	Ref.
640	36.42	137.68	Gas	8.29	±	0.18	6.8	Kusakabe et al. (2003)
641	36.42	137.20	Gas	2.57	±	0.08	140	Umeda et al. (2008a)
642	36.40	137.77	Water	1.38	±	0.09	0.79	Kusakabe et al. (2003)
643	36.40	137.75	Water	1.14	±	0.21	2.56	Kusakabe et al. (2003)
644	36.40	136.53	Water	4.40	±	0.39	14.1	Ohwada et al. (2007)
645	36.40	136.53	Water	4.84	±	0.11	17.4	Ohwada et al. (2007)
646	36.37	137.28	Gas	2.45	±	0.04	310	Umeda et al. (2008a)
647	36.34	136.69	Water	0.90	±	0.21	1.96	Ohwada et al. (2007)
648	36.34	138.38	Gas	5.71	±	0.04	9.1	Sano and Wakita (1985)
649	36.34	138.36	Gas	5.24	±	0.17		Sano and Wakita (1985)
650	36.33	138.00	Gas	0.97	±	0.02	11.4	Umeda et al. (2011b)
651	36.33	138.40	Gas	4.97	±	0.22	12.195	Sano et al. (1982)
652	36.33	136.45	Water	4.73	±	0.15	14.6	Ohwada et al. (2007)
653	36.33	136.45	Water	5.03	±	0.10	33.6	Ohwada et al. (2007)
654	36.32	136.66	Water	5.87	±	0.55	147.0	Ohwada et al. (2007)
655	36.31	137.83	Water	4.36	±	0.06	0.87	Umeda et al. (2011b)
656	36.29	138.83	Gas	0.92	±	0.04	630	Sano and Wakita (1985)
657	36.29	136.37	Water	4.86	±	0.10	6.24	Ohwada et al. (2007)
658	36.25	137.62	Water	4.93	±	0.22	3.32	Kusakabe et al. (2003)
659	36.25	138.00	Gas	3.44	±	0.06	2.04	Umeda et al. (2011b)
660	36.24	136.76	Gas	11.40	±	1.10	8.1	Nagao et al. (1981)
661	36.23	137.53	Gas	3.28	±	0.08	3.40	Kusakabe et al. (2003)
662	36.22	137.60	Gas	4.15	±	0.10	0.92	Kusakabe et al. (2003)
663	36.22	137.57	Gas	11.18	±	0.15	242	Kusakabe et al. (2003)
664	36.22	137.57	Gas	10.71	±	0.12	289	Scott et al. (1994)
665	36.22	137.52	Gas	6.14	±	0.28	50.0	Kusakabe et al. (2003)
666	36.22	137.52	Gas	7.29	±	0.18	30.3	Kusakabe et al. (2003)
667	36.22	138.73	Gas	1.29	±	0.13	9.1	Sano and Wakita (1985)
668	36.19	138.09	Water	1.16	±	0.02	0.75	Umeda et al. (2011b)
669	36.18	137.80	Water	2.94	±	0.08	24.7	Kusakabe et al. (2003)
670	36.18	137.62	Water	1.47	±	0.08	1.29	Kusakabe et al. (2003)
671	36.17	136.63	Gas	5.65	±	0.27	470	Nagao et al. (1981)
672	36.17	137.95	Water	2.27	±	0.06	4.14	Umeda et al. (2011b)
673	36.15	137.63	Gas	8.91	±	0.26	8.57	Kusakabe et al. (2003)
674	36.15	137.63	Gas	9.65	±	0.28	16	Sano and Wakita (1985)
675	36.14	138.01	Water	1.44	±	0.03	0.33	Umeda et al. (2011b)

No.	Latitude	Longitude	Feature	$^{3}He/^{4}He(x10^{-6})$	±	1σ	$^{4}He/^{20}Ne$	Ref.
676	36.11	138.24	Water	5.74	±	0.08	1.03	Umeda et al. (2011b)
677	36.10	138.01	Gas	1.48	±	0.03	29	Sano and Wakita (1985)
678	36.08	138.09	Gas	6.62	±	0.10	16.3	Umeda et al. (2011b)
679	36.05	137.67	Gas	4.86	±	0.11	2.33	Kusakabe et al. (2003)
680	36.05	138.11	Gas	7.44	±	0.12	20.6	Umeda et al. (2011b)
681	36.04	138.30	Gas	9.24			15	Ono et al. (1993)
682	36.04	138.30	Gas	9.24	±	0.07	12	Sano and Wakita (1985)
683	36.02	138.48	Gas	7.00			66	Ono et al. (1993)
684	36.02	138.48	Gas	7.22	±	0.15	63	Sano and Wakita (1985)
685	36.00	137.40	Gas	4.44	±	0.15	1.5	Sano et al. (1986)
686	36.00	137.40	Gas	4.89	±	0.15	2.7	Sano et al. (1986)
687	36.00	137.40	Gas	5.07	±	0.10	3.3	Sano et al. (1986)
688	36.00	137.40	Gas	5.92			7.9	Sano et al. (1998a)
689	36.00	137.40	Gas	5.26			5.5	Sano et al. (1998a)
690	36.00	137.40	Gas	2.86			0.52	Takahata et al. (2003)
691	36.00	137.40	Gas	2.17			0.39	Takahata et al. (2003)
692	35.99	137.40	Gas	4.81	±	0.18	2.5	Sano et al. (1984)
693	35.97	137.99	Gas	1.11	±	0.02	93.3	Umeda et al. (2011b)
694	35.93	138.11	Water	1.65	±	0.04	0.33	Umeda et al. (2011b)
695	35.92	137.45	Gas	8.61	±	0.11	32	Sano et al. (1984)
696	35.92	137.45	Gas	8.40	±	0.13	79	Sano et al. (1986)
697	35.92	137.45	Gas	8.79	±	0.15	55	Sano et al. (1986)
698	35.92	137.45	Gas	8.48	±	0.24	29	Sano et al. (1986)
699	35.92	137.45	Gas	8.95			39	Sano et al. (1998a)
700	35.92	137.45	Gas	8.29			15	Sano et al. (1998a)
701	35.92	137.45	Gas	6.12			0.99	Takahata et al. (2003)
702	35.92	137.45	Gas	7.29			2.22	Takahata et al. (2003)
703	35.92	137.45	Gas	7.99			5.56	Takahata et al. (2003)
704	35.92	137.33	Gas	2.60	±	0.16	0.77	Sano et al. (1984)
705	35.92	137.33	Gas	2.65	±	0.24	0.86	Sano et al. (1986)
706	35.92	137.33	Gas	1.44	±	0.15	0.36	Sano et al. (1986)
707	35.92	137.33	Gas	1.85	±	0.32	0.52	Sano et al. (1986)
708	35.92	137.33	Gas	2.98			1.8	Sano et al. (1998a)
709	35.92	137.33	Gas	3.57			66	Sano et al. (1998a)
710	35.91	137.57	Gas	8.13	±	0.27	83	Sano et al. (1984)
711	35.90	137.58	Gas	6.80	±	0.07	150	Sano et al. (1984)
712	35.90	137.58	Gas	7.83	±	0.06	120	Sano et al. (1986)
713	35.90	137.58	Gas	7.06	±	0.29	49	Sano et al. (1986)
714	35.90	137.58	Gas	7.25	±	0.11	43	Sano et al. (1986)
715	35.90	137.58	Gas	8.55			2.3	Sano et al. (1998a)
716	35.90	137.58	Gas	3.08			0.42	Sano et al. (1998a)
717	35.90	137.58	Gas	7.66			2.45	Takahata et al. (2003)

Table 1. (Continued)

No.	Latitude	Longitude	Feature	$^3He/^4He(x10^{-6})$	±	1σ	$^4He/^{20}Ne$	Ref.
718	35.90	137.56	Gas	8.44	±	0.20	92	Sano et al. (1986)
719	35.90	137.56	Gas	8.69	±	0.18	92	Sano et al. (1986)
720	35.90	137.56	Gas	8.33	±	0.35	91	Sano et al. (1986)
721	35.90	137.56	Gas	9.91			51	Sano et al. (1998a)
722	35.90	137.56	Gas	8.68			8.0	Sano et al. (1998a)
723	35.90	137.56	Gas	8.93			30.9	Takahata et al. (2003)
724	35.90	137.56	Gas	9.07			27.0	Takahata et al. (2003)
725	35.90	137.56	Gas	3.64			0.46	Takahata et al. (2003)
726	35.90	137.31	Gas	3.00	±	0.15	37	Sano et al. (1984)
727	35.90	137.31	Gas	2.86	±	0.07	110	Sano et al. (1986)
728	35.90	137.31	Gas	2.93	±	0.07	130	Sano et al. (1986)
729	35.90	137.31	Gas	2.98	±	0.08	99	Sano et al. (1986)
730	35.90	137.31	Gas	3.21			120	Sano et al. (1998a)
731	35.90	137.31	Gas	2.91			285	Sano et al. (1998a)
732	35.90	137.31	Gas	3.02			54.5	Takahata et al. (2003)
733	35.90	137.31	Gas	3.00			113	Takahata et al. (2003)
734	35.90	137.31	Gas	3.04			52.7	Takahata et al. (2003)
735	35.90	138.54	Gas	7.19	±	0.16	3.22	Umeda et al. (2011b)
736	35.89	137.52	Gas	9.44			4.5	Sano et al. (1998a)
737	35.89	137.48	Gas	10.02				Sano and Marty (1995)
738	35.89	137.48	Gas	10.23			67	Sano et al. (1998a)
739	35.89	137.48	Gas	10.34	±	0.11	253	Scott et al. (1994)
740	35.89	138.41	Gas	7.68	±	0.10	58.7	Umeda et al. (2011b)
741	35.88	138.32	Gas	6.50	±	0.10	297	Umeda et al. (2011b)
742	35.88	138.26	Gas	8.44	±	0.11	184	Umeda et al. (2011b)
743	35.87	137.60	Gas	7.13		0.08	4.6	Sano et al. (1986)
744	35.87	137.60	Gas	7.25		0.10	22	Sano et al. (1986)
745	35.87	137.60	Gas	7.00			1.2	Sano et al. (1998a)
746	35.87	137.60	Gas	2.88			0.40	Sano et al. (1998a)
747	35.87	137.60	Gas	4.48			0.58	Takahata et al. (2003)
748	35.87	137.60	Gas	7.13			1.57	Takahata et al. (2003)
749	35.87	137.60	Gas	7.35			1.40	Takahata et al. (2003)
750	35.84	137.46	Gas	7.63	±	0.18	4.3	Sano et al. (1984)
751	35.82	137.63	Gas	5.59			3.67	Sano et al. (1998a)
752	35.82	137.63	Gas	5.43			2.29	Takahata et al. (2003)
753	35.81	138.34	Gas	6.88	±	0.11	2150	Umeda et al. (2011b)
754	35.80	137.71	Gas	2.83	±	0.10	17	Sano et al. (1984)
755	35.80	137.71	Gas	2.90	±	0.15	9.3	Sano et al. (1986)
756	35.80	137.71	Gas	2.93	±	0.08	3.4	Sano et al. (1986)

No.	Latitude	Longitude	Feature	$^{3}He/^{4}He(x10^{-6})$	±	1σ	$^{4}He/^{20}Ne$	Ref.
757	35.80	137.71	Gas	2.87	±	0.17	3.2	Sano et al. (1986)
758	35.80	137.71	Gas	3.53			61	Sano et al. (1998a)
759	35.80	137.71	Gas	3.56			7.9	Sano et al. (1998a)
760	35.80	137.71	Gas	3.51			15.3	Takahata et al. (2003)
761	35.80	137.71	Gas	3.39			6.86	Takahata et al. (2003)
762	35.80	137.71	Gas	3.46			11.1	Takahata et al. (2003)
763	35.80	137.68	Gas	4.17	±	0.06	23	Sano et al. (1984)
764	35.80	137.68	Gas	4.26	±	0.06	41	Sano et al. (1986)
765	35.80	137.68	Gas	4.70	±	0.10	47	Sano et al. (1986)
766	35.80	137.68	Gas	4.80	±	0.07	96	Sano et al. (1986)
767	35.80	137.68	Gas	5.45			150	Sano et al. (1998a)
768	35.80	137.68	Gas	5.26			82	Sano et al. (1998a)
769	35.80	137.68	Gas	3.88			0.89	Takahata et al. (2003)
770	35.80	137.68	Gas	5.59			107	Takahata et al. (2003)
771	35.80	137.68	Gas	5.73			44.3	Takahata et al. (2003)
772	35.78	138.38	Gas	5.40	±	0.08	419	Umeda et al. (2011b)
773	35.76	138.32	Water	7.39	±	0.13	11.8	Umeda et al. (2011b)
774	35.72	137.68	Gas	2.39	±	0.04	71	Sano et al. (1984)
775	35.72	137.68	Gas	2.24			13.4	Takahata et al. (2003)
776	35.69	140.03	Gas	0.21	±	0.01	50.000	Sano et al. (1982)
777	35.66	138.57	Water	5.64	±	0.08	33	Umeda et al. (2011b)
778	35.65	138.41	Water	0.96	±	0.03	0.52	Umeda et al. (2011b)
779	35.65	134.98	Gas	3.15	±	0.07		Matsumoto et al. (2003)
780	35.63	134.82	Gas	6.23	±	0.11		Matsumoto et al. (2003)
781	35.62	138.36	Gas	3.55	±	0.06	1.82	Umeda et al. (2011b)
782	35.62	140.50	Gas	0.35	±	0.01	14.085	Sano et al. (1982)
783	35.60	135.08	Gas	0.55	±	0.01		Matsumoto et al. (2003)
784	35.59	139.75	Gas	0.14	±	0.01	100.000	Sano et al. (1982)
785	35.57	135.15	Water	1.74	±	0.06	20.4	Umeda et al. (2008a)
786	35.57	135.17	Gas	1.75	±	0.04		Matsumoto et al. (2003)
787	35.55	138.31	Gas	4.05	±	0.05	17.6	Umeda et al. (2011b)
788	35.55	134.50	Gas	7.50	±	0.18	62	Nagao et al. (1981)
789	35.55	134.48	Gas	5.39	±	0.10	59	Sano and Wakita (1985)
790	35.54	139.73	Gas	<0.3			23	Nagao et al. (1981)
791	35.54	135.20	Water	1.12			0.53	Sano et al. (2009)
792	35.53	139.72	Gas	0.86	±	0.32	0.72	Nagao et al. (1981)
793	35.52	135.28	Water	1.02			3.94	Sano et al. (2009)
794	35.51	138.39	Water	9.54	±	0.12	235	Umeda et al. (2011b)
795	35.49	135.57	Water	1.12			59.8	Sano et al. (2009)
796	35.47	133.03	Gas	5.94	±	0.28	200	Nagao et al. (1981)
797	35.46	136.28	Water	2.17			0.31	Sano et al. (2009)

Table 1. (Continued)

No.	Latitude	Longitude	Feature	$^3He/^4He(x10^{-6})$	±	1σ	$^4He/^{20}Ne$	Ref.
798	35.45	133.36	Gas	7.66	±	0.11	180	Nagao et al. (1981)
799	35.44	139.06	Water	1.92	±	0.02	0.72	Sakamoto et al. (1992)
800	35.44	138.33	Gas	2.82	±	0.10	100	Sano and Wakita (1985)
801	35.42	138.48	Gas	5.68	±	0.23	10	Sano and Wakita (1985)
802	35.42	133.90	Gas	8.35	±	0.30	150	Nagao et al. (1981)
803	35.40	137.28	Gas	1.10	±	0.03	59	Sano and Wakita (1985)
804	35.40	133.88	Gas	7.56			13	Ono et al. (1993)
805	35.40	133.88	Gas	7.38	±	0.02	6.3	Sano and Wakita (1985)
806	35.40	133.01	Water	4.32	±	0.04	29	Sano and Wakita (1985)
807	35.40	134.80	Gas	7.64		0.20		Matsumoto et al. (2003)
808	35.38	136.40	Water	1.74			0.47	Sano et al. (2009)
809	35.37	136.26	Water	1.61			0.31	Sano et al. (2009)
810	35.37	133.37	Water	5.98	±	0.08	4.71	Umeda and Ninomiya (2009)
811	35.37	137.25	Water	0.93			681	Nagao (2004)
812	35.37	137.22	Water	0.21			4.90	Nagao (2004)
813	35.37	137.22	Water	0.08			5.47	Nagao (2004)
814	35.37	137.22	Water	0.14			5.79	Nagao (2004)
815	35.37	137.22	Water	0.12			7.25	Nagao (2004)
816	35.37	137.22	Water	0.09			6.22	Nagao (2004)
817	35.37	137.22	Water	0.18			2.33	Nagao (2004)
818	35.37	137.22	Water	0.31			1.97	Nagao (2004)
819	35.37	137.22	Water	1.32			0.53	Nagao (2004)
820	35.37	137.22	Water	1.03			0.28	Nagao (2004)
821	35.37	137.22	Water	1.14			0.40	Nagao (2004)
822	35.37	137.22	Water	1.34			0.55	Nagao (2004)
823	35.37	137.22	Water	1.04			0.43	Nagao (2004)
824	35.37	137.22	Water	1.18			0.69	Nagao (2004)
825	35.37	137.22	Water	1.27			0.33	Nagao (2004)
826	35.37	137.22	Water	0.83			0.58	Nagao (2004)
827	35.37	137.22	Water	0.92			0.30	Nagao (2004)
828	35.37	137.22	Water	1.01			0.75	Nagao (2004)
829	35.37	137.22	Water	0.47			0.74	Nagao (2004)
830	35.37	137.22	Water	0.86			0.40	Nagao (2004)
831	35.37	137.22	Water	0.59			0.69	Nagao (2004)
832	35.37	137.22	Water	0.50			0.62	Nagao (2004)
833	35.37	137.22	Water	0.62			0.59	Nagao (2004)
834	35.37	137.22	Water	0.44			0.68	Nagao (2004)
835	35.37	137.22	Water	0.30			1.08	Nagao (2004)
836	35.37	137.22	Water	0.64			0.59	Nagao (2004)
837	35.37	137.22	Water	0.55			0.71	Nagao (2004)
838	35.37	137.22	Water	0.72			0.682	Nagao (2004)

No.	Latitude	Longitude	Feature	$^{3}He/^{4}He(x10^{-6})$	±	1σ	$^{4}He/^{20}Ne$	Ref.
839	35.37	137.22	Water	1.00			0.506	Nagao (2004)
840	35.37	137.22	Water	0.28			36.9	Nagao (2004)
841	35.37	137.22	Water	0.40			112	Nagao (2004)
842	35.37	137.22	Water	0.42			19.2	Nagao (2004)
843	35.37	137.22	Water	0.40			123	Nagao (2004)
844	35.37	137.22	Water	0.45			109	Nagao (2004)
845	35.37	137.22	Water	1.82			3.86	Nagao (2004)
846	35.37	137.22	Water	0.43			20.6	Nagao (2004)
847	35.37	137.22	Water	1.26	±	0.04		Morikawa (2001)
848	35.37	137.22	Water	0.98	±	0.04		Morikawa (2001)
849	35.37	137.22	Water	0.34	±	0.01		Morikawa (2001)
850	35.37	137.22	Water	0.29	±	0.01		Morikawa (2001)
851	35.37	137.22	Water	1.48	±	0.05		Morikawa (2001)
852	35.37	137.22	Water	0.63	±	0.02		Morikawa (2001)
853	35.37	137.22	Water	0.18	±	0.01		Morikawa (2001)
854	35.37	137.22	Water	1.37	±	0.05		Morikawa (2001)
855	35.37	137.22	Water	1.28	±	0.05		Morikawa (2001)
856	35.37	137.22	Water	1.08	±	0.04		Morikawa (2001)
857	35.37	137.22	Water	1.26	±	0.05		Morikawa (2001)
858	35.37	137.22	Water	0.12	±	0.00		Morikawa (2001)
859	35.37	137.22	Water	0.13	±	0.01		Morikawa (2001)
860	35.37	137.22	Water	0.56	±	0.02		Morikawa (2001)
861	35.37	137.22	Water	0.49	±	0.02		Morikawa (2001)
862	35.37	137.22	Water	0.28	±	0.01		Morikawa (2001)
863	35.37	137.22	Water	0.31	±	0.01		Morikawa (2001)
864	35.37	137.22	Water	1.25	±	0.04		Morikawa (2001)
865	35.37	137.22	Water	0.45	±	0.02		Morikawa (2001)
866	35.37	137.22	Water	0.44	±	0.02		Morikawa (2001)
867	35.37	137.22	Water	0.35	±	0.02		Morikawa (2001)
868	35.37	137.22	Water	0.23	±	0.01		Morikawa (2001)
869	35.37	137.22	Water	0.24	±	0.01		Morikawa (2001)
870	35.37	137.22	Water	0.24	±	0.01		Morikawa (2001)
871	35.37	137.22	Water	0.17	±	0.01		Morikawa (2001)
872	35.37	137.22	Water	0.38	±	0.01		Morikawa (2001)
873	35.37	137.22	Water	0.42	±	0.02		Morikawa (2001)
874	35.37	137.22	Water	0.40	±	0.01		Morikawa (2001)
875	35.37	137.22	Water	0.33	±	0.01		Morikawa (2001)
876	35.37	137.22	Water	0.33	±	0.01		Morikawa (2001)
877	35.37	137.22	Water	0.49	±	0.02		Morikawa (2001)
878	35.37	137.22	Water	0.50	±	0.02		Morikawa (2001)
879	35.37	137.22	Water	0.47	±	0.02		Morikawa (2001)
880	35.37	137.22	Water	0.50	±	0.02		Morikawa (2001)
881	35.37	137.22	Water	0.49	±	0.02		Morikawa (2001)
882	35.37	137.22	Water	0.44	±	0.02		Morikawa (2001)
883	35.37	137.22	Water	0.44	±	0.02		Morikawa (2001)
884	35.37	137.22	Water	0.45	±	0.02		Morikawa (2001)
885	35.37	137.22	Water	0.45	±	0.02		Morikawa (2001)

Table 1. (Continued)

No.	Latitude	Longitude	Feature	$^{3}He/^{4}He(x10^{-6})$	±	1σ	$^{4}He/^{20}Ne$	Ref.
886	35.37	137.22	Water	0.46	±	0.02		Morikawa (2001)
887	35.37	137.22	Water	0.14	±	0.01		Morikawa (2001)
888	35.37	137.22	Water	0.13	±	0.01		Morikawa (2001)
889	35.37	137.22	Water	0.14	±	0.01		Morikawa (2001)
890	35.37	137.22	Water	0.10	±	0.01		Morikawa (2001)
891	35.37	137.22	Water	0.11	±	0.00		Morikawa (2001)
892	35.37	137.22	Water	0.11	±	0.01		Morikawa (2001)
893	35.37	137.22	Water	0.09	±	0.00		Morikawa (2001)
894	35.37	137.22	Water	0.10	±	0.00		Morikawa (2001)
895	35.37	140.35	Gas	0.31	±	0.01	4.762	Sano et al. (1982)
896	35.37	140.28	Gas	0.22	±	0.01	4.167	Sano et al. (1982)
897	35.37	138.98	Water	6.94	±	0.07	5.1	Sakamoto et al. (1992)
898	35.36	133.28	Water	1.55	±	0.04	0.42	Umeda and Ninomiya (2009)
899	35.35	133.76	Gas	6.02	±	0.08	30	Sano and Wakita (1985)
900	35.35	133.30	Water	1.41	±	0.03	0.38	Umeda and Ninomiya (2009)
901	35.34	135.92	Water	0.17			3.37	Sano et al. (2009)
902	35.34	135.92	Water	0.19	±	0.01	2.4	Umeda et al. (2008a)
903	35.34	133.33	Water	1.71	±	0.03	0.43	Umeda and Ninomiya (2009)
904	35.33	133.36	Water	1.26	±	0.03	0.43	Umeda and Ninomiya (2009)
905	35.30	136.56	Gas	0.66	±	0.04	53	Umeda et al. (2006b)
906	35.30	133.41	Water	4.29	±	0.08	1.84	Umeda et al. (2011a)
907	35.30	133.26	Water	1.12	±	0.02	0.40	Umeda and Ninomiya (2009)
908	35.29	133.31	Water	1.28	±	0.02	0.40	Umeda and Ninomiya (2009)
909	35.29	138.42	Water	1.84	±	0.05	0.34	Umeda et al. (2011b)
910	35.29	133.23	Water	0.94	±	0.02	0.52	Umeda and Ninomiya (2009)
911	35.28	133.35	Water	6.20	±	0.10	4.80	Umeda and Ninomiya (2009)
912	35.26	133.47	Water	1.47	±	0.06	0.28	Umeda and Ninomiya (2009)
913	35.26	133.38	Water	3.12	±	0.04	0.60	Umeda and Ninomiya (2009)
914	35.25	133.16	Water	5.08	±	0.13	4.03	Umeda et al. (2011a)
915	35.25	133.24	Water	1.15	±	0.02	0.39	Umeda and Ninomiya (2009)
916	35.24	136.24	Water	1.50			0.26	Sano et al. (2009)
917	35.24	139.06	Gas	8.65	±	0.08	2.3	Sakamoto et al. (1992)
918	35.24	139.06	Water	7.93	±	0.08	2.0	Sakamoto et al. (1992)
919	35.24	139.02	Gas	9.63	±	0.12	19	Sakamoto et al. (1992)
920	35.24	139.02	Gas	8.79	±	0.10	7.9	Sakamoto et al. (1992)
921	35.24	139.02	Gas	9.25	±	0.08	12	Sakamoto et al. (1992)

No.	Latitude	Longitude	Feature	$^{3}He/^{4}He(x10^{-6})$	±	1σ	$^{4}He/^{20}Ne$	Ref.
922	35.24	139.02	Gas	9.11	±	0.08	10	Sakamoto et al. (1992)
923	35.24	139.02	Gas	9.25	±	0.09	26	Sakamoto et al. (1992)
924	35.24	133.41	Water	2.50	±	0.04	0.98	Umeda and Ninomiya (2009)
925	35.24	139.03	Gas	9.61	±	0.15	19	Sakamoto et al. (1992)
926	35.24	139.03	Gas	9.71	±	0.21	17	Sakamoto et al. (1992)
927	35.24	139.01	Gas	8.29	±	0.20	16	Sano and Wakita (1985)
928	35.23	139.02	Gas	8.51	±	0.40	23	Nagao et al. (1981)
929	35.23	133.13	Water	1.85	±	0.04	1.07	Umeda et al. (2011a)
930	35.23	139.10	Water	7.46	±	0.07	1.4	Sakamoto et al. (1992)
931	35.23	133.32	Water	1.61	±	0.04	0.31	Umeda and Ninomiya (2009)
932	35.23	133.41	Water	0.55	±	0.01	2.27	Umeda and Ninomiya (2009)
933	35.23	139.03	Gas	8.60			40	Craig et al. (1978)
934	35.23	133.09	Gas	7.02	±	0.11	41.9	Umeda et al. (2011a)
935	35.22	139.04	Gas	9.16	±	0.09	20	Sakamoto et al. (1992)
936	35.22	139.04	Gas	8.84	±	0.19	11	Sakamoto et al. (1992)
937	35.22	133.39	Water	1.70	±	0.04	0.31	Umeda and Ninomiya (2009)
938	35.22	139.03	Gas	7.62	±	0.07	7.1	Sano and Wakita (1985)
939	35.22	139.04	Water	6.73	±	0.07	0.85	Sakamoto et al. (1992)
940	35.20	133.73	Gas	3.96	±	0.22	10	Nagao et al. (1981)
941	35.20	133.38	Water	2.66	±	0.04	1.46	Umeda and Ninomiya (2009)
942	35.19	133.35	Water	1.08	±	0.03	0.35	Umeda and Ninomiya (2009)
943	35.19	133.18	Water	1.93	±	0.06	0.61	Umeda et al. (2011a)
944	35.19	133.11	Water	2.49	±	0.06	0.38	Umeda et al. (2011a)
945	35.18	136.66	Gas	0.71	±	0.04	150	Umeda et al. (2006b)
946	35.17	135.28	Gas	0.45	±	0.03	4100	Umeda et al. (2006b)
947	35.15	132.60	Gas	9.15	±	0.52	160	Nagao et al. (1981)
948	35.15	132.58	Gas	5.84	±	0.05	59	Sano and Wakita (1985)
949	35.15	136.52	Water	0.34	±	0.01	1.9	Umeda et al. (2007c)
950	35.14	136.17	Water	2.42			10.2	Sano et al. (2009)
951	35.14	132.56	Gas	8.58	±	0.44	180	Nagao et al. (1981)
952	35.12	135.82	Water	0.41			1.37	Sano et al. (2009)
953	35.12	132.62	Water	1.19			0.49	Sano et al. (2006)
954	35.12	135.95	Gas	1.82	±	0.10	45	Sano and Wakita (1985)
955	35.12	135.95	Gas	1.82	±	0.10	45.455	Sano et al. (1982)
956	35.12	133.05	Water	0.79	±	0.01	1.97	Umeda et al. (2011a)
957	35.10	139.07	Gas	7.60	±	0.14	6.1	Ohno et al. (2011)
958	35.10	135.90	Water	0.09	±	0.00	5.3	Umeda et al. (2008a)
959	35.10	132.35	Gas	9.12	±	0.57	120	Nagao et al. (1981)

Table 1. (Continued)

No.	Latitude	Longitude	Feature	$^3He/^4He(x10^{-6})$	±	1σ	$^4He/^{20}Ne$	Ref.
960	35.09	136.72	Gas	1.04	±	0.03	62	Umeda et al. (2008a)
961	35.08	132.35	Water	6.58			1.01	Sano et al. (2006)
962	35.08	138.98		7.20	±	0.22	1.4	Ohno et al. (2011)
963	35.07	132.80	Gas	3.62	±	0.08	18	Sano and Wakita (1985)
964	35.05	139.08	Gas	7.78	±	0.13	2.2	Ohno et al. (2011)
965	35.05	139.09	Gas	8.12	±	0.17	11	Ohno et al. (2011)
966	35.04	135.81	Water	1.43			0.31	Sano et al. (2009)
967	35.04	135.33	Gas	0.89	±	0.04	8.3	Sano and Wakita (1985)
968	35.04	138.94	Gas	9.58	±	0.20	9.5	Ohno et al. (2011)
969	35.03	136.74	Gas	0.83	±	0.03	41	Umeda et al. (2008a)
970	35.03	139.09	Gas	9.52	±	0.18	16	Ohno et al. (2011)
971	35.02	136.51	Gas	0.54	±	0.03	29	Umeda et al. (2006b)
972	35.01	135.85	Water	0.99			0.43	Sano et al. (2009)
973	35.01	136.11	Water	1.01	±	0.02	1.9	Umeda et al. (2008a)
974	35.00	139.08	Water	8.36	±	0.14	5.1	Ohno et al. (2011)
975	35.00	135.10	Gas	0.22	±	0.01		Matsumoto et al. (2003)
976	35.00	133.00	Water	1.58			36.8	Sano et al. (2006)
977	34.99	134.99	Gas	1.66	±	0.03	106.383	Sano et al. (1982)
978	34.98	138.92	Water	1.51	±	0.08	0.20	Ohno et al. (2011)
979	34.97	138.93	Gas	1.67	±	0.17	0.27	Nagao et al. (1981)
980	34.97	137.61	Gas	1.73	±	0.04	100	Sano and Wakita (1985)
981	34.96	138.98	Gas	9.69	±	0.17	12	Ohno et al. (2011)
982	34.96	139.09	Gas	8.71	±	0.14	1.8	Ohno et al. (2011)
983	34.95	134.70	Gas	1.06	±	0.04	19	Sano and Wakita (1985)
984	34.95	139.03	Gas	9.77	±	0.17	2.9	Ohno et al. (2011)
985	34.95	138.94	Gas	9.17	±	0.15	1.6	Ohno et al. (2011)
986	34.95	135.40	Gas	0.87	±	0.07		Matsumoto et al. (2003)
987	34.94	139.13	Water	10.29	±	0.13	14	Ohno et al. (2011)
988	34.94	139.03	Gas	1.46	±	0.04	0.32	Sano and Wakita (1985)
989	34.93	136.15	Gas	3.63	±	0.07		Matsumoto et al. (2003)
990	34.93	132.20	Water	3.27			2.05	Sano et al. (2006)
991	34.93	135.77	Water	1.14	±	0.05	6.1	Umeda et al. (2006b)
992	34.93	139.04	Water	1.76	±	0.04	0.28	Ohno et al. (2011)
993	34.93	139.11	Water	8.57	±	0.17	1.2	Ohno et al. (2011)
994	34.93	136.16	Water	3.23			114	Sano et al. (2009)
995	34.92	139.02	Gas	1.69	±	0.04	0.312	Ohno et al. (2011)
996	34.92	138.79	Gas	8.60	±	0.13	3.8	Ohno et al. (2011)
997	34.92	134.88	Gas	1.66	±	0.03	110	Sano and Wakita (1985)
998	34.92	136.17	Gas	3.15	±	0.06		Matsumoto et al. (2003)
999	34.91	138.93	Gas	1.69	±	0.11	0.27	Ohno et al. (2011)

No.	Latitude	Longitude	Feature	$^3He/^4He(x10^{-6})$	±	1σ	$^4He/^{20}Ne$	Ref.
1000	34.91	139.14	Water	10.02	±	0.29	27	Ohno et al. (2011)
1001	34.91	138.80	Gas	8.08	±	0.14	1.9	Ohno et al. (2011)
1002	34.90	139.13	Water	9.62	±	0.13	3.2	Ohno et al. (2011)
1003	34.90	135.42	Gas	1.53	±	0.06	110	Nagao et al. (1981)
1004	34.89	138.93	Water	7.76	±	0.11	2.0	Ohno et al. (2011)
1005	34.89	135.81	Water	1.58			0.33	Sano et al. (2009)
1006	34.88	135.38	Gas	1.43			97.3	Sano et al. (2009)
1007	34.88	132.23	Water	7.05			1.45	Sano et al. (2006)
1008	34.88	135.68	Water	1.64			0.47	Sano et al. (2009)
1009	34.88	135.67	Water	1.72			0.44	Sano et al. (2009)
1010	34.87	135.59	Water	0.24	±	0.02	3.29	Morikawa et al. (2008b)
1011	34.87	132.27	Water	4.27			1.39	Sano et al. (2006)
1012	34.87	135.67	Water	1.79			0.75	Sano et al. (2009)
1013	34.87	135.38	Gas	1.71	±	0.06	45	Umeda et al. (2007c)
1014	34.87	135.42	Gas	1.29	±	0.04		Matsumoto et al. (2003)
1015	34.87	139.10	Gas	10.72	±	0.13	4.1	Ohno et al. (2011)
1016	34.86	135.23	Water	6.16				AIST (2004)
1017	34.86	135.67	Water	1.85			7.93	Sano et al. (2009)
1018	34.86	139.09	Water	9.59	±	0.34	12	Ohno et al. (2011)
1019	34.85	135.31	Water	4.39	±	0.04	67.8	Morikawa et al. (2008b)
1020	34.85	135.90	Water	1.39			0.31	Sano et al. (2009)
1021	34.84	135.19	Water	6.17	±	0.13	3.27	Morikawa et al. (2008b)
1022	34.84	135.57	Water	1.03	±	0.01	329	Morikawa et al. (2008b)
1023	34.84	135.47	Water	1.20	±	0.02	0.477	Morikawa et al. (2008b)
1024	34.84	135.47	Water	1.03	±	0.02	0.705	Morikawa et al. (2008b)
1025	34.82	135.24	Water	10.64				AIST (2004)
1026	34.82	139.07	Gas	10.84	±	0.06	28	Ohno et al. (2011)
1027	34.81	138.83		8.12	±	0.21	1.4	Ohno et al. (2011)
1028	34.81	135.34	Gas	5.66			0.52	Sano et al. (2009)
1029	34.80	139.06	Water	1.70	±	0.09	0.67	Sano and Wakita (1985)
1030	34.80	135.35	Gas	8.68			45	Ono et al. (1993)
1031	34.80	135.35	Gas	8.15	±	0.21	42	Sano and Wakita (1985)
1032	34.80	135.25	Gas	11.00	±	0.40	210	Nagao et al. (1981)
1033	34.80	135.25	Gas	9.24			60	Ono et al. (1993)
1034	34.80	135.25	Gas	9.22	±	0.11	37	Sano and Wakita (1985)
1035	34.80	135.25	Gas	9.22	±	0.11	37	Wakita et al. (1987)
1036	34.80	135.25	Gas	9.69	±	0.12	530	Wakita et al. (1987)
1037	34.80	135.22	Water	9.18	±	0.05		Morikawa et al. (2005)
1038	34.80	135.25	Gas	10.42	±	0.18		Matsumoto et al. (2003)
1039	34.80	135.35	Gas	8.04	±	0.46		Matsumoto et al. (2003)

Table 1. (Continued)

No.	Latitude	Longitude	Feature	$^3He/^4He(x10^{-6})$	±	1σ	$^4He/^{20}Ne$	Ref.
1040	34.80	134.90	Gas	0.93	±	0.04	130	Umeda et al. (2007c)
1041	34.80	135.49	Water	0.81	±	0.01	390	Morikawa et al. (2008b)
1042	34.80	135.59	Water	1.21	±	0.01	1.90	Morikawa et al. (2008b)
1043	34.80	135.56	Water	0.71	±	0.01	62.7	Morikawa et al. (2008b)
1044	34.80	135.25	Water	9.93			45.8	Sano et al. (2009)
1045	34.79	135.25	Water	10.74	±	0.07		Morikawa et al. (2005)
1046	34.79	138.76	Gas	4.16	±	0.08	1.4	Ohno et al. (2011)
1047	34.78	136.28	Water	1.16	±	0.02	23	Umeda et al. (2008a)
1048	34.78	138.95	Water	9.25	±	0.27	3.5	Ohno et al. (2011)
1049	34.77	139.36	Water	3.60			0.45	Sano et al. (1995)
1050	34.77	139.36	Water	3.72			0.46	Sano et al. (1995)
1051	34.77	139.36	Water	3.36			0.48	Sano et al. (1995)
1052	34.77	139.36	Water	3.25			0.42	Sano et al. (1995)
1053	34.77	139.36	Water	3.30			0.40	Sano et al. (1995)
1054	34.77	139.36	Water	3.74			0.41	Sano et al. (1995)
1055	34.77	139.36	Water	4.52			0.36	Sano et al. (1995)
1056	34.76	135.32	Water	4.20				AIST (2004)
1057	34.76	135.93	Water	2.86			0.6	Sano et al. (2009)
1058	34.76	138.80	Gas	8.72	±	0.08	3.5	Ohno et al. (2011)
1059	34.75	139.41	Gas	2.39	±	0.01	0.37	Sano et al. (1988)
1060	34.75	139.41	Gas	2.65	±	0.03	0.38	Sano et al. (1988)
1061	34.75	139.41	Gas	2.38	±	0.03		Sano et al. (1988)
1062	34.75	139.41	Gas	4.30	±	0.06	0.49	Sano et al. (1988)
1063	34.75	139.41	Gas	4.89	±	0.06	0.58	Sano et al. (1988)
1064	34.75	139.41	Gas	5.60	±	0.06	0.81	Sano et al. (1988)
1065	34.75	139.41	Gas	5.05	±	0.11	0.62	Sano et al. (1988)
1066	34.75	139.41	Gas	4.96	±	0.07	0.62	Sano et al. (1988)
1067	34.75	139.41	Gas	5.66	±	0.11	0.79	Sano et al. (1988)
1068	34.75	139.41	Gas	5.57	±	0.06	0.82	Sano et al. (1988)
1069	34.75	139.41	Gas	5.81	±	0.08	0.85	Sano et al. (1988)
1070	34.75	139.41	Gas	6.64	±	0.06	0.97	Sano et al. (1988)
1071	34.75	139.41	Gas	6.64	±	0.13	1.1	Sano et al. (1988)
1072	34.75	139.41	Gas	6.62	±	0.11	1.1	Sano et al. (1988)
1073	34.75	139.41	Gas	7.03	±	0.06	1.5	Sano et al. (1988)
1074	34.75	139.41	Gas	7.03	±	0.04	1.8	Sano et al. (1988)
1075	34.75	139.41	Gas	7.06			1.60	Sano et al. (1991)
1076	34.75	139.41	Gas	7.01			1.50	Sano et al. (1991)
1077	34.75	139.41	Gas	7.11			1.80	Sano et al. (1991)
1078	34.75	139.41	Gas	7.27			1.50	Sano et al. (1991)
1079	34.75	139.41	Gas	7.39			1.60	Sano et al. (1991)
1080	34.75	139.41	Gas	7.31			1.60	Sano et al. (1991)
1081	34.75	139.41	Gas	7.56			1.80	Sano et al. (1991)
1082	34.75	139.41	Gas	7.55			1.80	Sano et al. (1991)

No.	Latitude	Longitude	Feature	$^3He/^4He(x10^{-6})$	±	1σ	$^4He/^{20}Ne$	Ref.
1083	34.75	139.41	Gas	7.69			1.80	Sano et al. (1991)
1084	34.75	139.41	Gas	7.53			2.00	Sano et al. (1991)
1085	34.75	139.41	Gas	7.29			1.70	Sano et al. (1991)
1086	34.75	139.41	Gas	7.07			1.40	Sano et al. (1991)
1087	34.75	139.41	Gas	7.07			1.30	Sano et al. (1991)
1088	34.75	139.41	Gas	7.17			1.39	Sano et al. (1991)
1089	34.75	139.41	Gas	7.03			1.30	Sano et al. (1991)
1090	34.75	139.41	Gas	6.97			1.21	Sano et al. (1991)
1091	34.75	139.41	Gas	7.00			1.43	Sano et al. (1991)
1092	34.75	139.41	Gas	6.85			1.23	Sano et al. (1991)
1093	34.75	139.41	Gas	6.65			1.11	Sano et al. (1991)
1094	34.75	139.41	Gas	6.41			1.06	Sano et al. (1991)
1095	34.75	139.41	Gas	6.36			0.98	Sano et al. (1991)
1096	34.75	139.41	Gas	6.23			0.98	Sano et al. (1991)
1097	34.75	139.41	Gas	5.98			0.80	Sano et al. (1991)
1098	34.75	139.41	Gas	5.85			0.93	Sano et al. (1991)
1099	34.75	139.41	Gas	5.98			0.82	Sano et al. (1991)
1100	34.75	139.41	Gas	6.19			0.99	Sano et al. (1991)
1101	34.75	139.41	Gas	6.43			1.08	Sano et al. (1991)
1102	34.75	139.41	Gas	6.24			0.98	Sano et al. (1991)
1103	34.75	139.41	Gas	5.95			0.76	Sano et al. (1991)
1104	34.75	139.41	Gas	5.67			0.74	Sano et al. (1991)
1105	34.75	139.41	Gas	5.43			0.76	Sano et al. (1991)
1106	34.75	138.96	Water	0.93	±	0.02	0.65	Sano and Wakita (1985)
1107	34.75	138.83	Water	4.53	±	0.15	0.74	Sano and Wakita (1985)
1108	34.75	135.33	Gas	4.96			0.71	Sano et al. (1995)
1109	34.75	135.33	Gas	5.11			0.58	Sano et al. (1995)
1110	34.75	135.33	Water	1.44			0.253	Sano et al. (1998b)
1111	34.75	135.33	Water	1.20			0.308	Sano et al. (1998b)
1112	34.75	135.33	Water	1.36			0.267	Sano et al. (1998b)
1113	34.75	135.33	Water	3.50			0.262	Sano et al. (1998b)
1114	34.75	135.33	Water	2.20			0.338	Sano et al. (1998b)
1115	34.75	135.33	Water	2.55			0.246	Sano et al. (1998b)
1116	34.75	135.33	Water	1.39			0.341	Sano et al. (1998b)
1117	34.75	135.33	Water	2.74			0.246	Sano et al. (1998b)
1118	34.75	135.33	Water	1.41			0.260	Sano et al. (1998b)
1119	34.75	135.33	Water	1.11			0.320	Sano et al. (1998b)
1120	34.75	135.33	Water	2.28			0.285	Sano et al. (1998b)
1121	34.75	135.33	Water	1.51			0.357	Sano et al. (1998b)
1122	34.75	135.33	Water	1.47			0.274	Sano et al. (1998b)
1123	34.75	135.33	Water	1.17			0.325	Sano et al. (1998b)
1124	34.75	135.95	Gas	5.29	±	0.07		Matsumoto et al. (2003)
1125	34.74	135.59	Water	0.27	±	0.01	200	Morikawa et al. (2008b)
1126	34.73	139.40	Gas	1.48	±	0.01	0.32	Sano et al. (1988)
1127	34.73	139.40	Gas	1.48	±	0.01	0.32	Sano et al. (1988)

Table 1. (Continued)

No.	Latitude	Longitude	Feature	$^3He/^4He(x10^{-6})$	±	1σ	$^4He/^{20}Ne$	Ref.
1128	34.73	132.43	Water	2.81			2.18	Sano et al. (2006)
1129	34.72	135.31	Water	5.69	±	0.15		Morikawa et al. (2005)
1130	34.72	135.45	Water	4.08	±	0.19	1.29	Okada et al. (1994)
1131	34.72	135.45	Water	4.48	±	0.06	418	Okada et al. (1994)
1132	34.71	136.05	Water	1.69	±	0.03	1.1	Umeda et al. (2008a)
1133	34.71	136.05	Water	2.10			2.67	Sano et al. (2009)
1134	34.71	135.24	Water	7.43	±	0.18		Morikawa et al. (2005)
1135	34.71	135.38	Water	4.29	±	0.15		Morikawa et al. (2005)
1136	34.71	135.52	Water	0.59	±	0.01	63.4	Morikawa et al. (2008b)
1137	34.71	135.22	Water	7.73	±	0.12		Morikawa et al. (2005)
1138	34.70	135.20	Gas	2.88	±	0.10	0.8	Wakita et al. (1987)
1139	34.70	135.20	Gas	2.74	±	0.07	0.9	Wakita et al. (1987)
1140	34.70	133.08	Water	1.08			1.03	Sano et al. (2006)
1141	34.70	138.93	Gas	7.57	±	0.14	1.9	Ohno et al. (2011)
1142	34.70	135.20	Water	3.60	±	0.07		Morikawa et al. (2005)
1143	34.70	135.20	Water	3.78	±	0.10		Morikawa et al. (2005)
1144	34.69	135.55	Water	0.83	±	0.01	2.88	Morikawa et al. (2008b)
1145	34.69	135.36	Water	1.83	±	0.04	2.65	Morikawa et al. (2008b)
1146	34.69	135.55	Water	0.56	±	0.01	317	Okada et al. (1994)
1147	34.69	135.55	Water	0.51	±	0.02	260	Okada et al. (1994)
1148	34.69	135.16	Gas	3.77	±	0.06	43	Sano and Wakita (1985)
1149	34.69	135.56	Water	0.80	±	0.01	5.12	Morikawa et al. (2008b)
1150	34.68	135.62	Water	0.55	±	0.01	8.35	Morikawa et al. (2008b)
1151	34.68	132.02	Water	2.16			0.82	Sano et al. (2006)
1152	34.68	135.78	Gas	3.68	±	0.06	1020	Umeda et al. (2007c)
1153	34.67	135.18	Water	5.30	±	0.13		Morikawa et al. (2005)
1154	34.67	133.10	Water	2.33			8.53	Sano et al. (2006)
1155	34.67	135.48	Water	1.42	±	0.37	1.80	Okada et al. (1994)
1156	34.67	135.46	Water	1.34	±	0.01	4.03	Morikawa et al. (2008b)
1157	34.67	135.46	Gas	7.96	±	0.07	572	Morikawa et al. (2008b)
1158	34.66	135.58	Water	1.59	±	0.02	105	Morikawa et al. (2008b)
1159	34.65	135.45	Water	2.83	±	0.03	181	Morikawa et al. (2008b)
1160	34.65	132.53	Water	2.11			0.55	Sano et al. (2006)
1161	34.65	135.72	Water	0.35	±	0.01	1.4	Umeda et al. (2007c)
1162	34.65	138.87	Gas	8.39	±	0.08	18	Ohno et al. (2011)
1163	34.62	135.53	Water	1.16	±	0.02	0.694	Morikawa et al. (2008b)

No.	Latitude	Longitude	Feature	$^{3}He/^{4}He(x10^{-6})$	±	1σ	$^{4}He/^{20}Ne$	Ref.
1164	34.60	138.23	Gas	0.10	±	0.00	6.667	Sano et al. (1982)
1165	34.60	135.00	Gas	0.67	±	0.04		Matsumoto et al. (2003)
1166	34.59	135.02	Gas	0.68	±	0.04	0.79	Umeda et al. (2008a)
1167	34.58	135.54	Water	3.44	±	0.03	160	Morikawa et al. (2008b)
1168	34.55	135.70	Water	4.06				AIST (2004)
1169	34.55	135.46	Water	3.02	±	0.03	154	Morikawa et al. (2008b)
1170	34.52	135.57	Water	3.61	±	0.03	0.577	Morikawa et al. (2008b)
1171	34.50	135.60	Water	1.71	±	0.16	0.450	Okada et al. (1994)
1172	34.50	135.60	Water	2.72	±	0.04	472	Okada et al. (1994)
1173	34.50	135.55	Water	4.43	±	0.49	1.90	Okada et al. (1994)
1174	34.50	135.55	Water	4.60	±	0.06	3050	Okada et al. (1994)
1175	34.48	132.28	Water	0.30			4.21	Sano et al. (2006)
1176	34.48	133.20	Water	1.37			0.27	Sano et al. (2006)
1177	34.48	135.39	Water	4.84	±	0.04	38.2	Morikawa et al. (2008b)
1178	34.47	132.12	Water	2.60			23	Sano et al. (2006)
1179	34.47	135.36	Water	1.53	±	0.36	0.440	Okada et al. (1994)
1180	34.47	135.36	Water	2.00	±	0.14	1.70	Okada et al. (1994)
1181	34.47	135.58	Water	4.81	±	0.04	13.5	Morikawa et al. (2008b)
1182	34.46	135.59	Water	5.36	±	0.10	0.806	Morikawa et al. (2008b)
1183	34.46	135.60	Water	3.51	±	0.05	0.827	Morikawa et al. (2008b)
1184	34.45	135.56	Water	8.15	±	0.11	11.8	Okada et al. (1994)
1185	34.45	135.20	Water	1.82				AIST (2004)
1186	34.45	133.02	Water	1.37			0.28	Sano et al. (2006)
1187	34.45	135.57	Water	2.94				AIST (2004)
1188	34.44	135.53	Water	6.72				AIST (2004)
1189	34.43	135.60	Gas	7.98			12	Ono et al. (1993)
1190	34.43	131.87	Water	6.64			5.92	Sano et al. (2006)
1191	34.43	131.77	Water	6.12			25.8	Sano et al. (2006)
1192	34.43	133.20	Water	1.81			0.39	Sano et al. (2006)
1193	34.43	135.57	Gas	8.33			33.2	Sano et al. (2009)
1194	34.43	132.75	Water	1.64	±	0.02		Takahata and Sano (2000)
1195	34.43	132.75	Water	1.98	±	0.02		Takahata and Sano (2000)
1196	34.43	132.75	Water	1.66	±	0.02		Takahata and Sano (2000)
1197	34.43	132.75	Water	1.66	±	0.02		Takahata and Sano (2000)
1198	34.43	132.75	Water	1.61	±	0.02		Takahata and Sano (2000)
1199	34.42	135.58	Water	6.21	±	0.15	0.987	Morikawa et al. (2008b)
1200	34.42	135.58	Gas	7.44	±	0.09	13.2	Morikawa et al. (2008b)
1201	34.42	135.58	Gas	7.77	±	0.10	16.2	Morikawa et al. (2008b)
1202	34.42	131.90	Water	3.26			0.35	Sano et al. (2006)

Table 1. (Continued)

No.	Latitude	Longitude	Feature	$^3He/^4He(x10^{-6})$	±	1σ	$^4He/^{20}Ne$	Ref.
1203	34.42	132.83	Water	0.08			12.1	Sano et al. (2006)
1204	34.42	135.58	Gas	6.45	±	0.48	30	Nagao et al. (1981)
1205	34.42	135.56	Gas	6.54	±	0.21	31	Sano and Wakita (1985)
1206	34.42	135.51	Gas	10.40	±	0.30	780	Nagao et al. (1981)
1207	34.42	135.51	Gas	8.90	±	0.24	610	Wakita et al. (1987)
1208	34.42	135.58	Gas	7.31		0.22		Matsumoto et al. (2003)
1209	34.41	135.54	Water	6.96	±	0.06	4.32	Morikawa et al. (2008b)
1210	34.41	135.68	Water	7.00				AIST (2004)
1211	34.40	135.91	Gas	2.42	±	0.11	6.3	Sano and Wakita (1985)
1212	34.40	135.60	Water	6.70	±	0.04	14.2	Morikawa et al. (2008b)
1213	34.39	135.88	Water	6.16				AIST (2004)
1214	34.39	135.60	Water	2.10				AIST (2004)
1215	34.38	136.22	Gas	5.48	±	0.14	0.86	Umeda et al. (2007c)
1216	34.37	135.41	Water	3.08				AIST (2004)
1217	34.37	135.45	Water	1.35	±	0.02	46.1	Morikawa et al. (2008b)
1218	34.37	135.45	Water	1.92	±	0.02	246	Morikawa et al. (2008b)
1219	34.37	135.41	Water	3.03	±	0.05	2.12	Morikawa et al. (2008b)
1220	34.37	132.13	Water	1.45			0.25	Sano et al. (2006)
1221	34.37	135.86	Gas	6.44	±	0.80	12	Nagao et al. (1981)
1222	34.37	135.60	Water	2.12	±	0.05	1.48	Morikawa et al. (2008b)
1223	34.37	135.87	Gas	3.35		0.07		Matsumoto et al. (2003)
1224	34.35	131.25	Water	1.59			0.27	Sano et al. (2006)
1225	34.35	132.07	Water	1.99			4.47	Sano et al. (2006)
1226	34.34	135.38	Water	0.63				AIST (2004)
1227	34.34	135.38	Water	0.63	±	0.04	4.84	Morikawa et al. (2008b)
1228	34.32	139.21	Gas	8.60	±	0.08	86.207	Marty et al. (1989)
1229	34.31	135.58	Water	2.52				AIST (2004)
1230	34.29	135.75	Water	1.40				AIST (2004)
1231	34.28	131.10	Water	4.18			3.17	Sano et al. (2006)
1232	34.27	135.37	Water	1.02				AIST (2004)
1233	34.27	135.43	Water	1.54				AIST (2004)
1234	34.27	135.07	Water	1.68				AIST (2004)
1235	34.27	131.07	Water	2.23			16.4	Sano et al. (2006)
1236	34.27	135.75	Water	0.54	±	0.02	5.5	Umeda et al. (2007c)
1237	34.27	135.13	Water	0.41	±	0.02	5.9	Umeda et al. (2007c)
1238	34.27	135.88	Water	4.06				AIST (2004)
1239	34.24	135.18	Water	2.38				AIST (2004)
1240	34.24	135.49	Water	1.34				AIST (2004)
1241	34.23	136.31	Water	4.79	±	0.11	5.1	Umeda et al. (2006b)

No.	Latitude	Longitude	Feature	$^3He/^4He(x10^{-6})$	±	1σ	$^4He/^{20}Ne$	Ref.
1242	34.23	135.21	Water	2.38				AIST (2004)
1243	34.23	131.17	Water	4.41			1.64	Sano et al. (2006)
1244	34.20	135.23	Gas	5.25	±	0.12	110	Sano and Wakita (1985)
1245	34.20	133.82	Gas	3.10	±	0.10	618	Umeda et al. (2006b)
1246	34.19	135.16	Water	3.64				AIST (2004)
1247	34.18	134.73	Gas	1.09	±	0.11		Matsumoto et al. (2003)
1248	34.18	133.82	Gas	2.91	±	0.06		Matsumoto et al. (2003)
1249	34.18	136.03	Water	3.22				AIST (2004)
1250	34.17	135.59	Water	5.88				AIST (2004)
1251	34.15	135.75	Gas	5.17		0.22		Matsumoto et al. (2003)
1252	34.15	134.58	Gas	1.53			251	Sano et al. (2006)
1253	34.15	131.45	Water	1.12			0.48	Sano et al. (2006)
1254	34.14	134.09	Gas	0.94	±	0.04	5	Umeda et al. (2006b)
1255	34.13	134.08	Gas	1.12	±	0.06		Matsumoto et al. (2003)
1256	34.13	130.93	Water	3.91			2.9	Sano et al. (2006)
1257	34.12	134.67	Gas	1.16			36	Sano et al. (2006)
1258	34.12	136.00	Gas	3.53	±	0.06		Matsumoto et al. (2003)
1259	34.10	131.10	Water	3.04			3.89	Sano et al. (2006)
1260	34.08	135.43	Water	1.68				AIST (2004)
1261	34.08	135.20	Water	1.40				AIST (2004)
1262	34.08	131.68	Water	2.80			37.7	Sano et al. (2006)
1263	34.05	135.96	Water	6.25	±	0.17	2.9	Umeda et al. (2006a)
1264	34.03	131.95	Water	2.33			2.48	Sano et al. (2006)
1265	34.03	131.27	Water	2.55			3.0	Sano et al. (2006)
1266	34.00	135.81	Water	4.20				AIST (2004)
1267	33.98	135.80	Gas	4.74	±	0.12	3.9	Umeda et al. (2007c)
1268	33.98	132.94	Water	0.53		0.01	1.12	Doğan et al. (2006)
1269	33.97	134.36	Gas	1.38	±	0.04	5.9	Wakita et al. (1987)
1270	33.97	135.30	Water	1.34				AIST (2004)
1271	33.96	133.83	Gas	0.44	±	0.02	28	Umeda et al. (2006b)
1272	33.96	135.56	Water	2.00	±	0.04	5.4	Umeda et al. (2007c)
1273	33.96	135.56	Water	1.68				AIST (2004)
1274	33.95	132.18	Water	1.23			0.30	Sano et al. (2006)
1275	33.94	135.94	Gas	3.08	±	0.07	63.9	Umeda et al. (2006a)
1276	33.93	135.73	Gas	3.80			93	Sakata et al. (1987)
1277	33.93	135.73	Water	3.64				AIST (2004)
1278	33.93	133.02	Water	1.13	±	0.08	0.344	Doğan et al. (2006)
1279	33.92	133.82	Water	1.08	±	0.04	47.0	Doğan et al. (2006)
1280	33.92	133.82	Gas	0.66	±	0.03		Matsumoto et al. (2003)
1281	33.90	133.31	Gas	3.47	±	0.13	1.02	Doğan et al. (2006)
1282	33.90	133.31	Gas	3.10	±	4.14	830	Wakita et al. (1987)
1283	33.89	133.82	Gas	0.80	±	0.03	40	Umeda et al. (2006b)
1284	33.89	133.16	Gas	4.98	±	0.07	83.1	Doğan et al. (2006)

Table 1. (Continued)

No.	Latitude	Longitude	Feature	$^{3}He/^{4}He(x10^{-6})$	±	1σ	$^{4}He/^{20}Ne$	Ref.
1285	33.89	133.16	Water	2.77	±	0.04	47.9	Doğan et al. (2006)
1286	33.89	133.16	Gas	4.13	±	0.13	43.5	Doğan et al. (2006)
1287	33.89	133.16	Gas	4.30	±	0.17	39.6	Doğan et al. (2006)
1288	33.89	133.16	Gas	2.36	±	0.02	17	Wakita et al. (1987)
1289	33.87	135.90	Water	1.59	±	0.06	2.6	Umeda et al. (2006a)
1290	33.87	132.83	Water	1.83	±	0.04	13.4	Doğan et al. (2006)
1291	33.87	132.83	Water	1.85	±	0.04	3.30	Doğan et al. (2006)
1292	33.87	132.83	Gas	1.77	±	0.07	14	Wakita et al. (1987)
1293	33.87	132.83	Gas	1.78	±	0.03	16	Wakita et al. (1987)
1294	33.87	132.83	Gas	1.77	±	0.03	10	Wakita et al. (1987)
1295	33.86	133.04	Water	1.44	±	0.03	6.42	Doğan et al. (2006)
1296	33.84	135.35	Water	5.60				AIST (2004)
1297	33.83	135.76	Gas	6.68			66.0	NEDO (1994)
1298	33.83	135.80	Water	2.38				AIST (2004)
1299	33.82	135.75	Gas	6.56	±	0.28	63	Nagao et al. (1981)
1300	33.82	135.75	Gas	5.88	±	0.09	34	Sano and Wakita (1985)
1301	33.82	135.76	Gas	5.84	±	0.08	59	Sano and Wakita (1985)
1302	33.82	135.75	Gas	5.68	±	0.35		Matsumoto et al. (2003)
1303	33.82	135.77	Gas	6.33	±	0.29		Matsumoto et al. (2003)
1304	33.81	135.77	Gas	5.65	±	0.37	1.0	Nagao et al. (1981)
1305	33.81	133.62	Water	1.76	±	0.59	3.42	Doğan et al. (2006)
1306	33.80	135.88	Gas	4.42	±	0.06		Matsumoto et al. (2003)
1307	33.80	135.87	Gas	4.46	±	0.10	48.7	Umeda et al. (2006a)
1308	33.78	133.31	Water	2.93	±	0.04	41.7	Doğan et al. (2006)
1309	33.76	132.80	Gas	2.02	±	0.07	84	Umeda et al. (2006b)
1310	33.75	135.32	Gas	1.95	±	0.05	0.26	Sano and Wakita (1985)
1311	33.75	132.80	Water	2.21	±	0.04	3.63	Doğan et al. (2006)
1312	33.74	135.58	Water	1.14	±	0.07	4.0	Umeda et al. (2006a)
1313	33.74	135.58	Water	1.20				AIST (2004)
1314	33.73	135.51	Gas	2.12	±	0.05	0.37	Sano and Wakita (1985)
1315	33.73	133.11	Gas	2.73	±	0.04	7.16	Doğan et al. (2006)
1316	33.73	133.11	Gas	2.94	±	0.21	155.8	Doğan et al. (2006)
1317	33.73	135.91	Water	3.47	±	0.05	26.4	Umeda et al. (2006a)
1318	33.72	132.78	Gas	2.55	±	0.18	65.3	Doğan et al. (2006)
1319	33.72	132.78	Gas	3.44	±	0.28	46.9	Doğan et al. (2006)
1320	33.72	132.78	Gas	3.35	±	0.20	13.4	Doğan et al. (2006)
1321	33.72	132.78	Gas	4.54	±	0.14	14.5	Doğan et al. (2006)
1322	33.69	135.43	Water	4.48				AIST (2004)
1323	33.67	135.35	Gas	6.80	±	0.14		Matsumoto et al. (2003)
1324	33.66	135.98	Gas	3.35	±	0.04	8.5	Umeda et al. (2006a)
1325	33.65	133.67	Water	2.00	±	0.20	0.8	Doğan et al. (2006)

No.	Latitude	Longitude	Feature	$^3He/^4He(x10^{-6})$	±	1σ	$^4He/^{20}Ne$	Ref.
1326	33.64	132.71	Gas	3.14	±	0.04	184	Doğan et al. (2006)
1327	33.62	135.92	Gas	3.82	±	0.11	1.9	Umeda et al. (2007c)
1328	33.59	133.50	Water	1.75	±	0.24	0.6	Doğan et al. (2006)
1329	33.57	134.06	Water	2.37	±	0.24	1.1	Doğan et al. (2006)
1330	33.57	134.06	Gas	1.45	±	0.03	2.2	Wakita et al. (1987)
1331	33.56	133.13	Water	0.90	±	0.04	44.5	Doğan et al. (2006)
1332	33.56	134.31	Water	1.05	±	0.05	9.8	Umeda et al. (2006b)
1333	33.54	132.66	Water	0.24	±	0.01	3.67	Doğan et al. (2006)
1334	33.54	135.49	Water	1.82				AIST (2004)
1335	33.54	135.79	Gas	3.13	±	0.05	6.2	Umeda et al. (2006a)
1336	33.54	135.80	Water	2.66				AIST (2004)
1337	33.53	133.42	Water	0.76	±	0.08	9.9	Doğan et al. (2006)
1338	33.49	132.66	Water	0.63	±	0.03	0.957	Doğan et al. (2006)
1339	33.48	135.78	Gas	3.01	±	0.07		Matsumoto et al. (2003)
1340	33.48	132.67	Water	0.29	±	0.01	12	Umeda et al. (2006b)
1341	33.47	135.79	Water	3.36				AIST (2004)
1342	33.44	133.28	Water	1.29	±	0.31	0.8	Doğan et al. (2006)
1343	33.39	133.01	Water	0.24	±	0.01	7.11	Doğan et al. (2006)
1344	33.37	132.80	Water	1.83	±	0.06	5.2	Umeda et al. (2006b)
1345	33.37	132.80	Water	1.99	±	0.03	14.4	Doğan et al. (2006)
1346	33.33	131.49	Gas	7.79	±	0.12	3.1	NEDO (1990c)
1347	33.33	131.46	Gas	9.41	±	0.25	55	NEDO (1990c)
1348	33.33	133.07	Water	0.35	±	0.01	1.75	Doğan et al. (2006)
1349	33.32	131.43	Gas	9.14	±	0.21	43	NEDO (1990c)
1350	33.32	131.48	Gas	6.46	±	0.32	1.5	NEDO (1990c)
1351	33.32	131.45	Water	6.72			1.57	Sturchio et al. (1996)
1352	33.32	131.46	Gas	8.41	±	0.25	58.824	Marty et al. (1989)
1353	33.32	131.46	Gas	9.00	±	0.47	75	Nagao et al. (1981)
1354	33.32	131.49	Gas	6.17	±	0.10	0.98	NEDO (1990c)
1355	33.32	131.50	Gas	3.83	±	0.05	0.41	NEDO (1990c)
1356	33.32	131.48	Gas	5.79	±	0.21	1.4	NEDO (1990c)
1357	33.32	131.47	Gas	7.51	±	0.22	12	NEDO (1990c)
1358	33.32	131.47	Gas	8.57	±	0.23	3.8	NEDO (1990c)
1359	33.32	131.48	Gas	8.03	±	0.20	7.9	NEDO (1990c)
1360	33.31	131.48	Gas	6.84	±	0.25	0.86	NEDO (1990c)
1361	33.31	131.48	Gas	6.69	±	0.20	1.2	NEDO (1990c)
1362	33.31	131.48	Gas	7.73	±	0.20	2.6	NEDO (1990c)
1363	33.31	131.42	Water	8.50			4.81	Sturchio et al. (1996)
1364	33.31	131.47	Gas	8.14	±	0.20	9.3	NEDO (1990c)
1365	33.31	131.47	Gas	5.76	±	0.15	0.77	NEDO (1990c)
1366	33.31	131.47	Gas	6.93	±	0.11	0.95	NEDO (1990c)
1367	33.31	131.46	Water	8.44			216	Sturchio et al. (1996)
1368	33.31	131.47	Gas	7.59	±	0.28	1.7	NEDO (1990c)
1369	33.31	131.46	Gas	9.21	±	0.26	14	NEDO (1990c)
1370	33.31	131.47	Gas	9.96	±	0.27	42	NEDO (1990c)
1371	33.31	131.47	Gas	9.03	±	0.23	19	NEDO (1990c)

Table 1. (Continued)

No.	Latitude	Longitude	Feature	$^3He/^4He(x10^{-6})$	±	1σ	$^4He/^{20}Ne$	Ref.
1372	33.31	130.37	Water	2.80			2.590	Mahara and Kitaoaka (2009)
1373	33.31	131.49	Gas	9.23	±	0.19	1.9	NEDO (1990c)
1374	33.30	134.13	Gas	2.32	±	0.06	7.9	Wakita et al. (1987)
1375	33.30	130.29	Water	5.01			5.601	Mahara and Kitaoaka (2009)
1376	33.30	131.48	Gas	8.98	±	0.15	5.5	NEDO (1990c)
1377	33.30	131.29	Gas	9.94	±	0.28	15	NEDO (1990c)
1378	33.29	131.47	Gas	7.55	±	0.14	4.3	NEDO (1990c)
1379	33.29	130.20	Water	1.55			0.267	Mahara and Kitaoaka (2009)
1380	33.29	131.43	Gas	6.48			1.0	NEDO (1990c)
1381	33.29	131.47	Water	8.95			4.64	Sturchio et al. (1996)
1382	33.29	132.94	Water	1.51	±	0.20	1.5	Doğan et al. (2006)
1383	33.29	131.46	Gas	9.76	±	0.16	92	NEDO (1990c)
1384	33.29	131.48	Gas	8.67	±	0.24	7.8	NEDO (1990c)
1385	33.29	131.50	Gas	2.53	±	0.12	0.35	NEDO (1990c)
1386	33.29	131.46	Water	8.78			100	Sturchio et al. (1996)
1387	33.29	131.48	Gas	6.74	±	0.19	1.7	NEDO (1990c)
1388	33.28	131.49	Gas	9.00	±	0.17	2.5	NEDO (1990c)
1389	33.28	131.49	Gas	8.04	±	0.16	4.4	NEDO (1990c)
1390	33.28	131.48	Gas	9.35	±	0.33	10	NEDO (1990c)
1391	33.28	131.49	Gas	9.39	±	0.32	54	NEDO (1990c)
1392	33.28	131.37	Gas	8.99	±	0.21	24	NEDO (1990c)
1393	33.27	130.29	Water	3.23			2.669	Mahara and Kitaoaka (2009)
1394	33.27	130.27	Water	3.36			1.185	Mahara and Kitaoaka (2009)
1395	33.27	130.30	Water	3.16			1.147	Mahara and Kitaoaka (2009)
1396	33.27	131.37	Water	3.92			0.451	Sturchio et al. (1996)
1397	33.27	130.19	Water	1.69			0.471	Mahara and Kitaoaka (2009)
1398	33.26	131.38	Gas	7.76	±	0.26	17	NEDO (1990c)
1399	33.26	131.36	Gas	8.98	±	0.14	5.9	NEDO (1990c)
1400	33.26	131.36	Water	8.99			17.1	Sturchio et al. (1996)
1401	33.25	131.36	Gas	9.32	±	0.32	3.2	NEDO (1990c)
1402	33.24	131.27	Gas	9.25	±	0.20	6.6	NEDO (1990c)
1403	33.24	132.72	Gas	2.65	±	0.09	4.2	Umeda et al. (2006b)
1404	33.23	130.33	Water	3.95			2.800	Mahara and Kitaoaka (2009)
1405	33.23	132.71	Water	2.41	±	0.04	1.86	Doğan et al. (2006)
1406	33.22	132.98	Water	1.27	±	0.15	0.397	Doğan et al. (2006)
1407	33.22	132.62	Gas	4.05	±	0.06	56.7	Doğan et al. (2006)
1408	33.21	130.10	Water	2.16			115.135	Mahara and Kitaoaka (2009)
1409	33.21	131.49	Gas	7.96	±	0.29	2.1	NEDO (1990c)
1410	33.21	131.51	Gas	8.72	±	0.22	20	NEDO (1990c)
1411	33.21	130.29	Water	3.00			1.229	Mahara and Kitaoaka (2009)

No.	Latitude	Longitude	Feature	$^3He/^4He(x10^{-6})$	±	1σ	$^4He/^{20}Ne$	Ref.
1412	33.20	131.51	Gas	4.43	±	0.15	0.69	NEDO (1990c)
1413	33.20	132.85	Water	0.53	±	0.06	1.41	Doğan et al. (2006)
1414	33.20	130.19	Water	2.41			10.718	Mahara and Kitaoaka (2009)
1415	33.20	130.10	Water	1.08			0.341	Mahara and Kitaoaka (2009)
1416	33.19	130.32	Water	3.12			13.111	Mahara and Kitaoaka (2009)
1417	33.19	130.12	Water	1.02			0.341	Mahara and Kitaoaka (2009)
1418	33.17	130.16	Water	0.90			0.540	Mahara and Kitaoaka (2009)
1419	33.17	132.80	Water	0.53	±	0.03	1.22	Doğan et al. (2006)
1420	33.17	132.80	Gas	0.37	±	0.02	1.8	Umeda et al. (2006b)
1421	33.17	130.11	Water	1.17			0.276	Mahara and Kitaoaka (2009)
1422	33.16	130.16	Water	1.97			1.336	Mahara and Kitaoaka (2009)
1423	33.16	131.13	Gas	5.70	±	0.40	0.75	Nagao et al. (1981)
1424	33.15	131.14	Gas	7.86			2.4	NEDO (1990c)
1425	33.12	130.14	Water	2.05			2.231	Mahara and Kitaoaka (2009)
1426	33.10	130.07	Water	1.72			0.335	Mahara and Kitaoaka (2009)
1427	33.09	132.53	Gas	3.48	±	0.12	30	Umeda et al. (2006b)
1428	33.09	132.53	Gas	3.39	±	0.06	31.5	Doğan et al. (2006)
1429	33.09	131.25	Gas	8.16	±	0.20	83.333	Marty et al. (1989)
1430	33.09	131.25	Gas	10.50	±	0.30	64	Nagao et al. (1981)
1431	33.08	130.14	Water	1.45			0.376	Mahara and Kitaoaka (2009)
1432	33.03	133.09	Water	0.35	±	0.03	1.91	Doğan et al. (2006)
1433	32.98	132.94	Water	0.71	±	0.03	19.5	Doğan et al. (2006)
1434	32.96	132.62	Gas	2.23			58.2	Doğan et al. (2006)
1435	32.96	132.62	Water	1.61		0.06	2.85	Doğan et al. (2006)
1436	32.96	132.62	Water	1.44	±	0.04	6.1	Umeda et al. (2006b)
1437	32.95	130.85	Water	1.73				Mahara (1994)
1438	32.95	132.62	Water	1.39	±	0.04	7.51	Doğan et al. (2006)
1439	32.94	130.79	Water	1.71				Mahara (1994)
1440	32.92	130.92	Water	1.50				Mahara (1994)
1441	32.89	131.04	Gas	9.38			130	NEDO (1991c)
1442	32.89	131.04	Gas	8.84			7.1	NEDO (1991c)
1443	32.89	131.04	Gas	8.19			23	NEDO (1991c)
1444	32.88	131.04	Gas	4.70			12.6	NEDO (1995)
1445	32.88	131.03	Gas	6.34	±	0.14	35.714	Marty et al. (1989)
1446	32.88	130.89	Water	2.70				Mahara (1994)
1447	32.87	130.24	Water	3.00	±	0.03	6.17	Morikawa et al. (2008a)
1448	32.87	130.76	Water	2.69				Mahara (1994)
1449	32.87	131.04	Gas	1.40			0.3	NEDO (1991c)
1450	32.86	131.04	Gas	8.50			8.3	NEDO (1991c)
1451	32.86	130.84	Water	1.56				Mahara (1994)
1452	32.86	130.33	Water	4.66	±	0.04	33.5	Morikawa et al. (2008a)

Table 1. (Continued)

No.	Latitude	Longitude	Feature	$^{3}He/^{4}He(x10^{-6})$	±	1σ	$^{4}He/^{20}Ne$	Ref.
1453	32.85	130.82	Water	2.17				Mahara (1994)
1454	32.85	130.76	Water	2.52				Mahara (1994)
1455	32.85	130.26	Water	1.78	±	0.01	0.278	Morikawa et al. (2008a)
1456	32.85	130.72	Water	2.39				Mahara (1994)
1457	32.85	130.34	Water	5.76	±	0.04	23.7	Morikawa et al. (2008a)
1458	32.85	130.30	Water	3.56	±	0.03	0.451	Morikawa et al. (2008a)
1459	32.85	130.31	Water	2.18	±	0.03	0.279	Morikawa et al. (2008a)
1460	32.85	130.72	Water	2.03				Mahara (1994)
1461	32.84	130.28	Water	3.95	±	0.04	0.598	Morikawa et al. (2008a)
1462	32.83	130.28	Water	3.01	±	0.03	0.476	Morikawa et al. (2008a)
1463	32.83	130.84	Water	2.54				Mahara (1994)
1464	32.83	130.33	Water	2.03	±	0.03	0.273	Morikawa et al. (2008a)
1465	32.83	131.10	Water	1.64				Mahara (1994)
1466	32.81	130.75	Water	1.37				Mahara (1994)
1467	32.81	130.71	Water	1.42				Mahara (1994)
1468	32.81	130.28	Water	1.83	±	0.02	0.274	Morikawa et al. (2008a)
1469	32.81	130.34	Water	1.46	±	0.03	0.274	Morikawa et al. (2008a)
1470	32.81	130.34	Water	1.85	±	0.02	0.274	Morikawa et al. (2008a)
1471	32.80	130.29	Water	2.19	±	0.02	0.284	Morikawa et al. (2008a)
1472	32.80	130.19	Water	2.10	±	0.30	0.3	Kita et al. (1993)
1473	32.80	130.36	Water	2.95	±	0.03	0.379	Morikawa et al. (2008a)
1474	32.80	130.18	Water	1.73	±	0.02	0.256	Morikawa et al. (2008a)
1475	32.79	130.74	Water	1.62				Mahara (1994)
1476	32.79	130.35	Water	7.30	±	0.10	1.2	Kita et al. (1993)
1477	32.79	130.34	Water	6.40	±	0.30	1.0	Kita et al. (1993)
1478	32.79	130.35	Water	8.04	±	0.03	1.22	Morikawa et al. (2008a)
1479	32.79	130.14	Water	1.39	±	0.02	0.506	Morikawa et al. (2008a)
1480	32.79	130.78	Water	2.65				Mahara (1994)
1481	32.79	130.35	Water	8.07	±	0.04	1.31	Morikawa et al. (2008a)
1482	32.79	130.23	Water	1.51	±	0.02	0.265	Morikawa et al. (2008a)
1483	32.79	130.35	Water	7.33	±	0.03	0.914	Morikawa et al. (2008a)
1484	32.79	130.35	Water	7.95	±	0.06	1.17	Morikawa et al. (2008a)
1485	32.79	130.29	Water	1.58	±	0.02	0.271	Morikawa et al. (2008a)
1486	32.78	130.21	Water	1.50	±	0.01	0.270	Morikawa et al. (2008a)

No.	Latitude	Longitude	Feature	$^{3}He/^{4}He(x10^{-6})$	±	1σ	$^{4}He/^{20}Ne$	Ref.
1487	32.78	130.22	Water	1.50	±	0.01	0.273	Morikawa et al. (2008a)
1488	32.78	130.75	Water	2.91				Mahara (1994)
1489	32.78	130.37	Water	7.86	±	0.07	1.06	Morikawa et al. (2008a)
1490	32.78	130.37	Water	8.97	±	0.04	2.64	Morikawa et al. (2008a)
1491	32.78	130.37	Water	6.90	±	0.03		Morikawa et al. (2008a)
1492	32.78	130.38	Water	9.28	±	0.08	3.14	Morikawa et al. (2008a)
1493	32.78	130.37	Water	9.16	±	0.05	2.98	Morikawa et al. (2008a)
1494	32.78	130.74	Water	3.00				Mahara (1994)
1495	32.78	130.20	Water	1.73	±	0.02	0.289	Morikawa et al. (2008a)
1496	32.78	130.37	Water	9.51	±	0.07	3.52	Morikawa et al. (2008a)
1497	32.77	130.71	Water	2.30				Mahara (1994)
1498	32.77	130.69	Water	4.40				Mahara (1994)
1499	32.77	130.36	Water	9.90	±	0.08	24.9	Morikawa et al. (2008a)
1500	32.77	130.37	Water	10.17	±	0.06	115	Morikawa et al. (2008a)
1501	32.77	130.84	Water	2.90				Mahara (1994)
1502	32.77	130.79	Water	3.61				Mahara (1994)
1503	32.77	130.36	Gas	9.35			185	Notsu et al. (2001)
1504	32.77	130.36	Gas	9.49			191	Notsu et al. (2001)
1505	32.77	130.36	Gas	9.41			155	Notsu et al. (2001)
1506	32.77	130.36	Gas	9.42			52.3	Notsu et al. (2001)
1507	32.77	130.36	Gas	9.46			174	Notsu et al. (2001)
1508	32.77	130.36	Gas	9.18			150	Notsu et al. (2001)
1509	32.77	130.36	Gas	9.52			210	Notsu et al. (2001)
1510	32.77	130.36	Gas	9.70			203	Notsu et al. (2001)
1511	32.77	130.36	Gas	9.88			178	Notsu et al. (2001)
1512	32.77	130.36	Gas	9.76			183	Notsu et al. (2001)
1513	32.77	130.36	Gas	9.63			164	Notsu et al. (2001)
1514	32.77	130.36	Gas	9.53			107	Notsu et al. (2001)
1515	32.77	130.36	Gas	9.58			150	Notsu et al. (2001)
1516	32.77	130.36	Gas	9.53			176	Notsu et al. (2001)
1517	32.77	130.36	Gas	9.55			197	Notsu et al. (2001)
1518	32.77	130.36	Gas	9.39			210	Notsu et al. (2001)
1519	32.77	130.36	Gas	9.35			164	Notsu et al. (2001)
1520	32.77	130.36	Gas	9.23			180	Notsu et al. (2001)
1521	32.77	130.36	Gas	9.11			76	Notsu et al. (2001)
1522	32.77	130.36	Gas	9.03			158	Notsu et al. (2001)
1523	32.77	130.36	Gas	9.67				Sano and Marty (1995)
1524	32.77	130.37	Water	9.95	±	0.07	2.8	Morikawa et al. (2008a)
1525	32.77	130.22	Water	1.65	±	0.01		Morikawa et al. (2008a)
1526	32.77	130.19	Water	1.97	±	0.03	0.296	Morikawa et al. (2008a)

Table 1. (Continued)

No.	Latitude	Longitude	Feature	$^{3}He/^{4}He(x10^{-6})$	±	1σ	$^{4}He/^{20}Ne$	Ref.
1527	32.76	130.37	Gas	9.90	±	0.50	154	Kita et al. (1993)
1528	32.76	130.37	Gas	10.30	±	0.30	151	Kita et al. (1993)
1529	32.76	130.37	Gas	10.30	±	0.10	180	Kita et al. (1993)
1530	32.76	130.37	Gas	10.10	±	0.30	286	Kita et al. (1993)
1531	32.76	130.37	Gas	9.60	±	0.20	160	Kita et al. (1993)
1532	32.76	130.37	Gas	9.60	±	0.10	186	Kita et al. (1993)
1533	32.76	130.37	Gas	8.11	±	0.14	200.000	Marty et al. (1989)
1534	32.76	130.30	Gas	6.40	±	0.30	0.9	Kita et al. (1993)
1535	32.76	130.30	Gas	9.47	±	0.03	69	Scott et al. (1994)
1536	32.76	130.30	Gas	9.41	±	0.05	12	Scott et al. (1994)
1537	32.76	130.20	Water	1.78	±	0.02	0.284	Morikawa et al. (2008a)
1538	32.76	130.30	Gas	10.20	±	0.10	350	Kita et al. (1993)
1539	32.76	130.81	Water	2.92				Mahara（1994）
1540	32.76	130.36	Water	3.18	±	0.02	0.222	Morikawa et al. (2008a)
1541	32.76	130.36	Water	2.51	±	0.01	0.337	Morikawa et al. (2008a)
1542	32.76	130.36	Water	1.64	±	0.02	0.275	Morikawa et al. (2008a)
1543	32.76	130.78	Water	2.41				Mahara（1994）
1544	32.75	130.30	Gas	9.98				Sano and Marty (1995)
1545	32.75	130.30	Gas	6.57				Sano and Marty (1995)
1546	32.75	130.37	Water	5.00	±	0.04	0.497	Morikawa et al. (2008a)
1547	32.75	130.37	Water	4.95	±	0.03		Morikawa et al. (2008a)
1548	32.75	130.36	Water	4.93	±	0.03		Morikawa et al. (2008a)
1549	32.75	130.37	Water	3.86	±	0.03	0.614	Morikawa et al. (2008a)
1550	32.75	130.36	Water	4.99	±	0.04	0.550	Morikawa et al. (2008a)
1551	32.75	130.26	Water	1.38	±	0.02	0.209	Morikawa et al. (2008a)
1552	32.75	130.23	Water	2.69	±	0.03	0.343	Morikawa et al. (2008a)
1553	32.75	130.36	Water	7.49	±	0.05	1.06	Morikawa et al. (2008a)
1554	32.75	130.26	Water	1.62	±	0.02		Morikawa et al. (2008a)
1555	32.75	130.26	Water	1.71	±	0.02	0.281	Morikawa et al. (2008a)
1556	32.74	130.26	Gas	6.23	±	0.06	197	Morikawa et al. (2008a)
1557	32.74	130.26	Gas	7.20	±	0.40	84	Kita et al. (1993)
1558	32.74	130.26	Gas	7.30	±	0.50	100	Kita et al. (1993)
1559	32.74	130.26	Gas	7.50	±	0.40	215	Kita et al. (1993)
1560	32.74	130.26	Gas	7.50	±	0.20	185	Kita et al. (1993)
1561	32.74	130.26	Gas	7.10	±	0.10	61	Kita et al. (1993)
1562	32.74	130.26	Gas	7.10	±	0.10	129	Kita et al. (1993)

No.	Latitude	Longitude	Feature	$^3He/^4He(x10^{-6})$	±	1σ	$^4He/^{20}Ne$	Ref.
1563	32.74	130.26	Gas	7.10	±	0.40	144	Kita et al. (1993)
1564	32.74	130.26	Gas	7.00	±	0.20	110	Kita et al. (1993)
1565	32.74	130.26	Gas	7.40	±	0.40	128	Kita et al. (1993)
1566	32.74	130.26	Gas	7.40	±	0.20	294	Kita et al. (1993)
1567	32.74	130.26	Gas	7.10	±	0.10	128	Kita et al. (1993)
1568	32.74	130.26	Gas	7.10	±	0.10	150	Kita et al. (1993)
1569	32.74	130.26	Gas	7.30	±	0.30	420	Kita et al. (1993)
1570	32.74	130.26	Gas	7.10	±	0.40	415	Kita et al. (1993)
1571	32.74	130.26	Gas	7.40	±	0.20	646	Kita et al. (1993)
1572	32.74	130.26	Gas	7.00	±	0.10	336	Kita et al. (1993)
1573	32.74	130.26	Gas	7.10	±	0.10	421	Kita et al. (1993)
1574	32.74	130.26	Gas	6.03	±	0.19	116.279	Marty et al. (1989)
1575	32.74	130.26	Gas	7.06			117	Notsu et al. (2001)
1576	32.74	130.26	Gas	7.13			144	Notsu et al. (2001)
1577	32.74	130.26	Gas	4.41			0.714	Notsu et al. (2001)
1578	32.74	130.26	Gas	7.00			119	Notsu et al. (2001)
1579	32.74	130.26	Gas	7.01			147	Notsu et al. (2001)
1580	32.74	130.26	Gas	6.79			161	Notsu et al. (2001)
1581	32.74	130.26	Gas	7.17			166	Notsu et al. (2001)
1582	32.74	130.26	Gas	7.11			129	Notsu et al. (2001)
1583	32.74	130.26	Gas	6.73			4.74	Notsu et al. (2001)
1584	32.74	130.26	Gas	7.32			67.8	Notsu et al. (2001)
1585	32.74	130.26	Gas	6.86			15.1	Notsu et al. (2001)
1586	32.74	130.26	Gas	7.14			110	Notsu et al. (2001)
1587	32.74	130.26	Gas	7.10			94.6	Notsu et al. (2001)
1588	32.74	130.26	Gas	7.18			100	Notsu et al. (2001)
1589	32.74	130.26	Gas	6.96			90.3	Notsu et al. (2001)
1590	32.74	130.26	Gas	6.89			72.8	Notsu et al. (2001)
1591	32.74	130.26	Gas	6.92			58.7	Notsu et al. (2001)
1592	32.74	130.26	Gas	6.82			105	Notsu et al. (2001)
1593	32.74	130.26	Gas	6.68			24	Notsu et al. (2001)
1594	32.74	130.26	Gas	6.62			76.3	Notsu et al. (2001)
1595	32.73	130.37	Water	5.81	±	0.05	0.762	Morikawa et al. (2008a)
1596	32.73	130.26	Gas	6.58	±	0.05	53.8	Morikawa et al. (2008a)
1597	32.73	130.21	Water	2.01	±	0.07	1.89	Morikawa et al. (2008a)
1598	32.73	130.36	Water	1.42	±	0.02	0.295	Morikawa et al. (2008a)
1599	32.73	130.21	Water	2.09	±	0.02	0.299	Morikawa et al. (2008a)
1600	32.73	130.36	Water	6.26	±	0.18	0.654	Morikawa et al. (2008a)
1601	32.73	130.21	Water	2.51	±	0.06	0.618	Morikawa et al. (2008a)
1602	32.73	130.21	Gas	5.73	±	0.04	13.8	Morikawa et al. (2008a)
1603	32.73	130.21	Gas	5.60				Sano and Marty (1995)
1604	32.73	130.21	Water	5.30	±	0.05	3.83	Morikawa et al. (2008a)

Table 1. (Continued)

No.	Latitude	Longitude	Feature	$^3He/^4He(x10^{-6})$	±	1σ	$^4He/^{20}Ne$	Ref.
1605	32.73	130.21	Gas	5.73	±	0.04	106	Morikawa et al. (2008a)
1606	32.73	130.21	Gas	5.70	±	0.04	100	Morikawa et al. (2008a)
1607	32.73	130.21	Water	1.58	±	0.05	1.10	Morikawa et al. (2008a)
1608	32.73	130.36	Water	2.06	±	0.04	0.295	Morikawa et al. (2008a)
1609	32.73	131.35	Water	0.37				Ohsawa et al. (2011)
1610	32.72	130.21	Water	3.13	±	0.03	0.450	Morikawa et al. (2008a)
1611	32.72	130.21	Gas	6.00	±	0.20	90	Kita et al. (1993)
1612	32.72	130.21	Gas	6.00	±	0.30	84	Kita et al. (1993)
1613	32.72	130.21	Gas	5.80	±	0.10	115	Kita et al. (1993)
1614	32.72	130.21	Gas	6.10	±	0.30	107	Kita et al. (1993)
1615	32.72	130.21	Gas	5.40	±	0.10	106	Kita et al. (1993)
1616	32.72	130.21	Gas	5.90	±	0.10	24	Kita et al. (1993)
1617	32.72	130.21	Gas	5.56	±	0.12	90.909	Marty et al. (1989)
1618	32.72	130.21	Gas	5.85			141	Notsu et al. (2001)
1619	32.72	130.21	Gas	5.87			124	Notsu et al. (2001)
1620	32.72	130.21	Gas	5.75			23.3	Notsu et al. (2001)
1621	32.72	130.21	Gas	5.78			112	Notsu et al. (2001)
1622	32.72	130.21	Gas	5.77			92.3	Notsu et al. (2001)
1623	32.72	130.21	Gas	5.87			97.9	Notsu et al. (2001)
1624	32.72	130.21	Gas	5.75			97.9	Notsu et al. (2001)
1625	32.72	130.21	Gas	5.89			104	Notsu et al. (2001)
1626	32.72	130.21	Gas	6.05			93.6	Notsu et al. (2001)
1627	32.72	130.21	Gas	5.84			104	Notsu et al. (2001)
1628	32.72	130.21	Gas	5.67			95.2	Notsu et al. (2001)
1629	32.72	130.21	Gas	5.77			86.8	Notsu et al. (2001)
1630	32.72	130.21	Gas	5.80			84.2	Notsu et al. (2001)
1631	32.72	130.21	Gas	5.89			124	Notsu et al. (2001)
1632	32.72	130.21	Gas	5.80			102	Notsu et al. (2001)
1633	32.72	130.21	Gas	5.73			65	Notsu et al. (2001)
1634	32.72	130.21	Gas	5.88			76.2	Notsu et al. (2001)
1635	32.72	130.21	Gas	5.70			80	Notsu et al. (2001)
1636	32.72	130.21	Gas	5.67			130	Notsu et al. (2001)
1637	32.72	130.22	Water	1.51	±	0.02	0.268	Morikawa et al. (2008a)
1638	32.72	130.35	Water	2.13	±	0.03	0.279	Morikawa et al. (2008a)
1639	32.72	130.20	Water	1.60	±	0.02	0.280	Morikawa et al. (2008a)
1640	32.71	130.31	Water	1.52	±	0.02	0.292	Morikawa et al. (2008a)
1641	32.70	130.33	Water	1.57	±	0.02	0.305	Morikawa et al. (2008a)
1642	32.70	130.32	Water	1.89	±	0.03	0.266	Morikawa et al. (2008a)
1643	32.70	130.35	Water	3.03	±	0.03	0.388	Morikawa et al. (2008a)
1644	32.70	130.35	Water	4.67	±	0.05	12.3	Morikawa et al. (2008a)
1645	32.69	130.20	Water	1.66	±	0.02	0.275	Morikawa et al. (2008a)
1646	32.67	131.39	Gas	0.66	±	0.02	21.8	Umeda et al. (2007a)
1647	32.67	130.73	Water	0.96				Ohsawa et al. (2011)
1648	32.65	130.29	Water	2.05	±	0.03	0.594	Morikawa et al. (2008a)

No.	Latitude	Longitude	Feature	$^3He/^4He(x10^{-6})$	±	1σ	$^4He/^{20}Ne$	Ref.
1649	32.64	130.26	Water	2.72	±	0.02	1.45	Morikawa et al. (2008a)
1650	32.61	130.88	Water	0.37				Ohsawa et al. (2011)
1651	32.61	130.18	Water	3.56	±	0.03	0.927	Morikawa et al. (2008a)
1652	32.44	131.47	Gas	0.95	±	0.02	110	Umeda et al. (2007a)
1653	32.40	131.34	Gas	1.11	±	0.03	68	Umeda et al. (2007a)
1654	32.39	131.33	Water	1.00				Ohsawa et al. (2011)
1655	32.25	130.75	Water	4.92				Ohsawa et al. (2011)
1656	32.23	130.72	Water	1.97				Ohsawa et al. (2011)
1657	32.22	131.16	Water	0.90				Ohsawa et al. (2011)
1658	32.21	130.76	Water	10.20				Ohsawa et al. (2011)
1659	31.97	131.46	Gas	0.41	±	0.02	324	Umeda et al. (2007a)
1660	31.96	130.78	Gas	4.24				Sato et al. (1999)
1661	31.95	130.86	Gas	9.70				Sato et al. (1999)
1662	31.95	130.85	Gas	9.70				Sato et al. (1999)
1663	31.94	130.85	Gas	8.33	±	0.17	37	Sano and Wakita (1985)
1664	31.92	130.89	Gas	10.30	±	0.70	24	Nagao et al. (1981)
1665	31.91	130.89	Gas	9.94				Sato et al. (1999)
1666	31.90	130.88	Gas	8.49	±	0.18	25	Sano and Wakita (1985)
1667	31.90	130.86	Gas	6.69				Sato et al. (1999)
1668	31.90	130.84	Gas	4.96				Sato et al. (1999)
1669	31.89	130.86	Gas	7.73				Sato et al. (1999)
1670	31.89	130.83	Gas	5.12	±	0.21	8.3	Sano and Wakita (1985)
1671	31.89	130.80	Gas	6.47	±	0.24	2.3	Sano and Wakita (1985)
1672	31.89	130.91	Gas	8.22				Sato et al. (1999)
1673	31.84	131.24	Gas	1.09	±	0.03	781	Umeda et al. (2007a)
1674	31.82	130.73	Gas	7.40	±	0.07	4.2	Sano and Wakita (1985)
1675	31.80	131.45	Gas	0.42	±	0.01	637	Umeda et al. (2007a)
1676	31.70	130.78	Gas	3.47	±	0.12	27	Sano and Wakita (1985)
1677	31.67	131.39	Gas	0.29	±	0.01	308	Umeda et al. (2007a)
1678	31.62	131.39	Gas	0.13	±	0.01	709	Umeda et al. (2007a)
1679	31.55	130.63	Gas	6.62	±	0.12	4.098	Marty et al. (1989)
1680	31.45	131.29	Gas	0.44	±	0.01	158	Umeda et al. (2007a)
1681	31.21	130.63	Gas	9.84	±	0.56	7.0	Nagao et al. (1981)
1682	30.80	130.32	Gas	9.21	±	0.10	4.831	Marty et al. (1989)
1683	30.80	130.32	Gas	8.64	±	0.09	4.098	Marty et al. (1989)
1684	30.80	130.30	Gas	3.66	±	0.10	0.529	Marty et al. (1989)
1685	30.79	130.30	Gas	10.20	±	0.60	9.6	Nagao et al. (1981)
1686	30.79	130.30	Gas	10.60	±	0.50	16	Nagao et al. (1981)
1687	30.79	130.31	Gas	9.54	±	0.28	17	Sano and Wakita (1985)
1688	30.79	130.31	Gas	9.91			11.0	Sato et al. (2002)
1689	30.79	130.31	Gas	10.21			8.1	Sato et al. (2002)
1690	30.79	130.31	Gas	10.11			10.1	Sato et al. (2002)
1691	30.79	130.31	Gas	9.86			9.8	Sato et al. (2002)
1692	30.79	130.31	Gas	10.99			21.6	Sato et al. (2002)
1693	30.79	130.31	Gas	9.34			11.6	Sato et al. (2002)
1694	30.79	130.31	Gas	10.57			14.7	Sato et al. (2002)

Table 1. (Continued)

No.	Latitude	Longitude	Feature	$^3He/^4He(x10^{-6})$	±	1σ	$^4He/^{20}Ne$	Ref.
1695	30.79	130.31	Gas	10.47			14.2	Sato et al. (2002)
1696	30.79	130.28		7.32	±	0.23	3.1	Sano and Wakita (1985)
1697	30.77	130.30	Water	1.41	±	0.04	0.45	Sano and Wakita (1985)
1698	26.12	127.66	Gas	0.07	±	0.00	340	Sano and Wakita (1985)
1699	24.81	141.33	Gas	7.53			23.9	Ohsawa and Yusa (2001)
1700	24.81	141.33	Gas	7.81	±	0.39		Sumino et al. (2004)
1701	24.81	141.33	Gas	7.62	±	0.11		Sumino et al. (2004)
1702	24.81	141.33	Gas	7.87	±	0.13		Sumino et al. (2004)
1703	24.80	141.34	Gas	7.63	±	0.38		Sumino et al. (2004)
1704	24.79	141.35	Gas	7.48	±	0.38		Sumino et al. (2004)
1705	24.79	141.35	Gas	7.55	±	0.27		Sumino et al. (2004)
1706	24.79	141.35	Gas	7.25	±	0.50		Sumino et al. (2004)
1707	24.79	141.35	Gas	7.69	±	0.13		Sumino et al. (2004)
1708	24.79	141.34	Gas	6.27	±	0.15	67.568	Marty et al. (1989)
1709	24.79	141.31	Gas	7.25			4.48	Ohsawa and Yusa (2001)
1710	24.79	141.34	Gas	7.73	±	0.42		Sumino et al. (2004)
1711	24.79	141.34	Gas	7.38	±	0.11		Sumino et al. (2004)
1712	24.79	141.34	Gas	7.81	±	0.15		Sumino et al. (2004)
1713	24.79	141.30	Gas	7.01	±	0.35		Sumino et al. (2004)
1714	24.79	141.30	Gas	4.34	±	0.11		Sumino et al. (2004)
1715	24.79	141.30	Gas	7.22	±	0.42		Sumino et al. (2004)
1716	24.79	141.30	Gas	7.24	±	0.14		Sumino et al. (2004)
1717	24.79	141.30	Gas	7.27	±	0.07		Sumino et al. (2004)
1718	24.78	141.32	Gas	6.41	±	0.12	65.359	Marty et al. (1989)
1719	24.78	141.33	Gas	6.93	±	0.35		Sumino et al. (2004)
1720	24.78	141.33	Gas	6.08	±	0.14		Sumino et al. (2004)
1721	24.78	141.33	Gas	7.13	±	0.38		Sumino et al. (2004)
1722	24.78	141.33	Gas	7.07	±	0.08		Sumino et al. (2004)
1723	24.78	141.33	Gas	7.36	±	0.14		Sumino et al. (2004)
1724	24.75	141.29	Gas	1.90	±	0.10		Sumino et al. (2004)
1725	24.75	141.29	Gas	1.72	±	0.07		Sumino et al. (2004)
1726	24.75	141.29	Gas	1.48	±	0.20		Sumino et al. (2004)
1727	24.75	141.29	Gas	2.04	±	0.04		Sumino et al. (2004)
1728	24.75	141.29	Gas	1.85	±	0.07		Sumino et al. (2004)

The Japanese Islands are located in a tectonically complex zone in the north-central western Pacific where four plates, the Pacific, Eurasian/Amurian, North American/Okhotsk, and the Philippine Sea plates, converge with one another. They are composed of two major tectonic blocks, Northeast Japan and Southwest Japan separated by the Itoigawa-Shizuoka tectonic line (ISTL). NE Japan is considered to be compressed along west-northwest direction by the convergence of the Pacific plate at a convergence rate of 60-80 mm/yr along the Japan Trench. The oceanic Philippine Sea plate (PHS) is descending beneath the SW Japan at a convergence rate of 20-60 mm/yr in a west-northwest direction along the Nankai Trough (e.g., Seno et al., 1993; Heki and Miyazaki, 2001) (Figure 1). NE Japan is a well-defined island arc, characterized by a deep trench, frontal and rear arcs (Kaizuka, 1972). Geophysical data vary systematically across this island arc system (Yoshii, 1977). In contrast, the volcanic front in SW Japan is not as clear as in NE Japan, and terrestrial heat flow values are relatively high in the trench region (Yamano et al., 1984).

The observed $^3He/^4He$ and $^4He/^{20}Ne$ ratios are correlated, indicating that all samples are from a mixing region with three end members: primordial helium, radiogenic helium and atmospheric helium dissolved in water (air saturated water; ASW). Therefore, it is possible to correct atmospheric helium contamination as follows (Craig et al., 1978):

$$RA_{cor} = (RA_{obs} - r)/(1 - r) \tag{1}$$

$$r = (^4He/^{20}Ne)_{ASW}/(^4He/^{20}Ne)_{obs}, \tag{2}$$

where RA_{cor} and RA_{obs} denote the corrected and observed $^3He/^4He$ ratios normalized to the atmospheric $^3He/^4He$ ratio of 1.4×10^{-6}, and $(^4He/^{20}Ne)_{ASW}$ and $(^4He/^{20}Ne)_{obs}$ are the ASW (0.254 at 15 °C) and observed $^4He/^{20}Ne$ ratios, respectively. Figure 2 shows the spatial distribution of the $^3He/^4He$ ratios corrected for air contamination throughout the Japanese Islands. Samples collected near volcanic fronts in NE Japan have $^3He/^4He$ ratios similar to the mean $^3He/^4He$ ratio of arc-related volcanism of 5.4±1.9 RA (Hilton et al., 2002). It has been recognized that a general trend of abrupt decrease in the $^3He/^4He$ ratios occurs in the fore-arc regions where the observed values are lower than the atmospheric value because crustal and atmospheric helium components are dominant without magmatic helium input (e.g., Sano and Wakita, 1985; Giggenbach et al., 1993). This trend has also been observed in well gases in NE Japan (Sano and Wakita, 1985), New Zealand (Giggenbach et al., 1993) and British Columbia (Clark and Phillips, 2000).

In the case of SW Japan, the high $^3He/^4He$ ratios observed near Quaternary volcanoes in the Chugoku region and on Kyusyu Island indicate a significant contribution of primordial ^3He carried by ascending magmas from the upper mantle in the present-day subduction system. However, there are several regions, such as Tottori City, around Osaka Bay, laterally away from active volcanoes where samples also have $^3He/^4He$ ratios in the range of 5.4 RA (Figure 2b). In the fore-arc side of Kyushu Island, $^3He/^4He$ ratios in gas samples are significantly lower than the atmospheric value, similar to other fore-arcs. In contrast, there is a significant proportion of mantle-derived helium in Shikoku Island and Kii Peninsula gas samples, which are different than other fore-arcs. It should be noted that a relatively older, cold lithosphere (Kyushu-Palau Ridge) is subducting beneath southern Kyushu, compared to the younger, warm oceanic lithosphere (Shikoku Basin) subducting below Shikoku Island and the Kii Peninsula.

Hyndman and Peacock (2003) indicated that large amounts of aqueous fluids could be released upward from dehydration reactions in a subducted slab below the fore-arc mantle wedge for a warm subduction zone. Actually, non-volcanic low-frequency tremors (LFTs), which lack sharp onsets and last from a few minutes to a few days with a predominant frequency of 1 to 10 Hz (Obara, 2002), occur in a E-W trending zone beneath Shikoku Island and the Kii Peninsula, but not beneath Kyushu Island.

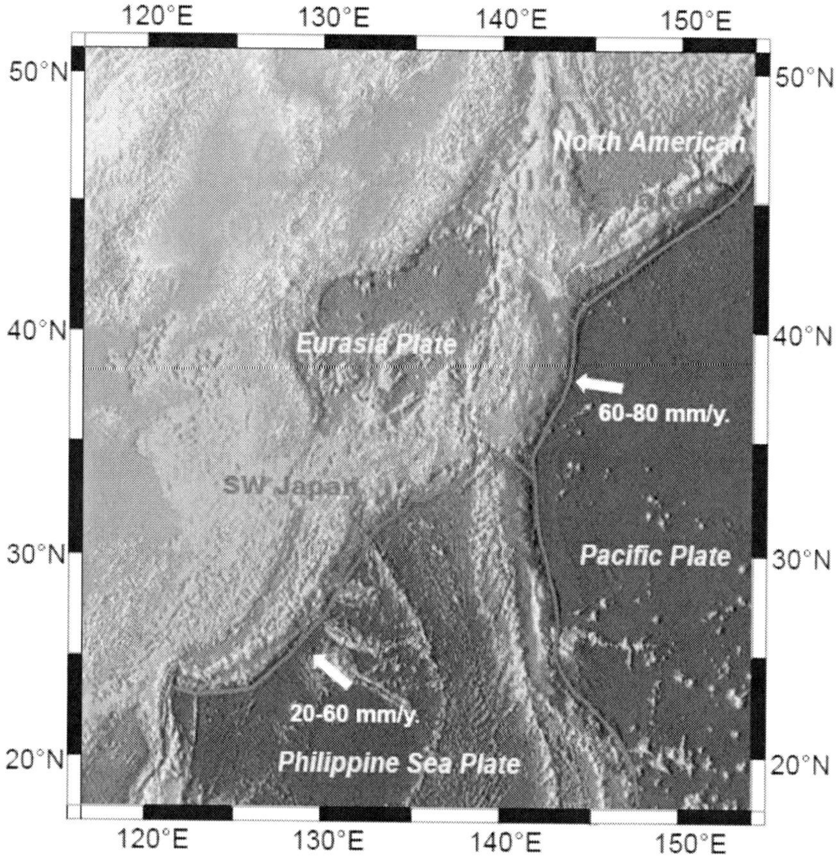

Figure 1. Tectonic setting in and around Japan. The solid lines indicate plate boundaries.

The LFTs are believed to be due to fracture dislocations in the presence of aqueous fluids generated by dehydration of the subducting PHS under high temperatures and pressures at a depth of around 30 km (e.g., Obara and Hirose, 2006). Therefore, the transport of mantle helium from the serpentinized fore-arc mantle is controlled by the higher advection process rather than by diffusion process, resulting in elevated $^3He/^4He$ ratios observed on Shikoku Island and the Kii Peninsula (Umeda et al., 2007). In the following sections, we review two manifestations of mantle and crustal helium discovered in seismically active regions.

3. The 2004 Mid-Niigata Prefecture Earthquake and the 2007 Niigataken Chuetsu-Oki Earthquake

Two shallow M_W 6.6 inland earthquakes (the mid-Niigata Prefecture Earthquake on October 23, 2004, and the Niigataken Chuetsu-oki Earthquake on July 16, 2007) occurred on the east side of the Itoigawa-Shizuoka Tectonic Line, Central Japan, where the Niigata sedimentary basin is located. The Niigata sedimentary basin locally contains a 6 or more km thick Neogene succession overlying a Paleozoic to Mesozoic basement (e.g., Takano et al., 2005).

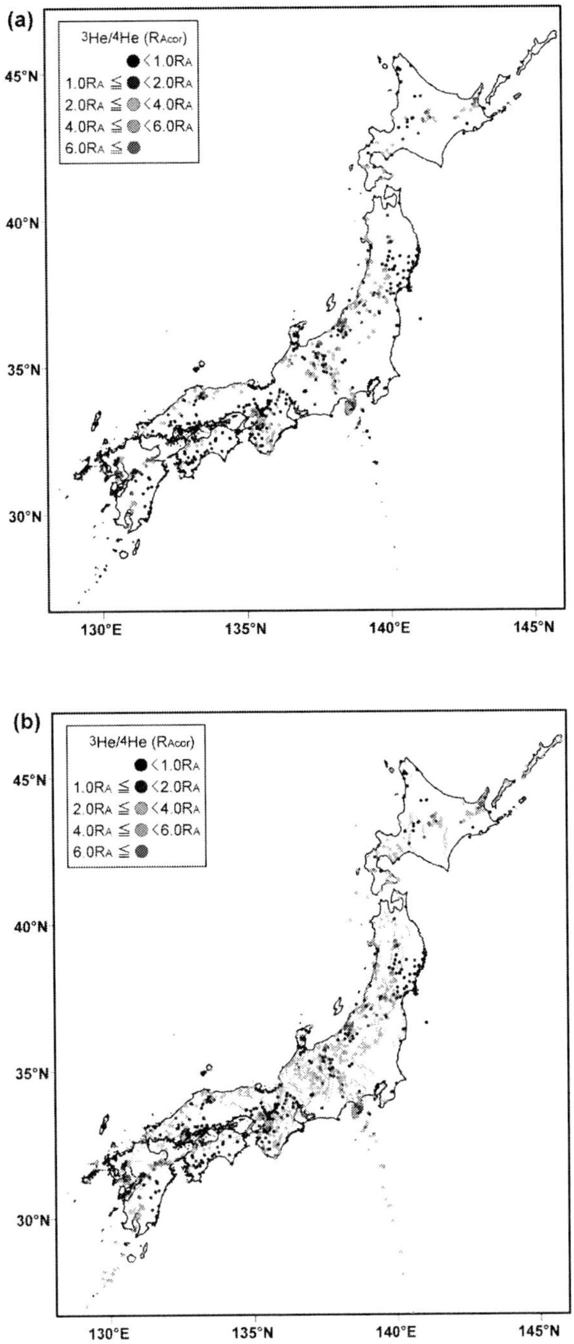

Figure 2. Geographic distribution of the air-corrected $^3He/^4He$ ratios in Japan (Figure 2a). Also shown in Figure 2b are Quaternary volcanoes (orange triangles) and active faults (gray lines).

The focal mechanisms for the mainshocks of both earthquakes were estimated to be west-dipping, high angle (dipping 50-60°W) reverse faults with a strike of approximately N30°E (e.g., Hirata et al., 2005; Kato et al., 2008). However, the high angle of these faults is less favorable for thrust-type aftershock events in the direction of the regional maximum compression which is almost horizontal to the WNW-ESE. Indeed, aftershocks within the sequence involved reverse slip on low angle (approximately 30°E) thrusts well-oriented for reactivation of WNW-ESE shortening (e.g., Sibson et al., 2007; Kato et al., 2008).

The spatial distribution of $^3He/^4He$ ratios in and around the source regions of the 2004 and 2007 earthquakes is shown in Figure 3. It should be noted that hot spring and natural gases with $^3He/^4He$ ratios about four times higher than the atmospheric value occur in the source regions where mainshock and aftershock epicenters were concentrated, and in an area between two regions of epicenter clusters along 37.5°N latitude. There appears to be a sharp decrease in $^3He/^4He$ ratios laterally away from the regions. Although temporal changes in the $^3He/^4He$ ratios in the samples from each well were unknown before and after the earthquakes, higher $^3He/^4He$ ratios were observed in the seismic source regions compared to other regions. In addition, high $^3He/^4He$ values are found in hot springs around active faults located on the eastern edge of the Niigata sedimentary basin.

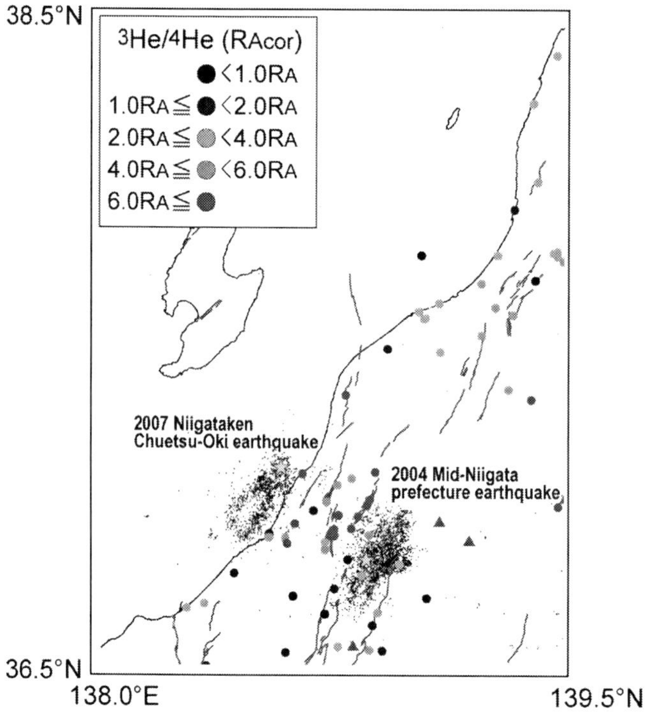

Figure 3. Geographic distribution of the air-corrected $^3He/^4He$ ratios in source regions of 2004 Mid-Niigata prefecture earthquake, and the 2007 Niigataken Chuetsu-Oki earthquake. Yellow stars and small black dots denote the epicenters of the mainshock and aftershocks (7 days after each mainshock), respectively.

Generally, $^3He/^4He$ ratios in either natural or hot spring gases that are lower than the atmospheric value have been interpreted to result from the accumulation of radiogenic ^4He derived from decay of U and Th in crustal rocks during geologic time (e.g., Sano et al., 1982). However, the CH_4 rich gases sampled in the Niigata sedimentary basin are characterized by

$^3He/^4He$ ratios several times higher than the atmospheric ratio. Wakita and Sano (1983) and Wakita et al. (1990) suggested that high helium isotopic values in natural gases can be attributed to an admixture of a biogenic CH_4 rich gas with low $^3He/^4He$ ratios, and a CH_4 rich gas with high $^3He/^4He$ ratios. It is thought that the latter gas is formed by inorganic reactions of CO_2 and H_2 in the volcaniclastics derived from magmatic activities in the Middle Miocene. The $^3He/^4He$ ratio for the ancient magma of Middle Miocene age is estimated to be 3.8 RA at the present time (Umeda et al., 2009).

We note that the $^3He/^4He$ ratios measured around the earthquake source regions are much higher than this value. Therefore, an additional ^3He source is required to explain the $^3He/^4He$ ratios in excess of 3.8 RA.

The results of many seismological and geophysical studies indicate that the 2004 mid-Niigata Prefecture Earthquake and the 2007 Niigataken Chuetsu-oki Earthquake were generated by reverse slip on NE-SW trending faults. Seismic tomography studies defined the detailed seismic velocity structure and the Vp/Vs ratio structure beneath the fault system that caused the 2004 mid-Niigata Prefecture Earthquake (e.g., Wang and Zhao, 2006). These results demonstrate that a zone with low seismic velocity and a Poisson's ratio anomaly is clearly imaged in the lower crust and the uppermost mantle beneath the source region. A conductive zone below the source region of the 2004 mid-Niigata Prefecture Earthquake was also imaged in the lower crust and the uppermost mantle (Ogawa and Honkura, 2004; Uyeshima et al., 2005). A heterogeneous distribution of anomalously low P- and S-wave velocities, and high electrical conductivity below the earthquake source region suggests the presence of aqueous fluids which are released due to dehydration of the subducting Pacific slab (Uyeshima et al., 2005; Wang and Zhao, 2006). This scenario is consistent with the hypothesis of a water-weakened lower crust and/or upper mantle beneath the NKTZ (Iio et al., 2002; Yamasaki and Seno, 2005). In addition, recent seismic tomography studies, also inferred the existence of a low velocity zone beneath the earthquake source region of the 2007 Niigataken Chuetsu-oki Earthquake from travel-time tomography (Nakajima and Hasegawa, 2008).

Hot springs in seismically active zones may have significantly higher $^3He/^4He$ ratios compared to non-active zones (e.g., Güleç et al., 2002). Active faults associated with extensional tectonics play an important role in the effective transfer of mantle fluid from the subcrustal lithosphere to the surface in non-volcanic regions (e.g., Kulongoski et al., 2005; Kennedy and van Soest, 2007). However, Kennedy et al. (1997) reported excess ^3He in groundwater from seismically active areas undergoing compression. Therefore, it is necessary to examine the possibility that reverse faulting in a compressional tectonic regime facilitates the transfer of mantle helium from the lower crust (below the seismogenic zone) to the Earth's surface.

As suggested above, geophysical findings indicate that the crust is overpressurized beneath the 2004 and 2007 earthquake source regions due to the upwelling of aqueous fluids produced by dehydration of the subducting Pacific slab. It is likely that aqueous fluids at high-temperature and high pressure (such as a supercritical fluid) contain volatiles with high $^3He/^4He$ values derived from the mantle wedge through which the fluids passed after being released from the subducted slab (Bureau and Keppler, 1999). Subsequently, aqueous fluids with high $^3He/^4He$ ratios may have lead to reactivation of the high-angle inverted normal faults and triggered the 2004 mid-Niigata Prefecture and the 2007 Niigataken Chuetsu-oki Earthquakes, while rising to the Earth's surface along permeable conduits. Near the Earth's surface, these fluids then mixed with geothermal fluids and natural gases, but without significant dilution by the crustal radiogenic helium, resulting in the emanation of high ^{3}He natural gases and hot spring gases in the source regions of the 2004 and 2007 earthquakes.

4. The 2011 Northern and Central Nagano Earthquakes

After the 2011 M_w 9.0 off the Pacific coast of Tohoku earthquake (hereafter referred to as "Tohoku-Oki earthquake"), increasd earthquake activity occurred within the inland crust, together with ordinary aftershocks. Several M ~6.0 shallow inland earthquakes struck as far as ~ 300 km from the source region of the mainshock (e.g., Hirose et al., 2011; Toda et al., 2011). This indicates that the static stress change caused by the fault slip of the earthquake considerably affected the seismicity pattern in inland areas. On the following day, March 12, 2011, northern Nagano was shook by the Mj 6.7 major earthquake near the Tokamachi fault (Figure 4a).

The epicenter was located at 36.985°N, 138.597°E, with a focal depth of 8 km. The focal mechanism of the mainshock indicates a reverse-fault type with a NE-SE compression axis, similar to previous earthquakes that occurred along the Tokamachi fault. On June 30, 2011, in Central Japan, the M_j 5.4 central Nagano earthquake occurred at a shallow depth of 5 km, central Japan. The epicenter was located at 36.188°N, 137.953°E, with a focal depth of 5 km. The focal mechanism, distribution of aftershocks and GPS observations with respect to the earthquake indicate WNW-ESE trending strike slip faulting. To the east of these events is the Gofukuji fault which forms part of the ISTL (Figure 4a). Because aftershocks occurred along a plane dipping eastward, the central Nagano earthquake is interpreted to be due to activity on the Gofukuji fault associated with dynamic stress/strain changes caused by the Tohoku-Oki earthquake (e.g., Miyazawa, 2011). In this section, the characteristics of temporal changes in the $^{3}He/^{4}He$ ratios before and after the 2011 northern and central Nagano earthquakes using both the data from Table 1 and unpublished data are discussed.

Most gas samples around the Tokamichi fault and the Gofukuji fault before the earthquakes have $^{3}He/^{4}He$ ratios higher than the atmospheric value, indicating a relatively higher contribution of mantle helium (Figure 4a).

The magmatic influence on the $^{3}He/^{4}He$ ratio in groundwater could be limited in the range from several kilometers to about a 10 kilometer radius from the centers of active volcanoes (e.g., Sakamoto et al., 1992). Since the sampled wells are located more than 10 km from the center of active volcanoes, the observed $^{3}He/^{4}He$ ratios suggest the effect of primordial ^{3}He carried in magma chambers beneath active volcanoes is insignificant. Accordingly, rather high $^{3}He/^{4}He$ ratios along the ISTL can be interpreted as leakage of mantle helium within the crust migrating upward along active faults. Several researchers have proposed that the continuous mantle-fluid flux model can explain the increased pore pressure as a result of vertical flux of deep crust or mantle fluids into seismic zone (e.g., Kennedy et al., 1997; Umeda et al., 2008).

The spatial variations in $^{3}He/^{4}He$ ratios observed after the 2011 northern and central Nagano earthquakes are shown in Figure 4b. Whether or not the earthquake occurred, the mantle helium components in samples collected adjacent to the Tokamachi and the Gofukuji faults are relatively higher than those away from active faults in non-volcanic regions.

Although the spatial variations in $^3He/^4He$ ratios observed before the earthquakes generally coincide with the variations after the earthquakes, considerable postseismic changes are observed in and around the source region of the 2011 earthquakes. Figure 4c displays the temporal changes in $^3He/^4He$ ratios, Δ (%), using the following equations:

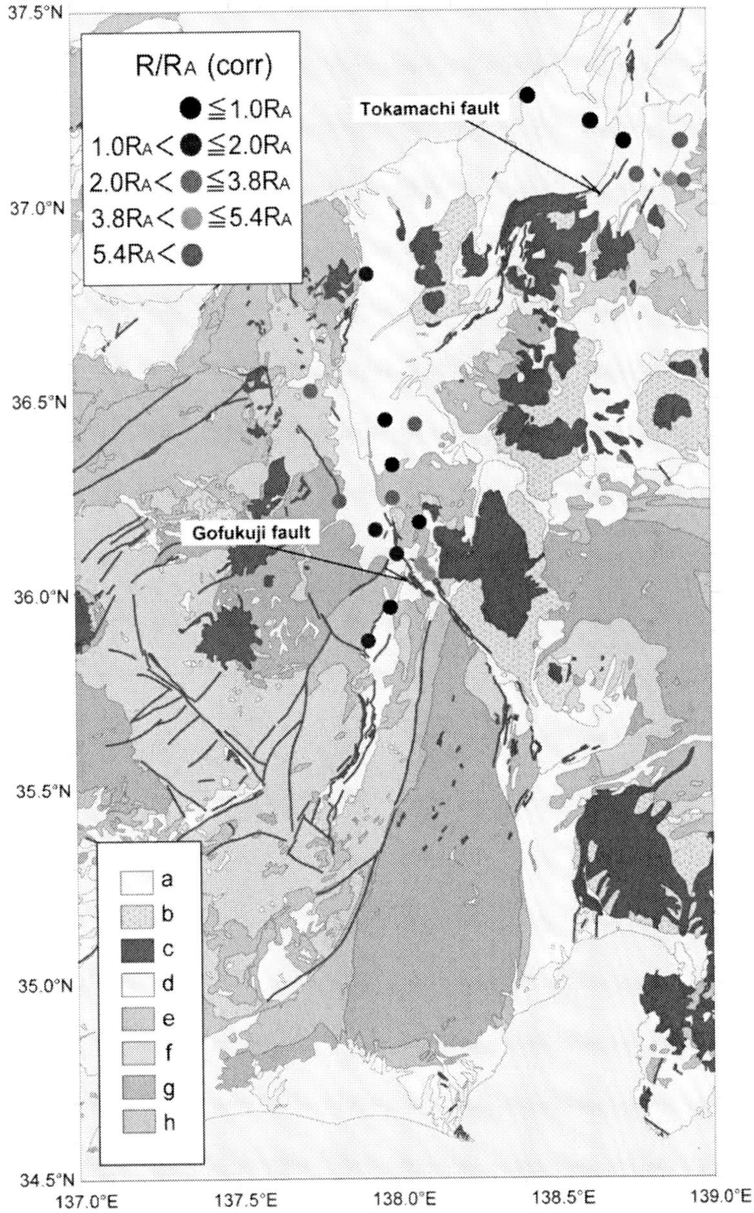

Figure 4a. Geographic distribution of the air-corrected $^3He/^4He$ ratios before the 2011 northern and central Nagano earthquakes. Also shown are the simplified geological map based on Geological Survey of Japan (1995). a, Quaternary sediments; b, debris; c, Quaternary volcanic rocks; d, Middle Miocene to Pliocene volcanic and sedimentary rocks; e, Middle to Late Miocene volcanic and sedimentary rocks; f, Miocene to Late Cretaceous igneous rocks; g, Mesozoic sedimentary rocks; h, Mesozoic metamorphic rocks.

$$\Delta\,(^3He/^4He) = ((^3He/^4He)_{after}/(^3He/^4He)_{before} - 1) \times 100 \qquad (3)$$

where $(^3He/^4He)_{before}$ and $(^3He/^4He)_{after}$ denote the corrected $^3He/^4He$ ratios before and after the earthquakes, respectively.

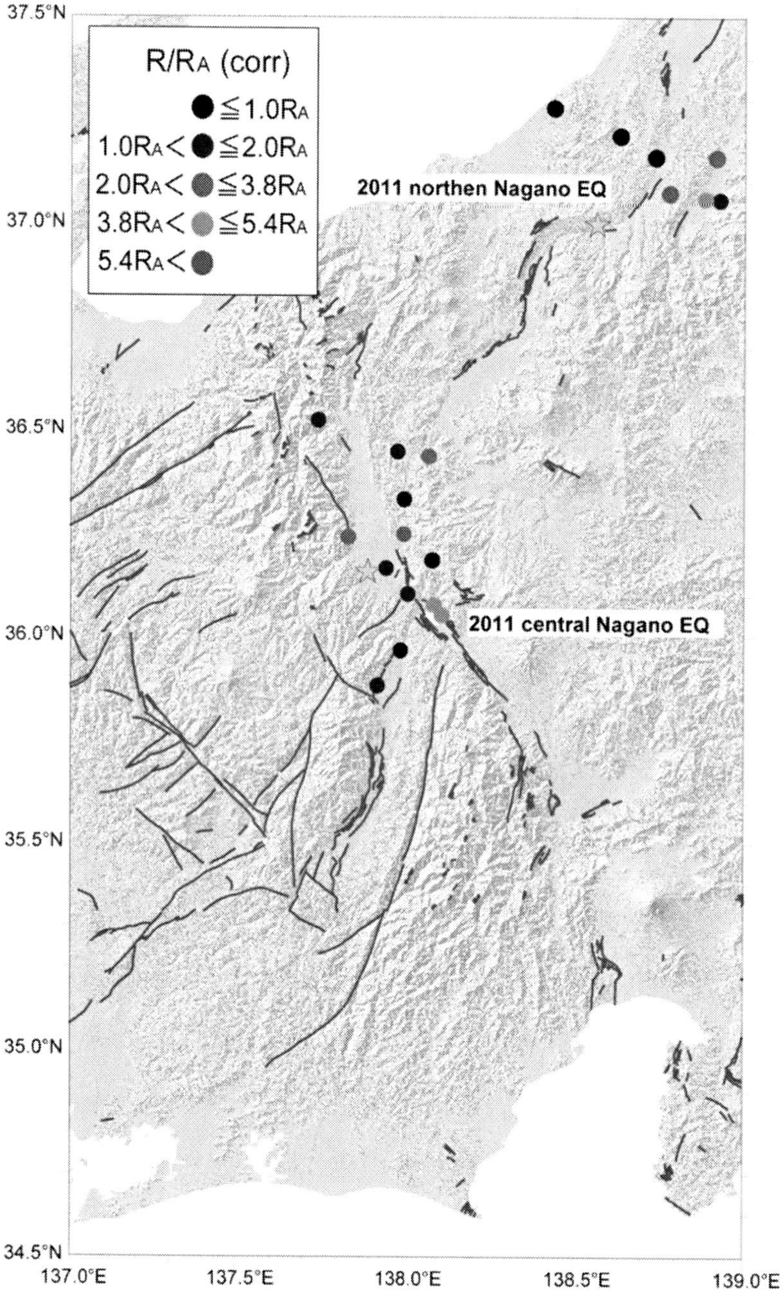

Figure 4b. Geographic distribution of of the air-corrected $^3He/^4He$ ratios after the 2011 northern and central Nagano earthquakes.

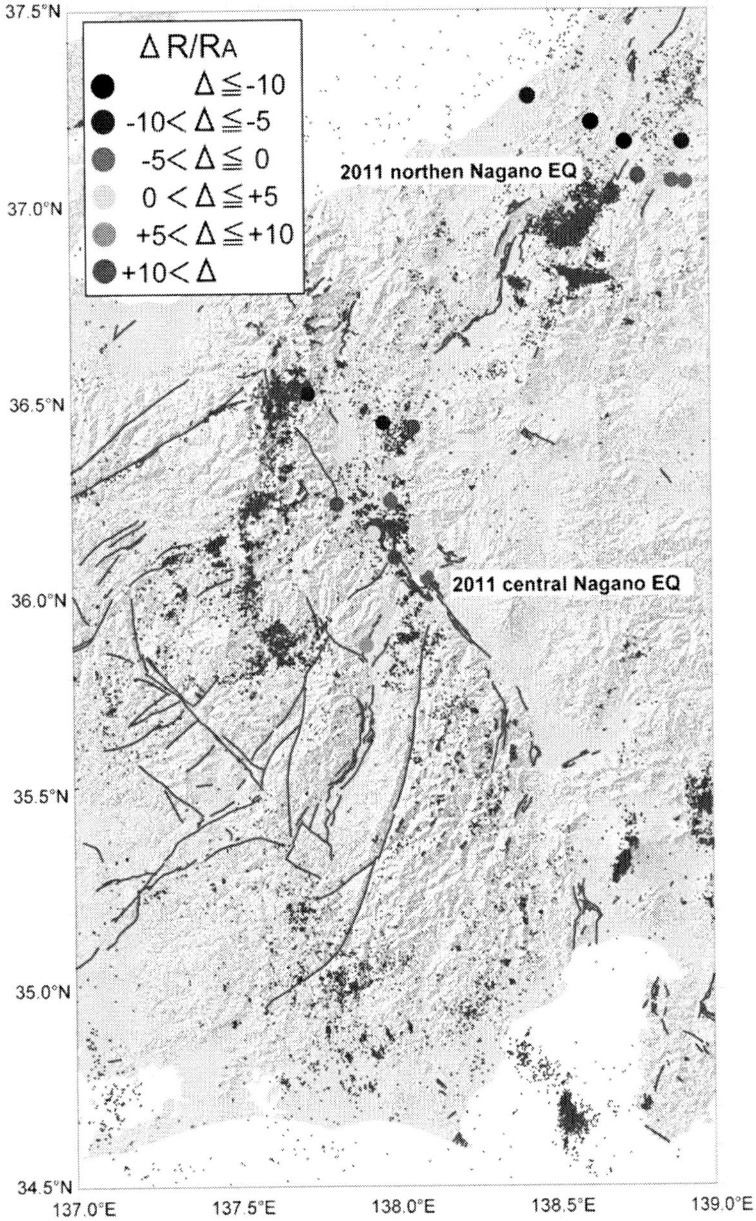

Figure 4c. Geographic distribution of temporal changes of $^3He/^4He$ ratios before and after the earthquakes. Also shown in blue is the hypocenter distribution of inland earthquakes at depths shallower than 30 km from 11 March 2011 to 1 August 2012.

It should be noted that most gas samples around the Gofukuji fault are characterized by postseismic $^3He/^4He$ ratio increases more than 10 %. In contrast, postseismic decreases less than 10 % occur around the Tokamachi fault. Episodic faulting and fracturing could release accumulated radiogenic helium from host rocks on geologically-short time scales (~ 1 k.y.) (Torgersen and O'Donnell, 1991), or enhance the transfer of mantle volatiles through permeable fault zones over long time scales (~ 1 m.y.) (Kulongoski et al., 2005), such that subsequent proximal fluid flow could explain spatial and temporal variations in $^3He/^4He$ ratios.

Generally, inland earthquakes are interpreted to arise from the frictional instability of existing faults induced by coupling of increased hydraulic pressure and tectonic shear stress. The influx of fluids into fault zones and development of high fluid pressure facilitate fault movement and earthquake nucleation by weakening the shear strength of the source fault. The recurrence of earthquake ruptures induced by fluid pressure increases would produce interconnected networks allowing fluid flow (Miller and Nur, 2000). Sibson et al. (1992) proposed that fault permeability is highest immediately after faulting, so that an earthquake rupture transecting overpressurized fluids in the lower crust induces postseismic discharge along the rupture zone owing to a pressure gradient. The 2011 central Nagano earthquake injected high pressure fluids with mantle helium into the seismogenic upper crust after the earthquake. These high pressure fluids mix with meteoric groundwater (recharge meteoric water), but without significant dilution by the crustal radiogenic helium, so that the elevated $^3He/^4He$ ratios of gases are observed near the seismic source fault. In contrast, the $^3He/^4He$ ratios observed around the seismic source region of the 2011 northern Nagano earthquake may indicate an insignificant contribution of mantle helium carried by upwelling fluids along the conduits related with the seismic source fault.

CONCLUSION

A great deal of effort has been made on obtaining information about $^3He/^4He$ ratios in gas samples from fumaroles, natural gas wells, hot spring and drinking wells in Japan. Generally, fumarolic and hot spring gases associated with Quaternary volcanoes have $^3He/^4He$ ratios higher than the atmospheric value of 1.4×10^{-6}, indicating a significant contribution of primordial ^3He carried by ascending magmas from the upper mantle. However, anomalously elevated $^3He/^4He$ ratios are observed in the seismically active zones such as the source regions of 2004 mid-Niigata Prefecture earthquake, 2007 Niigataken Chuetsu-oki earthquake. Temporal changes of $^3He/^4He$ ratios before and after the 2011 central Nagano earthquake indicate that most gas samples near the seismic source region are characterized by increase in postseismic $^3He/^4He$ ratios. In contrast, the postseismic ratios decrease more than 10 % around the source region of the 2011 northern Nagano earthquake.

Episodic faulting could release accumulated crustal (radiogenic) helium from host rocks, or enhance the transfer of mantle volatiles through permeable fault zones, such that subsequent fluid flow proximal to the source region could then explain the spatio-temporal variations in $^3He/^4He$ ratios.

ACKNOWLEDGMENTS

We used hypocenter data processed with close coordination of the Japan Meteorological Agency and the Ministry of Education, Culture, Sports, Science and Technology. The data have been provided by the National Research Institute for Earth Science and Disaster Prevention, Geospatial Information Authority of Japan, Hokkaido University, Hirosaki

University, Tohoku University, the University of Tokyo, Nagoya University, Kyoto University, Kochi University, Kyushu University, Kagoshima University, Japan Agency for Marine-Earth Science and Technology, National Institute of Advanced Industrial Science and Technology, Tokyo Metropolitan Government, Aomori Prefectural Government, Shizuoka Prefectural Government, Hot Springs Research Institute of Kanagawa Prefecture, and the City of Yokohama. This research was partially supported by Grants-in-Aid for Scientific Research (No. 24510250), Ministry of Education, Culture, Sports, Science and Technology (MEXT).

REFERENCES

Asamori, K.; K. Umeda; A.; T. Negi. *Manifestations of upwelling mantle flow on the earth's surface*; Jon M. Phillips; Ed.; The Earth's Core: Structure, Properties and Dynamics; Nova Science Publishers: New York, USA, 2011, pp.79–94.

Asamori, K.; K. Umeda; Y. Ogawa; T. Oikawa. *International Journal of Geophysics*, 2010, doi:10.1155/2010/738139.

Bureau, H.; H. Keppler. *Earth Planet. Sci. Lett.*1999, *165*, 187-196.

Clark, I. D.; R. J. Phillips. *J. Volcanol. Geotherm. Res.* 2000, *104*, 261-276.

Clarke, WB; Beg, MA; Craig, H. *Earth Planetary Sci. Lett.*, 1969, *6*, 213-220.

Craig, H.; Lupton, J. E.; Horibe, Y. *A mantle helium component in circum-Pacific volcanic gases: Hakone, the Marianas and Mt. Lassen;* Alexander, E.C., Ozima, M.; Eds.; Terrestrial Rare Gases. ; Jpn. Sci. Soc. Press: Tokyo, Japan, 1978; pp. 3-16.

Craig, H; Clarke, WB; Beg, MA. *Earth Planet. Sci. Lett.*, 1975, *26*, 125-132.

Doğan, T.; H. Sumino; K. Nagao; K. Notsu. *Chemical Geology,* 2006, *233*, 235-248.

Geological Survey of Japan(1995). Geological Map of Japan, 1, 1,000,000 3rd Edition [CD-ROM]; Geol. Surv. Jpn.: Tsukuba, Japan, 1995.

Giggenbach, W. F.; S. Matsuo. *Applied Geochemistry*, 1991, *6*, 125–141.

Giggenbach, W. F.; Y. Sano; H. Wakita. *Geochim. Cosmochim. Acta*, 1993, *7*, 3427-3455.

Güleç, N; Hilton, DR; Mutlu, H. *Chem. Geol.*, 2002, *187*, 129-142.

Heki, K.; Miyazaki, S. *Geophys. Res. Lett.* 2001, *28*, 2313-2316.

Hilton, DR; Fischer, TP; Marty, B. *Noble Gases in Cosmochemistry and Geochemistry;* D; Porcelli, CJ; Ballentine, R; Wieler; Ed.; Reviews in Mineralogy and Geochemistry 47; Mineral. Soc. Am: Washington DC., 2002, 319-370.

Hirata, N.; H. Sato; S. Sakai; A. Kato; E. Kurashimo, *Landslides*, 2005, *2*, 153-157.

Hirose, F.; Miyaoka, K.; Hayashimoto, N.; Yamazaki, T.; Nakamura, M. *Earth Planets Space,* 2011, *63*, 513-518.

Horiguchi, K.; J. Matsuda. *Geochemical Journal*, 2008, *42*, e1–e4.

Horiguchi, K.; S. Ueki; Y. Sano; N. Takahata; A. Hasegawa; G. Igarashi. *Island Arc*, 2010, *19*, 60–70.

Hyndman, RD; Peacock SM. *Earth Planet. Sci. Lett.*, 2003, *212*, 417-432.

Hirabayashi J.; T. Ohba; M. Yoshida; M. Kusakabe; G. L. Scott. *Programme and abstracts the Volcanological Society of Japan*, 1992, 157, 103-103.(in Japanese)

Iio, Y.; T. Sagiya; Y. Kobayashi; I. Shiozaki. *Earth Planet. Sci. Lett.*, 2002, *203*, 245-253.

Iwasaki, I.; T. Ozawa; M. Yoshida; T. Katsura; B. Iwasaki; M. Kamada., *Bulletin of Volcanology*, 1962, *24*, 23–48.

Jenden, P. D.; I. R. Kaplan; R. J. Poreda; H. Craig., *Geochimica et Cosmochimica Acta*, 1988, *52*, 851–861.

Kaizuka, S. *Kagaku*, 1972, *42*, 573–581.

Kato, A; S. Sakai; N. Hirata; E. Kurashimo; T. Iidaka; T. Iwasaki; T. Kanazawa. *J. Geophys. Res.*, 2006, *111*, 10.1029/2005JB004016.

Kennedy, BM; Kharaka, YK; Evans, WC; Ellwood, A; DePaolo, DJ; Thordsen, J; Ambats, G; Mariner, RH. *Science*, 1997, *278*, 1278-1281.

Kennedy, BM; van Soest, MC. *Science*, 2007, *318*, 1433-1436.

Kita, I.; K. Nagao. *Chishitsu News*,1996, *506*, 29–34. (in Japanese)

Kita, I.; K. Nagao; S. Taguchi; K. Nitta; H. Hasegawa. *Geochemical Journal*, 1993, *27*, 251–259. (in Japanese)

Kita, I.; K. Nagao; Y. Nakamura; S. Taguchi. *J. Geotherm. Res. Soc. Japan*, 1992, *14*, 115–128.

Kulongoski, JT; Hilton, DR; Izbicki, JA. *Geochimica et Cosmochimica Acta*, 2005, *69*, 3857-3872.

Kurz, MD; Jenkins, WJ; Hart SR; Clague, D. *Earth Planet. Sci. Lett.*, 1983, *66*, 388-406.

Kurz, MD; Jenkins, WJ; Hart, SR. *Nature*, 1982, *297*, 43-47.

Kusakabe, M.; M. Ohwada; H. Satake; K. Nagao; I. Kawasaki. *Society of Economic Geologists*, 2003, *10*, 75–89.

Mahara, Y. *Development of dating method groundwater –proposal for new survey method for ground water flow sing dissolved noble gases-;* CRIEPI Research Report U94027; Central Research Institute of Electric Power Industry: Abiko, Japan, 1994. (in Japanese)

Mahara, Y.; E. Nakata. *Proposal for the methods to characterize fossil seawater- Distribution of anions, cations and satable isotopes, and estimation on the groundwater residence time by measuring 36Cl in the Taiheiyou Coal Mine*; CRIEPI Research Report U03019, Central Research Institute of Electric Power Industry: Abiko, Japan, 2003. (in Japanese)

Mahara, Y.; K. Kitaoka. *Applied Geochemistry*, 2009, *24*, 438–446.

Mahara, Y.; T. Igarashi. *Applied Geochemistry*, 2003, *18*, 719–738.

Marty, B.; A. Jambon; Y. Sano. , *Chemical Geology*, 1989, *76*, 25–40.

Matsubayashi, O.; S. Matsuo; I. Kaneoka; M. Ozima (1978): "*Rare gas abundance pattern of fumarolic gases in Japanese volcanic areas*; E.C. Alexander Jr; M. Ozima; Ed.; Terrestrial Rare Gases; Springer Netherlands, 1978, Vol.3, pp 27–32.

Matsumoto, T.; T. Kawabata; J. Matsuda; K. Yamamoto; K. Mimura. *Earth and Planetary Science Letters*, 2003, *216*, 221–230.

Matsunami, T.; Wake, T.; Hayakawa, F.; Futamase, T.; Yokoyama, E.; Uchida, Y.; Sakou, T.; Saito, N. *Geothermal indications and hotsprings of Hokkaido, Japan, (c) Central Hokkaido;* Special report of the Geological Survey of Hokkaido, 7; Geological Survey of Hokkaido: Sapporo, Japan, 1979. (in Japanese)

Matsuo, S.; J. Ossaka; J. Hirabayashi; T. Ozawa; K. Kimishima, *1st Workshop on Volcanic Gases, Bulletin*, 1980, *31*, pp.3–12.

Miller, S. A.; Nur, A. *Earth Planet. Sci. Lett.*,2000, *183*, 133-146.

Miyazawa, M. *Geophys. Res. Lett.*, 2011, *38*, 10.1029/2011GL049795.

Morikawa, N.; *Geochemical study on groundwater flow research; Groundwater survey by dissolved noble gases in groundwater from Tono area;* JNC-TN7400 2001-003; Japan Atomic Energy Agency:Japan, 2001. (in Japanese)

Morikawa, N.; K. Kazahaya; E. Fourre; H. A. Takahashi; P. Jean-Baptiste; M. Ohwada; F. J. LeGuern; A. Nakama. *Journal of Volcanology and Geothermal Research*, 2008, *175*, 218–230.

Morikawa, N.; K. Kazahaya; H. Masuda; M. Ohwada; A. Nakama; K. Nagao; H. Sumino.*Geochemical Journal*, 2008, *42*, 61–74.

Morikawa, N.; K. Kazahaya; M. Yasuhara; A. Inamura; K. Nagao; H. Sumino. *Geophysical Research Letters*, 2005, *32*, doi:10.1029/2004GL021501.

Nagao, K. *Development of Geochemical Monitoring Methods by Using Sr and Noble Gas Isotopic Compositions;* JNC-TY7400 2004-001; Japan Atomic Energy Agency: Japan, 2004. (in Japanese)

Nagao, K.; N. Takaoka; H. Wakita; S. Matsuo;N. Fujii., *Geochemical Journal,* 1980, *14*, 63–69.

Nagao, K.; N. Takaoka; O. Matsubayashi. *Earth and Planetary Science Letters*, 1979, *44*, 82–90.

Nagao, K.; N. Takaoka; O. Matsubayashi., *Earth and Planetary Science Letters,* 1981, *53*, 175–188.

Nagao, K.; N. Takaoka; S. Matsuo; Y. Mizutani; O. Matsubayashi. *Geochemical Journal*, 1980, *14*, 139–143.

New Energy and Industrial Technology Development Organization; *Geothermal development promotion research report. Mt. Shibetsu region;* New Energy and Industrial Technology Development Organization: Tokyo, Japan, 1997. (in Japanese).

New Energy and Industrial Technology Development Organization; *National survey on geothermal resources-the third wide-area study on thermal water flows.Fiscal year 1988.Tokachi region.*; New Energy and Industrial Technology Development Organization: Tokyo, Japan, 1988. (in Japanese)

New Energy and Industrial Technology Development Organization; *National study on geothermal resources (the third) study on wide-ranged systems.Fiscal year 1990.Akitakoma region.Fluid geochemical study report.Summary;* New Energy and Industrial Technology Development Organization: Tokyo, Japan, 1991. (in Japanese)

New Energy and Industrial Technology Development Organization; *National survey on geothermal resources-the third wide area study on thermal water flows.Fiscal year 1990.Bandai region. Survey report on volcanic rock distribution and age.A summary*; New Energy and Industrial Technology Development Organization: Tokyo, Japan, 1991. (in Japanese)

New Energy and Industrial Technology Development Organization; *National survey on geothermal resources - the third wide-area study on thermal water flows.Fiscal year 1990.Aso region.Survey report on volcanic rock distribution and age.A summary* New Energy and Industrial Technology Development Organization: Tokyo, Japan, 1991. (in Japanese)

New Energy and Industrial Technology Development Organization; *Regional Report on Geothermal Development Promotion Survey (37. Hongu Area)*; New Energy and Industrial Technology Development Organization: Tokyo, Japan, 1994. (in Japanese)

New Energy and Industrial Technology Development Organization; *Regional Report on Geothermal Development Promotion Survey (38. Asosan-Seibu Area).* New Energy and Industrial Technology Development Organization: Tokyo, Japan, 1995. (in Japanese)

New Energy and Industrial Technology Development Organization; *Report Geothermal Development Promotion Survey. No.C-2. Wasabizawa area;* New Energy and Industrial Technology Development Organization: Tokyo, Japan, 1998. (in Japanese).

New Energy and Industrial Technology Development Organization; *Report Geothermal Development Promotion Survey. No.A-3. Himekawa area;* New Energy and Industrial Technology Development Organization: Tokyo, Japan, 1999. (in Japanese).

New Energy and Industrial Technology Development Organization; *Regional Report on Geothermal Development Promotion Survey. B-2. Ashiro Area.* New Energy and Industrial Technology Development Organization: Tokyo, Japan, 1999. (in Japanese).

New Energy and Industrial Technology Development Organization; *Regional Report on Geothermal Development Promotion Survey. B-3. Kimaishi Area.* New Energy and Industrial Technology Development Organization: Tokyo, Japan, 1999. (in Japanese).

New Energy and Industrial Technology Development Organization; *Regional Report on Geothermal Development Promotion Survey. A-4. Kunbetsudake Area.* New Energy and Industrial Technology Development Organization: Tokyo, Japan, 2001. (in Japanese).

New Energy and Industrial Technology Development Organization; *Regional Report on Geothermal Development Promotion Survey. B-5. Musadake Area.* New Energy and Industrial Technology Development Organization: Tokyo, Japan, 2001. (in Japanese).

New Energy and Industrial Technology Development Organization; *Regional Report on Geothermal Development Promotion Survey. C-3. Akinomiya Area.* New Energy and Industrial Technology Development Organization: Tokyo, Japan, 2001. (in Japanese).

New Energy and Industrial Technology Development Organization; *Third survey on national geothermal resources - survey on hydrological thermal water flow systems.Fiscal year 1989.Mt. Tokachi area.Geothermal maps;* New Energy and Industrial Technology Development Organization: Tokyo, Japan, 1990. (in Japanese)

New Energy and Industrial Technology Development Organization; *Third survey on national geothermal resources - survey on hydrological thermal water flow systems.Fiscal year 1989.Mt. Nasu area.Geothermal maps;* New Energy and Industrial Technology Development Organization: Tokyo, Japan, 1990. (in Japanese)

New Energy and Industrial Technology Development Organization; *Third survey on national geothermal resources - survey on hydrological thermal water flow systems.Fiscal year 1989.Mt. Tsurumitake area.Geothermal maps;* New Energy and Industrial Technology Development Organization: Tokyo, Japan, 1990. (in Japanese)

Notsu, K.; S. Nakai; G. Igarashi; J. Ishibashi; T. Mori; M. Suzuki; H. Wakita., *Journal of Volcanology and Geothermal Research*, 2001, *111*, 89–98.

Obara, K. *Science*, 2002, *296*, 1679-1681.

Obara, K.; H. Hirose. *Tectonophys.*, 2006, *417,* 33-51.

Ogawa, Y. ; Y. Honkura. *Earth Plant. Space*, 2004, *56*, 1285-1291.

Ohno, M.; H. Sumino; P. A. Hernandez; T. Sato; K. Nagao., *Journal of Volcanology and Geothermal Research*, 2011, *199*, 118–126.

Ohsawa, S.; T. Sakai; M. Yamada; T. Mishima; S. Yoshikawa; T. Kagiyama. *Journal of Hot Spring Sciences*, 2011, *60*, 410–417.

Ohsawa, S.; Y. Yusa. *J. Geotherm. Res. Soc. Japan*, 2001, *23*, 197–205.

Ohsawa, S.; Y. Yusa; K. Oue; K. Kitaoka. *Journal of Hot Spring Sciences*, 1998, *48*, 9–14. (in Japanese)

Ohwada, M.; H. Satake; K. Nagao; K. Kazahaya. *Journal of Volcanology and Geothermal Research*, 2007, *168*, 55–67.

Okada, T.; T. Itaya; M. Sato; K. Nagao. *The Island Arc*, 1994, *3*, 221–231.

Ono, A.; Y. Sano; H. Wakita; W. F. Giggenbach., *Geochemical Journal*, 1993, *27*, 287–295.

Ossaka, J.; T. Ozawa; T. Nomura; T. Ossaka; J. Hirabayashi; A. Takaesu; T. Hayashi . *Bulletin of Volcanology,* 1980, *43*, 207–216.

Ozima M; Podosek FA. *Noble Gas Geochemistry (Second Edition);* Cambridge University Press: Cambridge, 2002.

O'Nions, RK; Oxburgh, ER. *Earth Planet. Sci. Lett.*, 1988, *90*, 331-347.

Porcelli, D; Ballentine, CJ. *Rev. Mineral. Geochem.*, 2002, *47*, 411-480.

Sakamoto, M.; Y. Sano; H. Wakita. *Geochemical Journal*, 1992, *26*, 189–195.

Sakata, S., *Geochimica et Cosmochimica*, 1991, *55*, 1395–1405.

Sakata, S.; M. Takahashi; S. Igari; N. Suzuki.*Chemical Geology*, 1989, *74*, 241–248.

Sakata, S.; M. Takahashi; S. Igari;Y. Matsuhisa; K. Hoshino.*Bull. Geol. Surv. Japan*,1996, *47*, 619–642. (in Japanese)

Sakata, S.; M. Takahashi;K. Hoshino. *Journal of the Japanese Association for Petroleum Technology*, 1986, *51*, 228–237. (in Japanese)

Sakata, S.; S. Maki; K. Yazaki; S. Nagata. *Chikyukagaku(Geochemistry)*, 1987, *21*, 65–74. (in Japanese)

Sano, Y.; A. Kameda; N. Takahata; J. Yamamoto; J. Nakajima. *Chemical Geology*, 2009, *266*, 50–56.

Sano, Y.; H. Wakita. *Geochemical journal*, 1988, *22*, 293–303.

Sano, Y.; J. Hirabayashi; T. Oba; T. Gamo. *Applied Geochemistry*, 1994, *9*, 371–377.

Sano, Y.; K. Notsu; J. Ishibashi; G. Igarashi; H. Wakita. *Earth and Planetary Science Letters*, 1991, *107*, 95–100.

Sano, Y.; Marty.*Chemical Geology*, 1995, *119*, 265–274.

Sano, Y.; N. Takahata; G. Igarashi; N. Koizumi; N. C. Sturchio. *Chemical Geology*, 1998, *150*, 171–179.

Sano, Y.; N. Takahata;T. Seno. *Pure and Applied Geophysics*, 2006, *163*, 745–757.

Sano, Y.; T. Gamo; K. Notsu; H. Wakita. *Journal of Volcanology and Geothermal Research*, 1995, *64*, 83–94.

Sano, Y.; T. Tominaga; Y. Nakamura; H. Wakita., *Geochemical Journal*, 1982, *16*, 237–245.

Sano, Y.; Y. Nakamura; H. Wakita., *Chemical Geology*, 1985, *52*, 1–8.

Sano, Y.; Y. Nakamura; H. Wakita; A. Urabe; T. Tominaga., *Sience*, 1984, *224*, 150–151.

Sano, Y.; Y. Nakamura; H. Wakita; K. Notsu; Y. Kobayashi. *Journal of Geophyical Research*, 1986, *91*, 12291–12295.

Sano, Y.; Y. Nakamura; K. Notsu; H. Wakita.*Geochimica et Cosmochimica Acta*, 1988, *52*, 1305–1308.

Sano, Y.; Y. Nishio; S. Sasaki; T. Gamo; K. Nagao. *Journal of Geophyical Research*, 1998, *103*, 23863–23873.

Sano, Y; Takahata, N; Igarashi, G; Koizumi, N; Sturchio, N. *Chem. Geol.*, 1998, *150*, 171-179.

Sano, Y; Wakita, H. *J. Geophys. Res.*, 1985, *90*, 8729-8741.

Sato, M.; T. Mori; K. Notsu; H. Wakita. *Bulletin of the Volcanological Society of Japan*, 1999, *44*, 279–283.(in Japanese)

Sato, M.; T. Mori; Y. Shimoike; K. Nagao; K. Notsu. *Earth Planets Space*, 2002, *54*, 257–263.

Scott, G.; M. Kusakabe; J. Hirabayashi; T. Ohba; K. Nagao; I. Kita.*Journal of Geological Society of the Philippines*, 1994, *59*, 89–110.

Seno, T.; S. Stein; A. E. Gripp. *J. Geophys. Res.*,1993, *98*, 17941-17948.

Shimoike, Y; Notsu, K. *J. Volcanol. Geotherm. Res.*, 2000, *101*, 211-221.

Sibson, R. H. *Earth Planet. Sci. Lett.*, 2007, *257*, 188-199.

Sibson, RH. *Tectonophys.,* 1992, *211*, 283-293.

Sturchio, N. C.; S. Ohsawa; Y. Sano; G. Arehart; K. Kitaoka; Y. Yusa. *Geothermics*, 1996, *25*, 215–230.

Sumino, H.; K. Notsu; S. Nakai; M. Sato; K. Nagao; M. Hosoe; H. Wakita. *Chemical Geology*, 2004, *209*, 153–173.

Takahata, N.; R. Yokochi; Y. Nishio; Y. Sano. *Geochemical Journal*, 2003, *37*, 299–310.

Takahata, N.; Y. Sano.*Applied Radiation and Isotopes*, 2000, *52*, 985–992.

Takano, O.; M. Tateishi; M. Endo. *Sed., Geol.,* 2005, *176*, 247-279.

Takaoka, N. *Bulletin of the Volcanological Society of Japan*, 1985, *30*, 185–195. (in Japanese)

Takaoka, N.; T. Konda. *Helium isotope ratios in hot springl gases* ; "Goshosan" Association of Yamagata genaral academic research: Yamagata, Japan, 1989; pp.33-36. (in Japanese)

Takaoka, N.; Y;Mizutani *Earth and Planetary Science Letters,* 1987, *85*, 74–78.

The National Institute of Advanced Industrial Science and Technology; *Compile geological data regarding the geological disposal of radioactive waste, 2003 Report;* The National Institute of Advanced Industrial Science and Technology, Research Center for Deep Geological Environments: Tsukuba, Japan, 2004.(in Japanese).

Toda, S.; Stein, R. S.; Lin J. *Geophys. Res. Lett.* 2011, *38*, 10.1029/2011GL047834.

Torgersen, T. *J. Geophys. Res.*, 1993, *98*, 16257-16269.

Torgersen, T.; O'Donnell J. *Geophys. Res. Lett.* 1991, *18*, 951-954.

Ueda, A.; K. Nagao; T. Shibata; T. Suzuki. *Geochemical Journal*, 2010, *44*, 545–560.

Umeda, K.; A. Ninomiya. *Geochemistry Geophysics Geosystems*, 2009, *10*, 10.1029/2009GC002501.

Umeda, K.; A. Ninomiya; G. F. McCrank. *Geochemistry Geophysics Geosystems*, 2008, *9*, 10.1029/2008GC002272.

Umeda, K.; A. Ninomiya; K. Asamori.*Release of mantle helium and its tectonic implications*; Lucas A. Becker; Ed.; Helium: Characteristics, Compounds, and Applications; Nova Science Publishers :NewYork, USA, 2011, pp.81–101.

Umeda, K.; A. Ninomiya; K. Shimada; J. Nakajima. *Helium isotope variations along the Niigata-Kobe tectonic zone, central Japan*, J. E. Anderson; R.W. Coateseds.; Ed.; The Lithospere: Geochemistry, Geology and Geophysics; Nova Science Publishers :NewYork, USA, 2008, pp.141–169.

Umeda, K.; A. Ninomiya; T. Negi. *Journal of Geophyical Research,* 2009, *114*, 10.1029/2008JB005812.

Umeda, K.; G. F. McCrank; A. Ninomiya.*Journal of Geophyical Research*, 2007, *112*, 10.1029/2007JB005031.

Umeda, K.; K. Asamori; A. Ninomiya; S. Kanazawa; T. Oikawa.*Journal of Geophyical Research*, 2007, *112*, .10.1029/2006JB004590.

Umeda, K.; K. Asamori; T. Negi;T. Kusano. *Journal of Geophysical Research*, 2011, *116*, 10.1029/2010JB007963.

Umeda, K.; S. Kanazawa; C. Kakuta; K. Asamori; T. Oikawa.*Geochemistry Geophysics Geosystems*, 2006, *7*, 10.1029/2005GC001210.

Umeda, K.; Y. Ogawa; K. Asamori; T. Oikawa. *Journal of Volcanology and Geothermal Research,* 2006, *149*, 47–61.

Umeda, K.; Y. Sakagawa; A. Ninomiya; K. Asamori.*Geophysical Research Letters*, 2007, *34*, 10.1029/2006GL028975.

Umeda, K; Kusano, T; Asamori, K; McCrank, G. F. *J. Geophys. Res.*, 2012, *117*, 10.1029/2012JB009409.

Urabe, A.; T. Tominaga; Y. Nakamura; H. Wakita., *Geochemical Journal*, 1985, *19*, 11–25.

Uto, K.; Hayakawa, Y.; Aramaki, S.; Ossaka, J. *Geological Map of Kusatsu-shirane Volcano*, Geological Survey of Japan, 1983.(in Japanese)

Uyeshima, M.; Y. Ogawa; Y. Honkura; S. Koyama; N. Ujihara; T. Mogi; Y. Yamaya; M.. Harada; S. Yamaguchi; I. Shiozaki; T. Noguchi; Y. Kuwaba; Y. Tanaka; Y. Mochido; N. Manabe; M. Nishihara; M. Saka; M. Serizawa *Earth Planet. Space*, 2005, *57*, 441-446.

Wakita H.; Y. Sano. *Nature,* 1983, *305*, 792-794.

Wakita, H.; N. Fujii; S. Matsuo; K. Notsu; K. Nagao; N. Takaoka. *Science*, 1978, *200*, 430–432.

Wakita, H.; Y. Sano; A. Urabe; Y. Nakamura. *Appl. Geochem.*, 1990, *5*, 263-278.

Wakita, H.; Y. Sano; M. Mizoue. *Journal of Geophysical Research*, 1987, *92*, 12539–12546.

Wang, Z.; D. Zhao. *Earth Planet. Sci. Lett.*, 2006, *244*, 16-31.

Yamano, M.; Honda, S.; Uyeda, S. *Mar. Geophys. Res.* 1984, *6*, 187–203.

Yamasaki, T.; T. Seno. *Earth Planet. Sci. Lett.*, 2005, *232*, 13-27.

Yoshii, T. *Kagaku*, 1977, *47*, 170–176. (in Japanese).

In: Helium: Occurrence, Applications and Biological Effects ISBN: 978-1-62808-612-6
Editor: Simon B. Davis

Chapter 6

WEAK SUPERFLUIDITY: NONSTATIONARY JOSEPHSON EFFECT IN He-4 AND POSSIBLE APPLICATIONS

A. M. Tskhovrebov and L. N. Zherikhina *
P.N. Lebedev Physical Institute of the Russian Academy of Sciences, Moscow, Russia

ABSTRACT

Unusual occurrences of nonstationary Josephson effect in superfluid He-4 are considered. Nonstationary Josephson effect is analyzed in the context with parametric amplifying of acoustic oscillations on kinetic nonlinearity of superfluid Josephson element. Mechanism of quantum fluctuations compression, following parametric rating, is taken into account. Appearance of Josephson fluxon in superfluid medium parted by a wall with narrow channels serving as weak links is analyzed by analogy with Josephson vortexes in the case of weak superconductivity. Possible applications of the effects above are discussed. Parametric amplifying of acoustic oscillations on conditions of quantum fluctuations compression is proposed to be used for making a high frequency ($f >> 1000$ Hz) gravitational waves detector. In this case if one have in mind that the efficiency of gravitational waves generation by a quadrupole radiator is proportional to the sixth power of frequency it seems interesting to use such a detector in a gravitational analog of Hertz experiment. Since superfluid Josephson vortexes are of quantum nature, it may be possible to model entangled states, which underline the matter wave q-bit system.

* E-mail: zherikh@sci.lebedev.ru.

INTRODUCTION ("HISTORY" OF JOSEPHSON EFFECT IN SUPERFLUID HELIUM)

Following theoretical prediction [1-4] of specific interference phenomena, accompanying the coherent tunneling of charge carriers in dissipationless systems, and then experimental detection of stationary [5-8] and non-stationary [9, 10] Josephson effects in superconductors, search attempts of similar effects in superfluid helium began [11-13].

At the first stage these search reduced to experiment setting [14-17], that reproduced observing conditions of Shapiro steps [9] in superfluid He^4. As analog of tunnel junction a narrow aperture - quantum throttle was used, the diameter and length of which were selected at the level of correlation length of superfluid condensatc [18]. Thc role of EM microwave irradiation carried out ultrasonic acoustic action. However to register surely the presence of stable differences of helium columns on each side of quantum throttle, corresponding to fixed pressure differentials (the direct analogs of "electric" Shapiro steps) in these experiments appeared to fail [19-22].

The next more successful search stage of Josephson effect in helium concerned investigations of He-3 superfluity [23, 24]. In contrast to first attempts on He-4, where the scheme of "passive" non-stationary Josephson effect was reproduced practically (external acoustic exposure has to "instigate" the stable pressure differential, proportional to the sound frequency - the direct analog of "electric" Shapiro steps [9]), here the analogy with "active" non-stationary effect was used, when the pressure differential, applied to quantum throttle, results in sound generation, at frequency proportional to this difference (the direct analog of microwave generation on Josephson tunneling barrier [10]). The complication of these experiments has been linked with the fact that in contrast to He-4 where the superfluidity is formed lower than the transition-point T_λ = 2,17 K, [25] the observing conditions of superfluidity of He-3 require the temperatures of the order of 1 mK [26, 27], that in its turn needs utilization of ultra-low-temperature cryogenics [28], including either system with adiabatic compression of solid He-3 on Pomeranchuk effect, or the dilution refrigerator He-3 - He-4 with a stage of nuclear demagnetization. In experiments with He-3 pressure differential produced by means of electrostatically deformable diaphragm was applied to the throttle, playing a part of Josephson junction. Authors happened to register sound, which frequency turned out to be proportional to pressure differential with "correct" proportionality factor according to the theory supposition $f_{He3} = \dfrac{m_{He3}}{\pi \rho_{He3} \hbar}(p_2 - p_1) \approx 187 \boldsymbol{kHz / Pa}$. "The quantum whistle" was heard with the help of SQUID-system, similar to that used earlier in the experiments on gravitational wave detection [29]. Such super high sensitive system perceived the sound generation in He-3, picking up acoustic vibrations of superconductor diaphragm, which periodically more or less supplanted magnetic flux from the input turn of superconducting transformer of DC-SQUID. The matrix $(m \times n)$ of microscopic aperture was used as a multiple throttle cell, carrying out the role of tunnel Josephson junction[23, 24, 30, 31], hence the intensity of "quantum whistle" enhanced in $(m \times n)^2$ times compared with the common-mode acoustic generation for one throttle.

On the next stage the method of "active" non-stationary Josephson effect proved, itself in investigations with He-3, has been extended over the analogous search experiments in superfluid He-4 [32-34]. The peculiarity of these substantially more simple in cryogenic relation experiments consisted in stabilizing the temperature as near as possible to the λ – point. In fact the stabilizing system provided the tuning out of operating temperature from transition point of He-4 into the superfluid state at the level of $T\text{-}T_\lambda = 3$ mK. So small tuning out ensured the value of correlation length of superfluid condensate of about 100 mkm [35], that makes it possible to reduce demands for trifle of aperture sizes of throttled circuit. One should note that the formula for frequency of "quantum whistle in superfluid He-3 differs from analogous value for He-4 $f_{He4} = \frac{m_{He4}}{2\pi\rho_{He4}\hbar}(p_2 - p_1) \approx 69\boldsymbol{kHz/Pa}$ by factor 2, due to the fact that atoms of He-4 below-temperature of superfluid transition do not form Cooper pairs in contrast to He-3.

In the course of subsequent works at first in He-3 [36], and then in He-4 [37] the effects of quantum interference of matter waves, analogous to stationary Josephson effect, realized in DC-SQUID, were observed. The resultant change of pressure differential in diametrical points of ring channel, divided by the pair of multiple-aperture throttles, was registered in these experiments with the help of non-stationary Josephson effect. The pressures cycling, corresponding to the growth of angular momentum of superfluid condensate in the channel on quantum equal to $N_{He}\ \hbar$, were proved in the form of periodical frequency variation of "quantum whistle" (where N_{He} - the number of helium atoms in the channel). SHeQUID - quantum interferometer on matter waves in superfluid helium, enabling in its construction the possibility of turning the channel plain relating to the axis of daily Earth rotation, allowed to register the effect of Earth rotation at first in experiments with He-3 [36], and further in He-4 [37]. These experiments provide a base for formulation a creation problem of gyroscope, the principle effect of which is analogous to the device of laser fiber-optical systems on Sagnac effect [38-40]. One should note that in earlier works [40, 41] the effect of daily Earth rotation was registered successfully. The interferential dependence of analog of Josephson kinetic inductance, corresponding to passing of superfluid He-3 or He-4 through the system of nanothrottles, was used in those experiments.

Quantum Compression on Josephson Effect in Superfluid Helium

The effect of quantum fluctuations squeezing [43-50] is lately often proposed to be used to enhance measurement precision and hence radical increase of sensitivity of recording systems. Practically it permits one to avoid indirectly the quantum-limit of accuracy of measurements, which is given by the Heisenberg uncertainty relation [51, 52]. The effect of quantum squeezing is caused by periodic changing of "static" parameters of initial undisturbed oscillating system (parametric modulation of capacity or inductance in electric *LC* circuit, correspondingly masses M or stiffness K in mechanical oscillator). The result of this like influence is namely squeezing effect, which consists in periodic oscillations of standard deviations $\Delta x = \Delta x(t)$, $\Delta p = \Delta p(t)$ of mutually additional quantum-mechanical variables, providing complete description of system, that are included in its Hamiltonian (impulse and coordinate, defining mass displacement in mechanical oscillator; voltage and flux of magnetic field, stored in *LC*-circuit inductance). These standard deviations usually fluctuate on the doubled base frequency of undisturbed system $\Omega_{mod} = 2\omega_0$ (correspondingly $\omega_0 = \sqrt{(K/M)}$, or $\omega_0 = 1/\sqrt{(LC)}$, or…), so that minimal value of amplitude of quantum-mechanical fluctuations $\Delta x(t_x)=min\{\Delta x(t)\}$, $\Delta p(t_p)=min\{\Delta p(t)\}$ can dissatisfy Heisenberg uncertainty relation $\Delta x\Delta p\geq\hbar/2$, if the latter is treated, as the limitation of their production from below (their - that is minimal values).Therefore if we fix variables at definite time points t_x and t_p answering to minimums of corresponding fluctuations ("the indeterminancy nodes"), then the precision of such gated measurements may substantially overbid quantum confinement $\Delta x(t_x)<\Delta x_{min}=\sqrt{(\hbar/(2M\omega_0))}$, $\Delta p(t_p)<\Delta p_{min}=\sqrt{(\hbar\omega_0 M/2)}$, where Δx_{min} and Δp_{min} – minimum uncertainties without squeezing effect.

For realization of measurement process in the mode of quantum squeezing it is necessary to include used detector into an oscillatory system, which resonance frequency $\omega_0=\omega_0(t)$ is parametrically modulated through periodic change of reactive parameters of oscillator with frequency twice exceeding its average resonant value $\Omega_{mod}=2<\omega_0(t)>$. In this respect the operation of quantum squeezing can be considered as the further evolution of widespread in radio electronics methods of parametric signal amplification (or more concretely, as the development of circuit of degenerate parametric amplifier).

For systems with «conventional» (that is observed in conditions of superconductivity) Josephson effect one can choose the Josephson inductance as the under control reactive parameter of the oscillator $L_J(\varphi)=\dfrac{\Phi_0}{2\pi I_{JC}\cos\varphi}$, where $\Phi_0=\pi\hbar/e\approx 2{,}07\times 10^{-15}Wb$ - magnetic flux quantum, I_{JC} - critical current of tunneling barrier, playing a part of amplitude of Josephson current. This inductance is kinetic by implication, that is patently not taking into consideration the contribution of magnetic energy. Definition of L_J is founded on taking into account the kinetic energy of Cooper pairs, tunneling through potential barrier in conditions of Josephson effect. Let's put the well known [1-2] dependence of Josephson current on phase difference upon tunneling barrier $I_J = I_{JC}\sin\varphi$ into Faraday induction law $-U = L\dfrac{dI}{dt}$. Then, meaning that the rate of phase changing in Josephson effect is proportional to the voltage drop across barrier $\dfrac{d\varphi}{dt}=\dfrac{2e}{\hbar}U$, one will have the kinetic inductance in the form of:

$$L_J=\frac{U}{\frac{d}{dt}(I_{JC}\sin\varphi)}=\frac{U}{I_{JC}\cos\varphi\frac{d\varphi}{dt}}=\frac{U}{I_{JC}\frac{2e}{\hbar}U\cos\varphi}=\frac{\hbar}{2eI_{JC}\cos\varphi}=\frac{(\pi\hbar/e)}{2\pi I_{JC}\cos\varphi}=\frac{\Phi_0}{2\pi I_{JC}\cos\varphi}.$$

Along with superconducting systems the parametric amplification in conditions of squeezing of quantum fluctuations can be realized in superfluid He either, where the required non-linear reactivity of oscillating system is generated due to quantum interference of matter waves, formed as a "superfluid analog" of Josephson element [11-13]. To determine non-linear reactivity, the role of which the effective mass (the mechanical analog of kinetic inductance) is executing, let's place the dependence of Josephson current density on phase difference upon "tunnel" barrier in the 2nd Newton's law. The rate of phase changing appears to be proportional to pressure differential, falling on Josephson barrier, which is the capillary or the array of capillaries, having a diameter comparable with superfluity correlation length

$\zeta_{He4}(T) = \dfrac{\zeta_{He4}(0)}{(1 - T/T_\lambda)^{2/3}}$ (where T_λ=2,17K, $\zeta_{He4}(0)$=0,3 nm [35]). As a result we get the expression for non-linear effective mass of helium inside the Josephson capillary:

$$M = \frac{F}{a} = \frac{s(P_2 - P_1)}{\dfrac{1}{n_{He4}}\dfrac{dj_J^{He4}}{dt}} = \frac{s(P_2 - P_1)}{\dfrac{1}{n_{He4}}\dfrac{d}{dt}\left(j_C^{He4}\, sin\left(\dfrac{P_2 - P_1}{\hbar n_{He4}}t\right)\right)} = \frac{s(P_2 - P_1)}{\dfrac{(P_2 - P_1) j_C^{He4}}{n_{He4}^2 \hbar} cos\phi \dfrac{d\varphi}{dt}} = \frac{\hbar n_{He4}^2 s}{j_C^{He4}\, cos\left(\dfrac{P_1 - P_2}{\hbar n_{He4}}t\right)},$$

which in the form of dependence on phase difference falling «on barrier» is represented as $M = M(\varphi) = \dfrac{\hbar n_{He4}^2 s}{j_C^{He4}\, cos\,\varphi}$, where s - the sectional area of «barrier» capillary.

The non-stationary Josephson effect in He-4 is observed generally at temperatures just below T_λ=2,17K, when liquid He-4 transfers into the non-dissipative superfluid state. Near the above mentioned phase transition within millikelvin deviation from λ-point the correlation length achieves values of the order of 10÷100 mkm, that allows using in experiments "comparatively thick" capillaries. Let's estimate the depth of relative modulation of effective mass

$$\left|\frac{\delta m}{m}\right| = \frac{|M(\varphi)|_{max}}{m_{He4} n_{He\$} s\ell} = \frac{\hbar n_{He4}^2 s}{j_C^{He4}} \Big/ \left(m_{He4} n_{He\$} s\ell\right) = \frac{\hbar}{\upsilon_{JC}^{He4} m_{He4} \ell},$$

where ℓ - capillary length, υ_{JC}^{He4} - critical Josephson velocity of He4 atoms in capillary. Consequently, selecting ℓ at the level of 10 micrometers one will have the depth of self-reactance modulation of the order of 10^{-3}, what tend to be quite enough for signal amplification. The oscillating system in proposed amplifier is formed as of waveguide channel, made up by internal cavity of Josephson capillary, at ends of which the acoustic waves experience the reflection. Wave's penetration outside of capillary is prevented by the jump of wave resistance, arising at the joint of narrow outlet of capillary with wide basin of superfluid helium. The base frequency of amplified acoustic vibrations will lie in the MGHz diapason ($C_{He4} \approx$ 220 m/sec - the sound velocity in helium when $T<T_\lambda$). To lead the high frequency acoustic signal on "the input" of such parametric amplifier it is possible by focusing it with the parabolic mirror over the area of allocation of the throttle.

Parametric modulation allows along with amplification to secure conditions of squeezing of quantum fluctuations of matter waves, that define the fundamental limits of sensitivity of the system. Expected ultra-high sensitivity of such acoustic systems can be used in particular for detection of "high frequency" (f>> 1000Gts) gravitational waves. The possibility to use such high frequency sound detector as a receiver in "Hertz gravitational experiment" is of certain interest since the generation efficiency of gravitational waves in quadrupole radiator grows proportional to frequency of the sixth degree [53].

Weak Superfluidity in Distributed Systems

In classical mechanics the transition from oscillations in lumped system to distributed one is formed within wave field theory. The physics of weak superconductivity also allows the transition from "conventional" point-contact Josephson effect to distributed system [54, 55] with coherent tunneling of Cooper pairs. In this case besides "common" waves in extended Josephson junction could arise [56, 57] soliton like vortex of Cooper condensate phase in which a magnetic flux quant $\Phi_0=\pi\hbar/e\approx 2{,}07\times 10^{-15}$ Вб is captured. A set of equations, taking into consideration the presence of distributed source of Josephson current in extended transition (the analog of Thomson equations of telegraphy), by using a substitution $U=\frac{\hbar}{2e}\frac{\partial\varphi}{\partial t}$ is transformed into the 2nd orders equation of the sine-Gordon type, as followers:

$$\begin{cases}\frac{\partial I}{\partial x}=C_X\frac{\partial U}{\partial t}+J_CW\sin\varphi\\ \frac{\partial U}{\partial x}=L_X\frac{\partial I}{\partial t},\end{cases}\Rightarrow\begin{cases}\frac{\partial I}{\partial x}=\frac{\varepsilon\varepsilon_0W\hbar}{2ed}\frac{\partial^2\varphi}{\partial t^2}+J_CW\sin\varphi\\ \frac{\hbar}{2e}\frac{\partial^2\varphi}{\partial x\partial t}=\frac{\mu\mu_0 d}{W}\frac{\partial I}{\partial t}\end{cases}\Rightarrow\frac{\partial^2\varphi}{\partial x^2}-\varepsilon\varepsilon_0\mu\mu_0\frac{\partial^2\varphi}{\partial t^2}=\frac{2\pi\mu\mu_0 d}{\Phi_0}J_C\sin\varphi,$$

where J_C - the density of Josephson critical current, W - electrodes width (in the X direction) of extended tunnel junction, d - barrier thickness, $C_X=\frac{\varepsilon\varepsilon_0 W}{d}$ and $L_X=\frac{\mu\mu_0 d}{W}$ - linear densities of capacity and inductance of the line, formed by electrodes. In the steady case $\frac{\partial\varphi}{\partial t}\approx 0$ for small phase values $\sin\varphi\approx\varphi$ the sine-Gordon equation is reduced up to $\frac{\partial^2\varphi}{\partial x^2}=-\frac{2\pi d\mu\mu_0}{\Phi_0}J_C\varphi$, and the solution there develops in the form $\varphi(x)\sim e^{-x/\lambda_J}$, where $\lambda_J=\sqrt{\frac{\Phi_0}{2\pi d\mu\mu_0 J_C}}$ - the characteristic length of penetration of magnetic field on the edge of junction, simultaneously determines the soliton or Josephson vortex length.

Let us consider the transition from point-contact to distributed systems in conditions of weak superfluidity. For this purpose let us write the lagrangian of wave motions of superfluid helium, using an action (Ξ) as a generalized coordinate. Such an approach allows to identify "conventional" and generalized momentums $L_W=\frac{1}{2m}\sum_i\left(\frac{\partial\Xi}{\partial x_i}\right)^2-\frac{1}{2mc^2}\left(\frac{\partial\Xi}{\partial t}\right)^2$. Substituting given Lagrangian in Euler equation $\sum_i\left(\frac{d}{dx_i}\frac{\partial L_W}{\partial\left(\frac{\partial\Xi}{\partial x_i}\right)}\right)+\frac{d}{dt}\frac{\partial L_W}{\partial\left(\frac{\partial\Xi}{\partial t}\right)}-\frac{\partial L_W}{\partial\Xi}=0$, we should get the wave equation $\frac{1}{m}\left(\sum_i\frac{\partial^2\Xi}{\partial x_i^2}-\frac{1}{c^2}\frac{\partial^2 S\Xi}{\partial t^2}\right)=0$ or $\nabla^2\Xi-\frac{1}{c^2}\frac{\partial^2\Xi}{\partial t^2}=0$. Now let;s define the

Josephson energy of helium flowing through cross-section of square s under the action of pressure differential (P_2-P_1):

$$E_J^{He4} = \int F\upsilon dt = \int s(P_1 - P_2)\frac{\dot{j}_J^{He4}}{mn}dt = \int s(P_1 - P_2)\frac{\dot{j}_C^{He4}}{mn}\sin\frac{(P_2 - P_1)t}{\hbar n}dt =$$

$$= -\int \frac{s\hbar \dot{j}_C^{He4}}{m}\sin\frac{(P_2 - P_1)t}{\hbar n}d\frac{(P_2 - P_1)t}{\hbar n} = \frac{s\hbar \dot{j}_C^{He4}}{m}\left(1 - \cos\left(\frac{2\pi\Xi}{\hbar}\right)\right).$$

After addition of Josephson contribution the Lagrangian develops an appearance:
$$L_{W+J} = L_W + E_J^{He4} = \frac{1}{2m}\sum_i\left(\frac{\partial\Xi}{\partial x_i}\right)^2 - \frac{1}{2mc^2}\left(\frac{\partial\Xi}{\partial t}\right)^2 - \frac{s\hbar \dot{j}_C^{He4}}{m}\cos\left(\frac{2\pi\Xi}{\hbar}\right)$$ (the non-alternating constant $\frac{s\hbar \dot{j}_C^{He4}}{m}$ is not taken into account here). Insertion of Lagrangian L_{W+J} into Euler equation results in soliton equation of the sine-Gordon type $\nabla^2 S - \frac{1}{c^2}\frac{\partial^2\Xi}{\partial t^2} - 4\pi s \dot{j}_C^{He4}\sin\left(\frac{2\pi\Xi}{\hbar}\right) = 0$. The stationary solution of the later for small values of action $\frac{1}{c^2}\frac{\partial^2\Xi}{\partial t^2} = 0,\ \frac{2\pi\Xi}{\hbar} \ll 1 \Rightarrow \frac{\partial^2\Xi}{\partial x^2} \approx \frac{8\pi^2 s\dot{j}_C^{He4}}{\hbar}\Xi \Rightarrow \Xi \sim \exp(-x/\lambda_J^{He4})$ shows the typical size of Josephson vortex in helium $\lambda_J^{He4} = \sqrt{\frac{\hbar}{8\pi^2 s\dot{j}_C^{He4}}}$.

In conditions of superconductivity models of distributed Josephson system are extended Josephson junction (one-dimensional tunnel channel) [58], two-dimensional Josephson net [59], 3D Josephson system [60], and in case of superfluidity these are correspondingly one and two-dimensional arrays of apertures and three dimensional regularized system of overlapping micro pores. Object with such spatially ordered micro pores is a photon crystal, which occurs in nature in the form of mineral hydrate of silicon dioxide of $SiO_2(H_2O)_N$, known as opal. Opal is colorless by itself, however spatially ordered micro spheres of $SiO_2(H_2O)_N$ with diameter 150÷900 nm, packed in the face-centered cubic lattice, generate bright color of crystal due to Bragg diffraction. Consequently, while transmitting of superfluid helium through an opal crystal one is to observe both highly synchronized "quantum whistle" and effects, related with Josephson vortexes motion. Quantum nature of quasi macroscopical Josephson vortexes will allow simulating of entangled states, that makes up the basis of construction of q-bite system on matter waves.

High Sensitive System SQUID - Magnetostrictor for Registration of Non-stationary Josephson Effect in Superfluid Helium

As it was mentioned above observation of quantum interference of matter waves, analogous to stationary Josephson effect, resulted in practical creation of SHeQUID [12-13] -

a superfluid analog of DC-SQUID, is based on registration of non-stationary effect. The resultant changes of pressure differential in diametrical points of the ring channel of interferometer, separated by a pair of multiple apertures arrays, appeared in the form of periodical variation of sound frequency - "quantum whistle". In its turn "quantum whistle" was heard with the help of a SQUID-system, similar to that widely applied earlier in the search experiments on gravitational wave detection [61]. Such super high sensitive system perceived the sound generation in helium, fixing acoustic vibrations of the superconductor diaphragm, which periodically in more or less degree supplanted magnetic flux from the input turn of superconducting transformer of DC-SQUID. The sensitivity of applied SQUID-system was at the level of $10^{-13}\div 10^{-15}$ m/$\sqrt{Hz}$. So high sensitivity is needed here due to the fact that under linear mode of non-stationary effect the amplitude of pressure variations in "quantum whistle" has to be considerably less then pressure difference, applied to the throttled system (by analogy with «electric» non-stationary Josephson effect, where the amplitude of generated microwaves voltage happens to be smaller then constant potential difference applied to transition). The smallness of static pressure difference, applied to the throttle element answers to smallness of sound quantum for sonic and ultrasonic frequencies, that essentially determines remarkably low level of generated amplitudes.

In most of experiment [23, 24, 36, 37, 41, 42] the membrane played two roles simultaneously, pumping of the required pressure differential and transmitting the resultant acoustic oscillations into the registering SQUID-system. Dividing functions of pressure "infuser" and high-sensitive "microphone", we suggest to use as the latter the system SQUID - magnetostrictor, which has been considered previously [62] as a basis of a compact gravitational wave detector. The detailed analysis of this system, performed in [63], showed that it should provide the registration of relative elongation of a probe magnetostrictor body at the level of $10^{-22}\div 10^{-24}$ 1/$\sqrt{Hz}$, that demonstrates its principal competitiveness as a modern gravitational wave detector.

In this case action of transducer of pressure and/or elongation into magnetic signal, which is expected to be used with SQUID, is based upon converse magnetostrictive effect - Villari effect (opened by E. Villari in 1865). This effect emerges in magnetization change, arising under the action of mechanical strain (that is pressure) and / or deformation of magnetostrictive sample. In its turn the "direct" magnetostrictive effect – "conventional" magnetostriction (opened by J. Joule in 1842), occurs in the form of change in size of magnetostrictive sample in the course of its magnetization.

To demonstrate the relation between the direct and inverse effects, let us write the expression for the variation of thermodynamic potential density $\wp$ of the magnetostrictor, depending on the variables U, T, S, P, H: $\delta\wp(U,T,S,P,H) = \delta\wp_0(U,T,S) + \frac{\ell(H)}{\ell_0}\delta P + B(P)\delta H$.

Here U, T, and S («thermal variables») are the internal energy, temperature, and entropy, respectively; $\boldsymbol{P}$ is the pressure, caused by the magnetostrictor deformation; H is the magnetic field strength; l and l_0 are the values of the magnetostrictor length in the presence of the field and without it; and B is the magnetic field induction inside the magnetostrictor. The last two terms characterize the contributions to $\delta\wp$ from the energy of deformation and magnetization. Then the partial derivatives of the density $\wp$ are calculated as $\frac{\partial\wp}{\partial P} = \frac{\ell(H)}{\ell_0}$,

$\frac{\partial \wp}{\partial H} = B(\ell)$ and from the equality of mixed derivatives it follows that $\frac{1}{\ell_0}\frac{\partial \ell}{\partial H} = \frac{\partial^2 \wp}{\partial H \partial P} = \frac{\partial^2 \wp}{\partial P \partial H} = \frac{\partial B}{\partial P}$. Then, expressing the elongation in terms of the pressure variation, in correspondence with Hook's law $\delta P = E\frac{\delta \ell}{\ell_0}$, we arrive at the quantitative relation between the direct (the constant Λ) and inverse (the constant $\Lambda^{(-1)}$) magnetostriction effects: $\Lambda = \frac{\partial P}{\partial B} = \frac{E}{\mu_0 \mu}\frac{\partial B}{\partial P} = \frac{E}{\mu_0 \mu}\Lambda^{(-1)}$ where E is the Young modulus, μ_0 and μ are the magnetic permeability of vacuum and the deformed body, respectively. So for alloy 54%Pt 46%Fe, which has far from record [64] magnetostrictive parameters $\Lambda/E \approx 6\times 10^{-3}\ T^{1}$ the inverse effect will be characterized by the constant at the level of $\Lambda^{(-1)} \approx \mu_0 \mu \Lambda/E \approx 7\times 10^{-6}\ T/Pa$ (here $\mu \approx 1000$).

Let's estimate limit possibilities of the method, based on the effect of reverse magnetostriction, in case of registration of magnetic response by the SQUID, on condition $\Delta\Phi = \delta\Phi$, that is when magnetic signal is equal to SQUID resolution. In practical aspect the converse magnetostrictive effect is characterized by the value of magnetostrictive sensitivity, which coincides with $\Lambda^{(-1)} = \frac{\partial B}{\partial P}$ and demonstrates the quantitative relationship between change of magnetic induction and elastic stress, causing this alteration in particular material. The change of induction B flux in magnetostrictive cylinder of section S will make $\delta\Phi = \Delta\Phi = S\Delta B = S\Lambda^{(-1)}\Delta P$, from where the limit sensitivity of system SQUID - magnetostrictor at registration of pressure is expressed in terms of resolution of the quantum interferometer in the form $\delta P = \delta\Phi/(S\Lambda^{(-1)})$. If $\langle \delta\Phi \rangle_{/\sqrt{Hz}} = 10^{-6}\Phi_0/\sqrt{Hz} = 2{,}07\times 10^{-21}\boldsymbol{Vb}/\sqrt{\boldsymbol{Hz}}$, $S = 0{,}003\ m^2$ and $\Lambda^{(-1)} \approx 7\times 10^{-6}\ T/Pa$, then the limit system sensibility on pressure registering will make up $\langle \boldsymbol{\delta P} \rangle_{/\boldsymbol{Hz}} = 10^{-13}\ \boldsymbol{Pa}/\sqrt{\boldsymbol{Hz}}$. To this pressure, which system in principle is able to fix, for magnetostrictor with typical value of Young's modulus E = 100 GPa answers extremely registered elongation $\langle \boldsymbol{\delta\ell/\ell} \rangle_{/\sqrt{Hz}} = 10^{-24}/\sqrt{\boldsymbol{Hz}}$.

The transmission factor of amplitude of acoustic displacements on the boundary helium magnetostrictor happens to be smaller than one $\kappa_{1\to 2} = \frac{2Z_1}{Z_1 + Z_2}\sqrt{\frac{c_1\cos\theta}{c_2\sqrt{1-(c_2/c_1)^2\sin^2\theta}}} \underset{|\theta|<<1}{\approx} \frac{2Z_1}{Z_1+Z_2}\sqrt{\frac{c_1}{c_2}}$ because of the reflection on the boundary, arising from the jump of wave resistances $Z_1 = c_1\rho_1$ // $Z_2 = c_2\rho_2$ (on the boundary helium – magnetostrictor $Z_1/Z_2 \approx 10^{-2}$, relation of sound velocities $c_1/c_2 \approx 10^{-1}$, θ - incidence angle of acoustic wave on magnetostrictor surface). However, even if to take into account the reflection losses, the acoustic response of system SQUID - magnetostrictor occurs to be rather high. More stiff construction of proposed registration system, as compared with the membrane mechanism of magnetic flux expulsion, would allow simplifying appreciably the facility of interferometer on matter waves as a whole.

Acoustic Superheterodyning on Effects of Weak Superfluidity and Quantum Gyroscopes on Matter Waves

The system SQUID - magnetostrictor can be used not only as a part of SHeQUID, but also to be applied as a super high sensitive acoustic receiver, able to register signals until hypersonic frequencies. Of course, magnetostrictive converter will not work further then at kHz frequency by itself. However for registration of giga- and even terahertz oscillations it is possible to transfer these signals into the lower-frequency region, accessible for considered system. As a non-linear element providing such superheterodyne transformation it is proposed to use the throttle, operating in superfluid He in condition of "active" or "passive" Josephson effect. In this case the dissipationless of superfluid state promotes guarantee of low noise level of such mixer, that safeguards preservation of high sensitivity of system SQUID - magnetostrictor in high frequency region with respect to low-one. In active mode the outer acoustic heterodyne is not needed. When superimposing of pressure differential ΔP the throttled circuit itself will start to generate acoustic vibrations $\omega = \frac{\Delta P}{\hbar n_{He4}}$, and the signal on difference frequency between high frequency acoustic signal $P=P_0 cos\omega_s t$, brought to throttle, and "quantum whistle" will be fixed already in low-frequency band by the system SQUID - magnetostrictor $\omega_{MgStr} \sim \left|\omega_S - \frac{\Delta P}{\hbar n_{He4}}\right|$. Indeed, writing Josephson current of superfluid helium $\dot{j}_j^{He4} = \dot{j}_C^{He4}\, sin\varphi = \dot{j}_C^{He4}\, sin\left(\varphi_0 + \frac{1}{\hbar n_{He4}}(\Delta Pt = \int P_0\, cos\omega_S tdt)\right)$, and then, having regard in its Fouriers decomposition $\dot{j}_j^{He4} = \dot{j}_C^{He4} \sum_{m=-\infty}^{\infty} J_n\left(\frac{P_0 / n_{He4}}{\hbar\omega_S}\right) sin\left(\left(\frac{\Delta P}{\hbar n_{He4}} + m\omega_S\right)t + \varphi_0\right)$ on the item with m=-1, we find out the component with required difference frequency $\left(\frac{\Delta P}{\hbar n_{He4}} - \omega_S\right)$. Note that the passive mode, using outer acoustic heterodyne, providing larger coherence of its harmonics in comparison with "quantum whistle", appears to be more preferable. The effect of frequency mixing is described in this case by Fourier decomposition of Josephson current, including double summation of row on $-\infty<m<\infty$ and $-\infty<k<\infty$ (m and k - repetition factors of signal and heterodyne harmonics correspondingly), summation indices of the component, corresponding to required difference frequency, being equal m=-1 и k=1. It is possible to guide the high frequency acoustic signal on mixer by focusing it over the area of throttle allocation with the help of a parabolic mirror. The operation principle of acoustic mixer described above practically allows to attribute it to the type of parametric one. Really the nonlinearity of reactive system parameters is used in its operation. It is known that mixers of this type perform the transformation of signal with power amplification. So in Up-converters on nonlinear crystals, developed in the 1980s for signal transfer from distant IR ranges to scope of photoelectron multipliers, tenfold intensification was achieved. Consequently, the acoustic mixer considered here can accomplish function of low-noise preamplifier also. Expected ultra-high sensitivity of such super heterodyne acoustic receiver with registration of difference signal by means of system SQUID - magnetostrictor can be used in particular for detection of "high frequency" (f>> 1000 Hz) gravitational waves [53] (see above).

The only possible practical application of interferometers on matter waves in superfluid He, on which directly was pointed out in [12, 13, 37, 41], is the creation of gyroscope at the base of SHeQUID, operating by analogy with circular laser systems on Sagnac effect [38-40]. The phase shift, corresponding to rotation of interferometer working ring with frequency Ω, to be measured, is equal to $\delta\varphi = \frac{1}{N\hbar}\oint \vec{p}d\vec{r} = \frac{M}{N\hbar}\oint \vec{v}d\vec{r} = \frac{m\Omega s}{\hbar}$, where $M=Nm$ – the total mass of matter, exerting wave properties. The origin the this phase shift is analogous to difference of Doppler frequencies, arising at waves traveling towards each other in a rotary ring resonator (the Sagnac effect $\delta f = \delta\varphi / t = \frac{4\pi\Omega s}{\lambda c} \Big/ \frac{\ell}{c} = \frac{4\pi\Omega s}{\lambda\ell}$). Appearance of phase shift of matter waves, corresponding to rotation of ring channel, has been observed in superfluid He [36, 37], in particular at appropriate turning of circulation plain of helium atoms the effect of Earth rotation was registered [36, 37]. In principle nothing interferes to substitute helium SHeQUID by "conventional" electronic DC-SQUID in order to create a gyroscope. In this case we are to put mass of electron instead of helium atom mass in the above formula for phase shift. Of course, as a consequence of such substitution resultant effect $\delta\varphi \sim m$, would diminish by 4 orders $m_e/m_{He4} \approx 10^{-4}$. It seems possible to compensate this decrease by increasing the area s of the working contour. However for both "conventional" DC-SQUID and SHeQUID expansibility of area of the working ring is limited. In the first case parameters of input ring of "conventional" DC-SQUID obey principal limitation: production of the input ring inductance on the critical current of Josephson tunnel junction $L_O I_{CJ}$ cannot considerably exceed the magnetic flux quantum $\Phi_0=2\pi\hbar/2e$=2,07×10^{-15} Vb. The cause of such limitation consists in demand of uniqueness of signal performance of interferometer, and comes to excluding of multiquantum hysteresis effects. These effects are possible, when magnetic flux created in input circuit by working currents $I \approx I_{CJ}$ and summered with measured magnetic flux exceeds fundamental quantum period equal to Φ_0 due to large inductance L_O In this case DC-SQUID behaves like the amplifier, embraced by strong positive feedback, operating in trigger conditions. Of course, it is impossible to use the interferometer in such conditions for measuring, however on similar "trigger" basis in 70-80 years of the last century the storage circuits of Josephson computer have been developed. In practice precisely because of limitation $2\pi L_0 I_{CJ} \lesssim \Phi_0$ ring inductance L_O can't be so large, in order to its dimensions corresponded to the common macroscopic scales. The solution of the problem of such limitation for SQUID is prompted by effects of mesoscopic quantum interference famous, beginning from classic Sharvin experiments [65]. The mesoscopic interference is observed in the form of periodical oscillations of resistance of metallic ring, become apparent with the increase of magnetic field flux, piercing this ring. Note that in absence of superconductivity in the role of oscillation period acts the flux quantum $\Phi_0=2\pi\hbar/2e$ exactly equal to quantum value, corresponding to the charge of Cooper pair $q_c=2e$ [66].The fact that without any cooper pairing "two-electron" quantum is fixed in experiment is explained by that nonzero magneto depended contribution to electron scattering, responsible for nonzero metal resistance, arises in second order of perturbation theory, that is as if a pair of electrons is involved. Unfortunately the mesoscopic interference in non-superconducting condition does not allow complete solving a problem of limitation $2\pi L_0 I_{CJ} \lesssim \Phi_0$ for SQUID technique: the Josephson critical current in normal state of course is equal to zero, however another "purely mesoscopic" limitation is superimposed on ring dimension in this case - its diameter should not substantially exceed the coherence length of electron waves, which even at helium temperatures both in metals [65, 66] and in semiconductor heterojunctions with electronic 2D-gas [67] does not exceed 1-10 mkm. We suggest that for imparting macroscopic dimensions to the working ring of quantum interferometer without loss in the interference effects to make the larger part of the ring from superconductor, where the wave-function phase will not be testing chaotic fails that restrict the length of spatial coherence. The experiments on mesoscopic [65-67] show the presence of Bohm - Aronov effect [68, 69], that not directly field unduction $\boldsymbol{B}$, but vector potential $\boldsymbol{A}$ integrated along trajectory of wave propagation affects the interfering electron waves. Hence, accumulated on superconducting site of the path magnetic contribution to wave phase, there will be applied to normal section of the ring (with microscopic dimensions, not exceeding the coherence length), where the quantum interference will arise the resistance variation. The same refers to "mechanical contribution" $\delta\varphi = \frac{1}{N\hbar}\oint \vec{p}d\vec{r} = \frac{M}{N\hbar}\oint \vec{v}d\vec{r} = \frac{m\Omega s}{\hbar}$, corresponding to circulation of electronic condensate. In practice in order to create such an interferometer it is necessary to include in the ring a pair of non-superconducting microscopic point-contacts connected by macroscopically extended segments of superconductor (by analogy with DC-SQUID). Feeding such ring by a low-current $I \ll \Phi_0/L$ and registering voltage drop on it (that is practically on its non-superconducting mesoscopic sites), it is possible to fix "mechanical" phase shift, without restriction on ring size inherent to "conventional" DC-SQUID. Measurement with large capture area of "rotation flow" s will allow to fix variation of the angular speed Ω of reference system with high precision and thus will provide the performing of function of mechanical gyroscope by purely electronic equipment.

REFERENCES

[1] B. D.Josephson. Possible new effects in superconductive tunneling. *Phys. Lett.* 1 (7), 251–253 (1962).

[2] B. D. Josephson. *Adv. Phys.* 14, 419 (1965).

[3] B. D.Josephson. The Discovery of Tunneling Supercurrents. *Science* 184, 527 (1974).

[4] *P. W.Anderson. How Josephson discovered his effect. Phys. Today* 23, 23–29 (1970).

[5] R. Holm, W.Meissner. *Zeitschrift für Physik* 74(11-12), 715-735 (1932).

[6] I. Giaver. *Phys. Rev. Lett.* 5, 464 (1960).

[7] P. W.Anderson, J. M. Rowe 11. *Phys. Rev. Lett.* 10, 230-232 (1963).

[8] R. C.Jak1evic, J. Lambe, A. H. Silver, J. E. Mercereau. *Phys. Rev. Lett.* 12, 159 (1964).

[9] S. Shapiro, A. R. Janus, S. Holly. *Rev. Mod. Phys.* 36, 223-225 (1964).

[10] I. K. Yanson, V. M. Svistunov, I. M. Dmitrenko. *JETP* 21, 650 (1965).

[11] P. W. Anderson. *Rev. Mod. Phys.* 38, 298 (1966).

[12] Y. Sato, R. E. Packard. *Phys. Today* 65, 31 (2012).

[13] Y. Sato, R. E. Packard. *Rep. Prog. Phys.* 75, 016401 (2012).

[14] P. L. Richards, P. W. Anderson. *Phys. Rev. Letters* 14, 540 (1965).

[15] B. M. Khorana, D. H. Douglass. *In Proc. 1-th Intern. Conf. Low Temp. Phys.* (Scotland 1968) 1,169 (Scotland, St. Andrews Univ. Press, 1968).
[16] P. L. Richards. *Phys. Rev. A* 2, 1534 (1970).
[17] P. Hulin, C. Laroche, A. Libchaber, B. Perrin. *Phys. Rev. A* 5, 1830 (1972).
[18] Yu. G. Mamaladze, O. D. Cheishvili. *JETP* 23, 112-117 (1966).
[19] D. L. Musinski, D. H. Douglass. *Phys. Rev. Lett.* 29, 1541 (1972).
[20] P. Leiderer, F. Pobell. *Phys. Rev. A* 7, 1130 (1973).
[21] I. Rudnick. *Phys. Rev. A* 8, 1969 (1973).
[22] D. R. Tilley, J. Tilley. Superfluidity and superconductivity. Dauville: Inderstche Printers & Publishers (1990).
[23] O. Avenel, E. Varoquaux. *Phys. Rev. Lett.* 55, 2704 (1985).
[24] S. V. Pereversev, A. Loshak, S. Backhaus, J. C. Davis, R. E. Packard. *Nature* 388, 449 (1997).
[25] P. L. Kapinza. *J. Phys. (USSR)* 5, 59 (1941).
[26] D. D. Osheroff. Nobel Lecture. http://www.nobelprize.org/nobel_prizes/physics/laureates/1996/osheroff-lecture.html
[27] D. Vollhardt, P. Wolfle. *The Superfluid Phases of Helium-3* (New York, Taylor and Francis, 1990).
[28] O. V. Lounasmaa. Experimental Principles and Methods Below 1 K (New York, Academic, 1974).
[29] J. Clarke. *Phys. Today* 39 (3), 36 (1986).
[30] O. Avenel, E. Varoquaux. *Phys. Rev. Lett.* 60, 416 (1988).
[31] J. C. Davis, R. E. Packard. *Rev. of Mod. Phys*. 74, 741-771 (2002).
[32] K. Sukhatme, Y. Mukharsky, T. Chui, D. Pearson. *Nature* 411, 280 (2001).
[33] E. Hoskinson, R. E. Parckard, T. Haard. *Nature* 433, 376 (2005).
[34] E. Hoskinson, R. E. Parckard. *Phys. Rev. Lett.* 94, 155303 (2005).
[35] R. P. Henkel, E. N. Smith J. D. Reppy. *Phys. Rev. Lett.* 23, 1276 (1969).
[36] R. W. Simmonds, A. Marchenkov, E. Hoskinson, J. C. Davis, R. E. Packard. *Nature* 412, 55 (2001).
[37] E. Hoskinson, Y. Sato, R. E. Packard. *Phys. Rev.* B 74 100509(R) (2006).
[38] G. Sagnac. *Comptes Rendus Acad. Sci.* 157, 708 (1913).
[39] E. J. Post. *Rev. Mod. Phys.* 39, 475 (1967).
[40] G. B. Malykin *Physics – Uspekhi* 43 1229 (2000)
[41] K. Schwab, N. Bruckner, R. E. Packard. *Nature* 386, 585 (1997).
[42] Y. Mukharsky Y, O. Avenel, E. Varoquaux. *Physica B* 284, 287 (2000).
[43] F. Mallet, M. A. Castellanos-Beltran, H. S. Ku, S. Glancy, E. Knill, K. D. Irwin, G. C. Hilton, L. R. Vale, K. W. Lehnert. *Phys. Rev. Lett.* 106, 220502 (2011).
[44] Rui-qin Fan, Hong-guang Yang, Zhi-ming Bai. *Applied Mathematics and Mechanics*. 30(9), 1161-1167 (2009).
[45] J. D. Teufel, T. Donner, M. A. Castellanos-Beltran, J. W. Hawlo, K. W. Lehnert. arXiv:0906.1212 (2009).
[46] C. Eichler, D. Bozyigit, C. Lang, M. Baur, L. Steffen, J. M. Fink, S. Filipp, A. Wallraff. *Phys. Rev. Lett.* 107,113601 (2011).
[47] E. Flurin, N. Roch, F. Mallet, M. H. Devoret, B. Huard. *Phys. Rev. Lett.* 109, 183901 (2012).

[48] J. R. Johansson, G. Johansson, C. M. Wilson, Franco Nori. *Phys. Rev. Lett.* 103, 147003 (2009).
[49] J. R. Johansson, G. Johansson, C. M. Wilson, Franco Nori. *Phys. Rev. A* 82, 052509 (2010).
[50] M. J. Everitt, T. D. Clark, P. B. Stiffell, A. Vourdas, J. F. Ralph, R. J. Prance, H. Prance. *Phys. Rev. A* 69, 043804 (2004).
[51] W. Heisenberg. Uber den anschaulichen Inhalt der quantentheoretischen Kinematik und Mechanik. *Zs. Phys.* 43, 172-198 (1927).
[52] H. P. Robertson. *Phys.Rev.* 46, 794-801 (1934).
[53] V. N. Rudenko. *Gravitation & Cosmology*. 10(№1-2),.41-49 (2004).
[54] R. A. Ferrell, R. E. Prange. *Phys.Rev.Lett.* 10, 479 (1963).
[55] И. О. Кулик. *Письма в ЖЭТФ* 2, 134 (1965).
[56] J. Clarke, A. B. Pippard, J. R. Waldram. *Physica* 55, 405 (1971).
[57] A. C. Scott, F. Y. F. Chu, D. W. McLaughlin. *Proc. IEEE* 61, 1443 (1973).
[58] A. S. Malishevskii, V. P. Silin, and S. A. Uryupin. *JETP Lett.* 69 (4), 348-352
[59] Э. Б. Сонин. *Письма в ЖЭТФ* 47(8), 415-418 (1988). E. B. Sonin, *JETP Letters* 47 (8), 496-499.
[60] A. I. Golovashkin, A. N. Zherikhin, L. N. Zherikhina, G. V. Kuleshova, A. M. Tskhovrebov. *JETP* 99 (2), pp 370-378 (2004)
[61] J. Clarke. *Phys. Today* 39(3), 36 (1986).
[62] A. I. Golovashkin, A. V. Gudenko, L. N. Zherikhina, O. M. Ivanenko, K. Yu.Mitsen, A. M. Tskhovrebov. *JETP Lett.* 60 (8) 612-616 (1994).
[63] A. I. Golovashkin, L. N. Zherikhina, A. M. Tskhovrebov, G. N. Izmailov. *Quantum Electronics,* 42 (12) 1140–1146 (2012).
[64] B. Kundys, Y. Bukhantsev, H. Szymczak, M. R. J. Gibbs, R. Zuberek. *J. Phys. D: Appl. Phys.* 35, 1095-1098 (2002).
[65] D. Yu. Sharvin, Yu. V. Sharvin. *JETP Lett.* 34, 272-275 (1981).
[66] B. Schwarzchild. *Physics Today,* January 1986, p.17.
[67] A. E. Hansem arXiv:cond-mat/9909246, 16 Sep.1999.
[68] W. Ehrenberg, R. E. Siday. *Proc. Phys. Soc. (London)* B62, 8-21 (1949).
[69] Y. Aharonov, D. Bohm. *Phys. Rev.* 115, 484 (1959).

In: Helium: Occurrence, Applications and Biological Effects ISBN: 978-1-62808-612-6
Editor: Simon B. Davis

Chapter 7

CLEAN ENERGY FROM HELIUM-3 FUELLED MUON-CATALYZED ANEUTRONIC FUSION

Cooroo Egan*

ABSTRACT

Helium-3 is a stable isotope of helium, terrestrially scarce, but with an estimated 6.5x10^8kg contained within the lunar regolith. Despite the difficulties inherent in commercial exploitation of these remote reserves, research progresses into its potential for use in fusion. It has now been posited that by employing existing and developing technologies, an energy efficient, helium-3 fuelled, muon-catalyzed fusion (μCF) reactor is able to be constructed. This reactor, which avoids the problems associated with maintenance of the extremely high temperatures normally required to catalyze fusion, is also aneutronic, offering total freedom from the need to deal with radioactive materials. The key processes of this radiation free, helium-3 powered μCF reactor are discussed, including a review of some of the technologies that are required, the issue of alpha-sticking in the reaction area, and the predicted muon-wave that simulations have suggested would partially negate the alpha-sticking while providing the primary agency for the predicted energy efficiency.

INTRODUCTION

Substantial progress has been made in the field of μCF since the principles involved were first understood. However, until recently, two problems have remained unresolved, preventing a μCF reactor from being considered a viable energy source. Firstly, the problem of negative muons (μ^-) sticking to created fusion products instead of being released to catalyze more reactions, known as alpha-sticking, has prevented reactions from reaching energy efficiency. Secondly, the problem of dealing with radiation, either from the fuels employed, or from the neutron bombardment associated with energy output, has also remained a deterrent. In order to overcome these two difficulties, a helium-3 fuelled μCF

* E-mail address: cooroo@ozlawyer.com. Affiliations: None at time of writing

reactor utilizing existing and currently developing technology has now been posited [1].[1] This proposed reactor is fundamentally different to previously considered μCF designs in that instead of utilizing a small number of μ^- and then seeking to establish a method of catalyzing enough reactions to compensate for their creation costs before alpha-sticking, it saturates a small fuel supply with a large amount of μ^- and then makes use of a simulation predicted effect, dubbed a muon-wave, to partially address the problem of alpha-sticking. This μ^- saturation approach, which allows for the possibility of using helium-3 as a fuel source, thus negating the need to deal with radioactive fuels such as tritium, has only recently become feasible by the implementation of the relatively new technology of tapered capillaries [1-4].

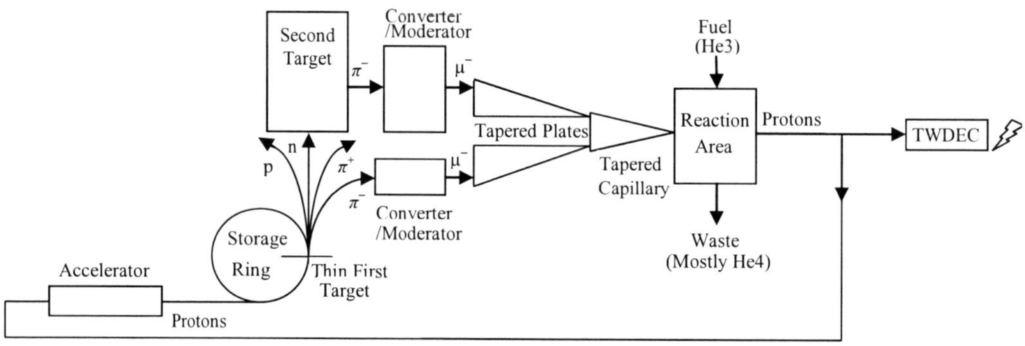

Figure 1 [1].[2] He-3 fuel is continuously pumped into a reaction area, where it fuses to release protons as well as create He-4 waste, which then flows out to be removed. A percentage of the protons from this are channeled into an accelerator before entering a storage ring for π^- production. Primary and secondary π^- are produced and immediately delivered to a moderator for cooling and phase space compression (PSC). The resulting high-intensity low-energy μ^- beam(s) is then channeled via tapered plates or tubes into a charged tapered capillary. Here they are focused and returned to the original target area, where the majority flow into and form part of a muon-wave, continuing to help catalyze fusion reactions for the remainder of their lifespan. Released protons not required for π^- production are channeled to a travelling wave direct energy converter (TWDEC) with a Direct-Current Direct Energy Converter installed downstream. These protons are converted directly to electricity, and a portion of this is spent accelerating protons bound for the storage ring and π^- production. The remaining electricity produced is a net gain from the process.

HELIUM-3 MUON-CATALYZED FUSION

μ^- are introduced to a supply of He-3, a stable isotope of helium with a nucleus comprised of 2 protons and 1 neutron. Due to their mass, when a μ^- orbits one of the protons of a helium atom, it orbits so closely that the positive charge of the proton orbited is effectively neutralized. Initially, the supply of He-3 is sufficiently saturated with μ^- to ensure that this neutralizing effect occurs for both protons within a substantial number of helium atoms, hence leaving them essentially chargeless. These chargeless helium atoms then experience no repulsion from other helium atoms, and through their natural thermal motion wander until close enough to a neighbouring atom for the nuclear forces to dominate. At this

[1] See figure 1.

[2] Diagram first used in a publication in Nucl. Instrum. Methods B.

point, if both atoms involved are He-3, the desired fusion effect occurs, creating helium-4 with surplus protons shooting out which can be harnessed and converted to usable energy.

He3 + He3 → He4 + 2p (12.86 MeV) [1, 5].

Meanwhile the μ^- involved are often released and free to repeat the process.

As with other potential μCF fuel sources, the tendency for a μ^- to become stuck to the helium-4 being created, known as alpha-sticking, is an issue, as is the relatively high probability as reactions continue, of existing or newly introduced μ^- sticking to helium-4 waste instead of the desired fuel. However, as will be discussed, the muon-wave addresses both these issues by allowing a substantial portion of these otherwise lost μ^- to be shaken loose, or reactivated through collisional detachment [1, 6-8].

Muon Generation

Current technology offers various methods for μ^- beam generation. This usually involves a pulsed beam hitting a target with sufficient energy to create negatively charged pions (π^-), which in a vacuum decay predominantly with a probability of 99.99% [1, 9] into a muon anti-neutrino and a μ^-, and attempt capture of the produced pions with a solenoid, having them decay in flight. The approach of Shin and Rafelski [10], incorporated in figure 1, has been suggested as a desirable method for three main reasons. Firstly, due largely to its ability to capitalize on an alternate source of π^- through the implementation of a secondary target, it is comparably energy efficient with respect to other available methods, of primary concern for any reactor. Secondly, it does not require radioactive materials such as tritium for beam targets, allowing the process to remain completely radiation free. Lastly, it requires only a proton beam, which can be siphoned from the fusion reactions directly, improving efficiency, but more importantly allowing the overall reaction process to be more self contained once ignition has been achieved.

Tapered Plates/Capillaries

Once a charged keV beam enters a capillary, the beam hits the inner wall, causing it to become charged. When the accumulated charge is sufficient to prevent the beam from touching the inner wall, it travels more or less parallel to the wall, and exits from the outlet still with its initial charge state. After this self-organized charge up of the inner wall has occurred, the outlet diameter of the tapered capillary can become the limiting factor [1-3]. While this suggests that the aim necessary to achieve the desired μ^- saturation level is able to be reached, testing in the area of guiding keV μ^- beams has not progressed since the initial proposal. Further investigation is required in order to determine the optimal angle, length and substance of the tapered plates used to narrow the muon beam to a width which can be channelled into the tapered capillary.

Unlike other charged beams, the short lifespan[3] and high energy cost of μ^- make it desirable to find a method of appropriately pre-charging the inner wall of the tapered capillary, using something with a lower energy cost and longer lifespan. This would ensure that μ^- are not wasted in charging the inner capillary wall. As the charge would dissipate in time, multiple tapered capillaries on a rotating system would possibly also be required in order to prevent μ^- being used to keep the charge at the desired level. The optimal setup for this would also require experimentation and refinement before implementation.

When a He-3 fuelled μCF reactor was first proposed, the smallest outlet diameter for a tapered capillary was 80nm [1, 2]. However, it is now possible to focus a charged beam to a 50nm diameter [4]. This substantially reduces the estimated target area during ignition, and subsequently the predicted quantity of μ^- required during ignition has also been reduced, with an estimated beam flux of 10^12 μ^-/second, nearly 1/3rd of previous estimates, now being adequate (see calculations and equations).[4] Further advances in this technology will result in even greater reductions on initial μ^- beam strength requirements.

Muon Slowing

As well as allowing increased focusing through capillary optics, the lower the μ^- energy, the smaller its penetration depth, and consequently its stopping range within a substance, in this case helium. As with other sections of the reactor, there are a variety of methods available to accomplish this slowing down of μ^-, known as muon cooling. The suggested method [1] investigated 'a moderator consisting of a large number of thin carbon foils placed perpendicularly to the axis of a high-field solenoid'. This would allow a 5keV beam, with a 1keV energy spread to be obtained; slowing μ^- quickly (with respect to their lifespan), accepting very large initial phase space volumes, while allowing a potentially high output/input ratio. Most importantly, it has an extremely large PSC factor, normally a significant limitation for potential muon beam fluxes.

This method of μ^- cooling is effective, and within the range of current technologies. However, due to each μ^- high energy cost and short lifespan, and this being the only section of the reactor where μ^- are not either being guided to, or in the reaction area, any time that can be saved in this process would translate to a direct increase in overall energy efficiency. As such it would be beneficial to evaluate alternatives which may have become available by the time of construction of a μCF reactor.

Muon-Wave

The graph in figure 2 is taken from a point in simulations some time after ignition is achieved, and shows the reaction area divided into 50 discs,[5] along with the corresponding quantities of μ^- in each disc at the time. There is a clear trend of increasing levels of μ^-

[3] Mean μ− lifespan at rest is 2.1969811μs ±0.0000022μs [9].

[4] μ^- beam strength requirements during ignition are likely the most difficult technological challenge facing the construction of a He-3 fuelled μCF reactor.

[5] With disc one located where He-3 fuel is introduced, and this disc flowing through the reaction area until it becomes disc 50, where the waste leaves or is extracted.

towards the downstream end, apart from a single distinct peak indicating the presence of the muon-wave. In order to understand how this spread eventuates, it is easiest to follow what happens in the reaction area from the very beginning.

Initially, there are a large number of μ^- introduced to the reaction area in essentially random locations. This is enough to ensure that a reasonable number of He-3 atoms attract two μ^-, become essentially chargeless, and are then able to interact with a random neighbouring He-3 atom allowing fusion to occur. The released protons from these leave the reaction area quickly, resulting in one helium isotope filling the space where two existed previously, causing a slight pressure drop. Meanwhile, fuel is constantly flowing or being pumped into the reaction area at the original pressure. Reactions continue to occur in essentially random locations throughout the reaction area while the flow downstream continues, resulting in discs further downstream having on average been involved in a greater number of fusion reactions, and hence likely to have a lower pressure.

At this early stage, the reaction area is still being saturated with μ^-, and there is now a pressure drop towards the downstream end. A helium atom which has attracted two μ^-, as before, now has an equal chance of being able to interact with any of its neighbouring atoms. However, due to the established pressure drop, there are now on average slightly more neighbouring helium isotopes to interact with upstream then there are downstream. This initially small difference means that μ^-, while still displaying a tendency to go with the flow downstream, also have a very slight tendency to react upstream against the flow. This in turn can cause slightly higher concentrations, or clumps of μ^- to form within discs, which in itself would probably not be enough to be significant, except that this tendency is self reinforcing.

With a small but consistent pressure drop from the upstream to the downstream of the reaction area, a small tendency for μ^- to react upstream results. However, a larger than average collection of μ^- in one disc allows the disc to be involved in a larger than average number of fusion reactions. This in turn results in there being a larger pressure difference on either side of the disc in question. A larger pressure difference means a higher chance of μ^- reacting upstream, resulting in a disc with even more μ^-. This reinforcing process, due to pressure, peaks when the upstream side of the muon-wave consists of mostly unreacted He-3, and the downstream side consists of mostly He-4 waste. After several generations of μ^-,[6] and with a carefully controlled flow rate, a second reinforcing effect to the muon-wave also begins to occur. The disc in question reaches a point where most, if not all of the helium within it has two μ^- attached,[7] with other μ^- in the area remaining unattached for a longer period of time. This saturation means that any μ^- which has been introduced upstream of the disc in question, as expected, attaches to a helium isotope and continues to flow downstream. However, now when flowing past the muon-wave, the chance of both not attracting a second μ^-, or becoming collisionally detached becomes negligible. In this way, the singleton μ^- becomes part of, and reinforces the muon-wave. There will always be a reasonable number of μ^- which detach from the muon-wave and continue to flow downstream.[8] Also, there are still μ^- introduced on the downstream side of the muon-wave which never flow into, and thus

---

[6] Multiple muon-waves may begin to form in this time, with each one effecting possible pressure differences further downstream. Sometimes two muon-waves may combine with one another, but it is generally more common for the one with the greatest pressure difference to reinforce itself most quickly, and thus become dominant.

[7] This is what allows for the collisional detachment and reactivation of otherwise lost μ^- (see alpha sticking and collisional detachment).

[8] Simulations suggest $<10\%$ [1].

never become part of it. However, so long as the flow rate is carefully monitored in order to ensure that it is not so great as to wash the muon-wave all the way downstream and out of the reaction area, it will continue to self reinforce, ensuring that the bulk of μ^- introduced will become, and remain a part of the muon-wave for the majority of their lifetime.

Not only does a muon-wave ensure that the bulk of He-3 flowing through the reaction area will be involved in fusion, but the μ^- flux required to maintain it relates directly to the helium flow. This means that it is no longer necessary to saturate the reaction area to achieve fusion reactions once one has been established, which substantially minimizes energy input requirements. The muon-wave also partially addresses the alpha-sticking problem. These two factors combined allow a particularly high number of reactions per μ^- created, thus allowing the overall reaction process to achieve the predicted energy efficiency.

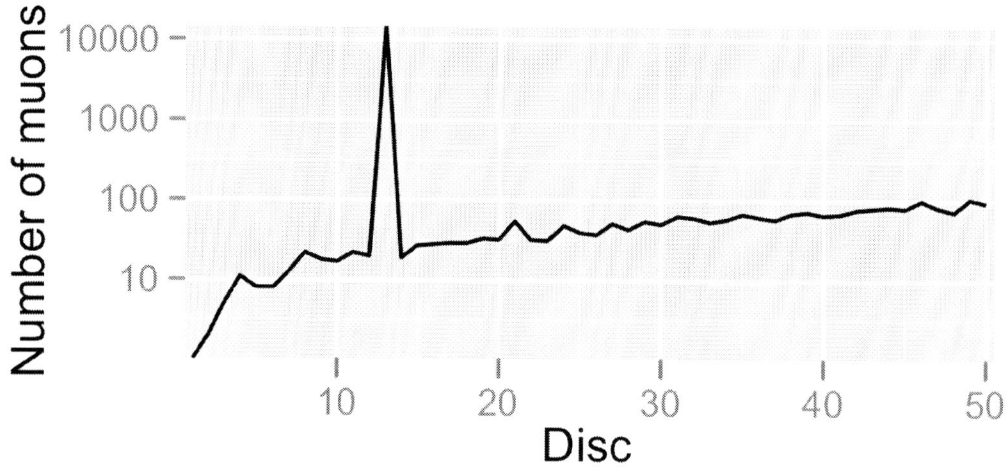

Figure 2. Graph from Monte Carlo simulations [1] depicting distribution of μ− in reaction area after the formation of a muon-wave.

Alpha Sticking and Collisional Detachment

In the outlined approach, fusion will only occur when two He-3 atoms are involved. However, this is not the only possible atom interaction. For example, He-4 waste in the reaction area can attract two μ^-, thus also becoming essentially chargeless. When these atoms interact with neighbouring atoms, instead of fusion, there is a chance for any μ^- involved to become reactivated via collisional detachment [1, 6-8], freeing them to potentially help catalyze future fusion reactions. Due to the high saturation level of μ^- in the muon-wave, the atoms in that area are constantly interacting with neighbouring atoms, so that while the actual probability of alpha-sticking occurring during fusion remains unchanged, any He-4 atom with μ^- attached in and around the muon-wave will have a high chance of this collisional detachment occurring. The overall detrimental effect of the alpha-sticking is thus substantially mitigated, with an estimated ~410 He-3 He-3 fusion reactions per μ^- being achievable once the muon-wave has been established [1].

VARIABLES

There are many untested variables that could have a direct effect on the efficiency and/or output of the proposed μCF reactor. These fall into three main categories; directly controllable, uncontrollable, and technologically dependant. Controllable variables are those which could be optimized given testing, including the temperature, pressure, and rate of the He-3 fuel being injected, the angle, length and substance of the tapered plates, and the size of the reaction area. Uncontrollable variables include those that still require testing, but are not able to be modified, such as the exact speed at which He-3 He-3 muon-catalyzed fusion occurs. The remaining type of variable depends on current technology levels, and could be improved upon given advances in different fields. These include the efficiency with which protons can be converted to electricity, the energy cost required per μ^-, and the remaining lifetime of μ^- reaching the reaction area after undergoing muon cooling. While conservative estimates were used wherever possible, 410 He-3 He-3 fusion reactions per introduced μ^- being achievable with current technologies is just an approximation, and would require considerable testing in order to confirm and refine.

HE-3 AVAILABILITY

While there are some possibilities for extracting He-3 from terrestrial sources [11, 12], these are extremely limited, with the natural abundance of the He-3 isotope as a proportion of terrestrial helium supplies being only 1.38ppm [13], for a total inventory of ~2x10^4kg [14, 15]. Manufacturing a source of He-3 via the radiolytic decay of tritium as is done for other fields such as medicine, is also not a feasible option for a reactor. However, He-3 sources exist extra-terrestrially [12-18]. While there are larger deposits elsewhere in the solar system, by far the most appealing given current technologies are those held within the lunar regolith, which, due to a lack of lunar atmosphere or magnetic field, has been accumulating He-3 from solar wind deposits for billions of years [12-16]. The amount of He-3 per unit area in the lunar regolith varies substantially from region to region. Current estimates suggest there exist 3.72x10^8kg on the lunar nearside, and 2.78x10^8kg on the lunar farside, giving a total estimated 6.5x10^8kg [14, 15], enough for ~1000 years of world energy supply [5].

Extracting He-3 from the lunar regolith would also result in the collection of substantial quantities of other substances. Gasses apart from helium recoverable from the heating of maria regolith[9] to ~700°C would include H_2, H_2O, N_2, CO_2, CH_4 and CO [5,16]. Isotropic separation by means of a superleak system[10] could then be applied to separate the He-4 from the He-3 [12, 16]. The large amounts of H_2 recovered could be used as a fuel, while the other recovered gasses could be directly used for the benefit of sustaining lunar life-support systems. This would drastically reduce the amount of supplies from Earth necessary to support a long term human mining endeavour on the lunar surface. Estimates for the projected

[9] Areas containing the mineral ilmenite ($FeTiO_3$), which is primarily found in the lunar basaltic marias, are expected to contain the largest concentrations of He-3 [12,16].

[10] When helium-4, sometimes referred to as helium I, is cooled to below ~2.1°K, it becomes a superfluid with negligible viscosity, referred to as helium II. A superleak separator makes use of this by introducing very fine filters to a mixture of helium which has been cooled to this point. The helium II passes through the filters, leaving behind any He-3 [1,12,18].

energy payback from such an endeavour vary based on various factors from operational lifetime of lunar mining equipment to achievable fusion efficiency, but are upwards of 270 times the energy invested [12, 16], making it a viable, and even appealing option from an economic standpoint. However, despite the promise of such an abundant source of clean energy, the initial difficulties involved in establishing a helium-3 mining presence on the lunar surface will be the immediate determent for its development.

CONCLUSION

Although testing would be required, the proposed He-3 fuelled μCF reactor is theoretically feasible, and could be constructed using existing technologies, producing a potent radiation-free energy source with He-4 as its only waste product. While technological advances in areas such as tapered capillaries will continue to make this option more obtainable and possibly more appealing as time progresses, practical implementation will be obstructed due to the difficulties involved in obtaining the required amounts of He-3 fuel needed to power the reactor. However, should the extraction of He-3 from extra-terrestrial sources such as the lunar regolith become a reality, its use in this reactor would become a prime option for power generation.

CALCULATIONS AND EQUATIONS

Initial estimates were given by first establishing the No. of atoms in 1000nm^3 of liquid He-3 = 11869.21 [1].

The volume of a cylinder with radius 40nm and height 55nm was then established: $\pi r^2 h = 88000\pi = 276460.15\text{nm}^3$.

The maximum number of He-3 atoms in $276{,}460\text{nm}^3$ was therefore given by 276.46 x 11869.21 = 3,281,361.8.

Initial estimate of required number of μ^- at any one time was 3x10^12 x 2x10^-6, or 6x10^6.

Initial estimate of starting number of protons was 3.281362x10^6 x 2, or 6.562724x10^6.

6x10^6 / 6.562724x10^6 then gave an estimated μ^-/proton ratio during ignition ~9:10.

For a cylinder with a radius of 25nm and height 55nm: $\pi r^2 h = 34375\pi = 107992.25\text{nm}^3$.

The maximum number of He-3 atoms in $107{,}992\text{nm}^3$ is therefore 107.99 x 11869.21 = 1,281,756.

Number of protons is 1.281756x10^6 x 2, or 2.563512x10^6.

2.563512x10^6 x 0.9 = 2.3071608x10^6.

2.3071608x10^6 / 2x10^-6 = 1.1535804x10^12, which is approximately 1/3rd of the initial μ^- beam strength requirement estimate of 3x10^12 [1].

REFERENCES

[1] Egan, C., A modified approach to muon-catalyzed fusion employing helium-3 as fuel. *Nucl. Instrum. Methods* B 287, 103-108 (September, 2012).

[2] Ikeda T., *et al.*, Glass capillary optics for producing nanometer sized beams and its applications. *Surf. Coat. Technol.*, doi:10.1016/j.surfcoat.2011.03.098, 5pp (2011).

[3] Ikeda T., *et al.*, Focusing of charged particle beams with various glass-made optics. *J. Phys. Conference Series* 88, 012031, 9pp (2007).

[4] Wang, H., *et al.*, Transmission of 16 keV Cu^- ions through Al_2O_3 nano-capillaries. *Nucl. Instrum. Methods* B 286, 351-354 (September 2012).

[5] Santarius, J. F., Lunar 3He and fusion power. Fusion Technology Institute, Univ. of Wisconsin, 48pp (28 September 2004).

[6] Petitjean, C., Progress in muon catalysed fusion. *Nucl. Phys.* A 543, 79c-97c (1992).

[7] Froelich, P., Larson, G., $\alpha\mu$ Stripping by ionization in dense deuterium/tritium mixture, and its implication for muon-catalyzed fusion. *J. Mol. Struct. (Theochem.)* 199, 189-200 (1989).

[8] Nagamine, K., Review of muon catalyzed fusion experiments – activities after EXAT98 and future perspectives. *Hyperfine Interact.* 138, 5-13 (2001).

[9] Beringer J., *et al.*, (Particle Data Group), *Phys. Rev.* D86(1), 010001 (July, 2012). (URL: *http://pdg.lbl.gov*).

[10] Shin, G. R., Rafelski, J., Pion production for MuCF. *Nucl. Instrum. Methods* A 287(3), 565-569 (February, 1990).

[11] Lupton, J. E., Craig, H., A major helium-3 source at 15^0S on the east pacific rise. *Science* 214, 13-18 (October, 1981).

[12] Wittenberg, L. J., Santarius, J. F., Kulcinski, G. L., Lunar source of He-3 for commercial fusion power. *Fusion Technol.* 10, 167, 37pp (March 1986).

[13] Andrew, A. M., Fusion power; centralized computing. *Kybernetes* 37(1), 17-19 (2008).

[14] WenZhe, F. A., YaQiu, J., Global inventory of helium-3 in lunar regoliths estimated by a multi-channel microwave radiometer on the chang-e 1 lunar satellite. *Chinese Sci. Bull.* 55(35), 4005-4009 (December 2010).

[15] Wenzhe, F., Ya-Qui, J., Quantitative estimation of helium-3 spatial distribution in the lunar regolith layer. *Icarus* 190(1), 15-23 (September 2007).

[16] Sviatoslavsky, I. N., Jacobs, M., Mobile helium-3 mining and extraction system and its benefits toward lunar base self-sufficiency. *WCSAR Technical Report*, 14pp (August 1988).

[17] Rood, R. T., Bania, T. M., Wilson, T. L., Detection of helium-3 in a planetary nebula. *Nature* 355, 618-620 (February 1992).

[18] Hampel, C. A., in *The Encyclopedia of the Chemical Elements* Hampel, C. A. Ed. (Reinhold Book Corp., New York, 1968), pp. 256-268.

INDEX

A

B

C

D

E

T

U

V

W

X

Y